Pvt. Ltd.

PEST MANAGEMENT IN VEGETABLES

VOLUME 1

SECOND EDITION

Editors
K P SRIVASTAVA
DHAMO K BUTANI

2009

Studium Press (India) Pvt. Ltd.

ISBN : Vol. 1 : 978-81-907577-4-4
ISBN : Vol. 2 : 978-81-907577-5-1
ISBN : SET : 978-81-907577-3-7

Published by:

STUDIUM PRESS (INDIA) PVT. LTD.
4735/22, 2nd Floor, Prakash Deep Building,
Ansari Road, Darya Ganj, New Delhi-110002.
Phone : +91-11-23240257, 65150445, 65150447
Fax : +91-11-23240273
E-mail : studiumpress@gmail.com

Printed at:
Salasar Imaging Systems
Delhi - 110035 (India)

Dedicated to

Dr. Mulchand G. Jotwani
(31.05.1926-04.06.1983)

*our friend,
philosopher & guide*

M. S. SWAMINATHAN RESEARCH FOUNDATION

FOREWORD

I am happy that the second revised edition of the book on Management of Vegetable Pests edited by K P Srivastava and Dhamo K Butani is now becoming available. The first edition of this book published in 1998 stimulated considerable interest in the cultivation and consumption of pesticide-free vegetables. The present revised edition gives more up-to-date information on the Scientific Management of vegetable Insect Pests, in such a manner that the productivity of vegetable crops can be enhanced in perpetuity without ecological or social harm. Thus Dr. Srivastava and the Late Dr. Dhamo Butani have laid the Foundation for an Evergreen Revolution Movement in Vegetable crops. I hope the book will be widely read and used.

M.S. SWAMINATHAN
Chairman

3rd Cross Street, Taramani Institutional Area, Chennai (Madras) - 600 113, INDIA
Phone: +91-44-22541698, 22542790 Fax: +91-44-22541319
E-mail: chairman@mssrf.res.in, swami@mssrf.res.in

PREFACE TO THE SECOND EDITION

The basic purpose of this new edition is to update the material in first edition which is now more than 10 years old. As a result, the chapters in that earlier edition have been revised and expanded and a new chapter in the 'Genetic engineering of vegetable crops for the insects pest management' added. The phenomenal growth of vegatable during the last ten years and the diversity of approached adopted have made this science truly and integrated one. The increased involvement of interdisplinary trend as well as the modern and newly evolved pest control methods made it necessited to revise the first edition.

Vegetable senario is changing very fast. The advent of plant molecular biology, genetic engineering and the introduction of gene into vegetable plant has resulted in a remarbable and rapid shift in vegetable practices.

The vegetables are the potential crops for improving nutrition and to provide food security. Vegetable being rich source of nutrients can play significant role for improving the nutritional intake. With the advent of modern technologies such as improved varieties, hybrid production, integrated pest management, protected cultivation; scenario of vegetable production in India is changing at a fast rate. The vegetable production has increased from 75 million tones in 1996-97 to 94 millions tones, which accounts for about 14.4 per cent of world production. The per capita consumption of vegetables has also increased from 95 to 175 grams. From 2002 onwards, the country has witnessed of 4.9 per cent increased in area for vegetable cultivation with 49% increase in production.

Vegetables provide more remuneration to farmers than cereals and pulses as the vegetables can be grown throughout the year. Many of the vegetables like spinach, potato, egg plant, pumkin, okra etc can be grown twice and even thrice in a year if irrigation facilities are available.

Vegetables play an important role in economy of India by way of exports of fresh, preserved and dehydrated vegetables. Fresh vegetables worth Rs 383.05 crore were exported in the year 1998-99. The vegetables such as onion, potato, okra, bitter gourd, green chilies and some other seasonal vegetables have a great export potential with the new technology of molecular biology and genetic engineering introduction of desired genes in a varieties is possible resulted in a remarkable and rapid shift to agricultural practices. The facctors controlling gene have been better understood, and it is hoped that new transgenic crops will provide more production of value added commodities.

Scientists at Flinders Medical centre in Adelaide plan to test the vaccine booster, or adjusted from natural sugar found in onion, garlic and chicory root which may help make influenze vaccines 10 times more powerful. In India also medicined value of dried vegetables under 'Natural Vegetable the Therapy' proved effective on number of ailment.

This revised edition provided adequate information for the postgraduate students, progressive farmers and plant protection workers about the production of vegetable crops in India.

Editors are greatly thankful to Padam Vibhusan Professor M.S. Swaminathan, Member Rajya Sabha who has been kind enough to spare his valuable time to write forward and evaluations of this book. The help rendered by Sri Arvind Jain, Sri Shray Jain of Studium Press (India) Pvt. Ltd., and Sri Arvind K. Mittal and Dinesh Gupta of Salasar Imaging Systems, New Delhi for designing, printing and publishing second edition of this book is also gratefully acknowledged.

K.P. Srivastava

NEW DELHI
December 2008

K.P. SRIVASTAVA

ABOUT THE EDITORS

Dr. Krishna Prasad Srivastava Principal Scientist (Vector-Entomologist) retired was born on 1st April, 1938, joined Division of Entomology, Indian Agricultural Research Institute (IARI), New Delhi in 1963; worked on insects genetics and host-plant resistance then switched over to insect pests of sorghum in 1967; became Assistant Entomologist (Sorghum) in1969 and was awarded Ph.D. degree in 1975. Promoted as S-2 in 1976, S-3 in 1983; working all along on sorghum pests. Invited to participate in International Sorghum Entomology Workshop at Nairobi (1980) and Texas A.M. University (1984). Awarded UNDP Fellowship to work at University of Kentucky, Lexington (USA). 1991, on virus vector relationship.

Dr. Krishna Prasad is working since 1991 in the Division of Plant Pathology, IARI, New Delhi, as Vector-Entomologist with special reference to aphids, beetles, jassids and whitefly. He has to his credit about 125 research papers, mostly on crop pests, besides chapters in various books, Dictionary of Agriculture (1990) and Dictionary of Botany (1991). He has also authored 'Soybean Pests in India and Their Management', 'Pest Management in Vegetable' (1998), 'Pest Management in Citrus' (1999), 'Dictionary of Entomology' (2008).

Dhamo Kessowdas Butani, B.Sc. (Agri.), M. Sc. (Agri.), D.Phil. (Agri.) – an eminent agricultural scientist, is a migrant from sind (Pakistan). He had specialized training (April-December, 1968) in biological control of insects at Paris and Antibes (France). Attended post-graduate certificate course in Nematology (1969-70) organized by AMU and International Agriculture Center, Wageningen (The Netherlands). Started his career as DAO (Sind) in 1944 and retired as Senior Entomologist (Fruits) from Indian Agricultural Research Institute, (IARI), New Delhi in 1983. He has experience of conducting and guiding research for ten years each, on insect pests of sugarcane, cotton and horticultural crops; besides teaching as Assistant Professor at Bihar Agricultural College, Sabour (1959-62) and as member of postgraduate faculty of IARI at New Delhi (1977-83).

Dr. Dhamo has to his credit around 250 research papers and review articles in English, French and Hindi; besides chapters in various books. He has also authored five books- Insects and fruits (1979); Laboratory Manual in Agricultural Entomology (1984); Insects in Vegetable (1984); Mango; Pest Problems (1993) and *Jantu Vigyan* (1994). After retirement, he has taken up as hobby, compiling dictionaries on biological science and those released so far, include, Dictionary of Science (1987), Hindi edition(1995), Dictionary of Medical Terms (1988)/ Medical Dictionary (1989), Dictionary of Biology (1990), Dictionary of Zoology(1990), Dictionary of Agriculture (1993). Biology Facts (1996) and Dictionary of Entomology (2008).

Dr. Butani wanted to contribute much more from his vast experience and knowledge but unfortunately his sad and sudden demise on 13.12.1998 inspired me to continue his desire.

CONTENTS

Volume I

Volume II

LIST OF ILLUSTRATION

1

INTRODUCTION

IN INDIA, varied agro-climatic conditions have made it possible to grow a wide variety of vegetables crops all the year round in one or the other part of the country. As many as 61 annual and 4 perennial vegetable crops are commercially cultivated in this country. The system of mixed, companion and inter-cropping in vegetable cultivation provide maximum output resulting more income per unit area of land to small farmers.

Production of vegetables in India till 1961-65 was about 2345 million tones at an average of 6.0 tonnes per hectare. It increased to 28.36 million tonnes in 1980, with an estimated yield of 50.99 million tonnes of vegetables in 1990-91. *(FAO Year Book, 1990).* India being the second largest producer of vegetables in the World next only to China, shares about 12% of the World output of vegetables from about 2.0% of cropped area in the country. Presently, India, produces 70 million tonnes of vegetables. In spite of such a large production, the per capita supply of vegetables in India is only 140 grammas per day.

Vegetables play an important role in balanced nutrition as these are valuable source of carbohydrate proteins, vitamins and minerals. Cabbage, carrot, amaranthus and spinach contain vitamin A. Cabbage, beet-root, bitter gourd and chillies are rich sources of vitamin C; beans and peas provide lot of protein that contain lysine while leafy vegetables supply calcium and iron. Vegetable crops not only provide nutritional security but are also capable of producing more biomass (five times the quantities of food per unit area) when compared to cereal crops.

Vegetables also assume importance as export earning crops. India has been exporting both fresh and processed vegetables and products are exported mainly to Gulf, Sri Lanka and South Asian countries. These include onion, okra, tomato, chillies, beans green cluster bean and garlic. Onion export was between 215830 and 360227 tonnes valued at 641.5 to 1180.3 million rupees while other fresh vegetables to the extent of 18230 to 41805 tonnes values at 15-87 to 23-45 crores during 1988-89 to 1990.

The factors like lack of widely adaptable high yielding varieties, lack of short duration superior varieties to fit in different farming systems, non-availability of seed of improved high yielding varieties, limited extension service, support several insect pests, diseases, unorganized marketing are the major constraints in the production of vegetable crops in our country.

Vegetables crops support several insect species. It is estimated that more than 40 per cent of yield loss is caused by insect pest attack in different vegetable crops.

Huge losses caused by number of diseases starting from placement of seed in the soil till the final produce is consumed, is a major constraint is the production. Disease are caused by fungi, bacteria, viruses, viroid, phytoplasma and other agents. Fungal disease of many kinds attacks vegetable crops during all stages of their growth. The influence of these diseases are highly destructive. Many factors are known to aggravate the disease incidence and economic losses are sustained. Bacterial diseases are responsible in reducing the quality and yield of produce. Some of the bacterial diseases are known to cause extensive loss to the crop, *e.g.*, bacterial wilt caused by *Pseudomonas solanacearum* can reduce the yield of egg plant and tomato upto 81 and 90.62 per cent respectively. Brown rot or bacterial wilt of potato may damage the crop in two different tubers both in fields and storage. As high as 90 per cent wilt incidence and 50 per cent tuber rot have been reported from India. Crop losses due to late blight caused by *Phytophthora infestans* in potato crop are more in hilly regions as compared to plains. At present, losses are higher in unsprayed crop which may go as high as 90-95 per cent. Potato crop harbours more than 30 viruses and allied pathogens. The loss in yield in potato was estimated to be about 40 per cent depending upon the virus, variety, season and quality of stocks.

Plant parasitic nematodes which take a heavy tall of vegetable crops every year throughout the World. Bulbs, rhizomes, roots, tubers or other subterranean parts of the plants are invaded by nematodes making the produce unsuitable for marketing. The infestation level of root-knot nematode alone has been reported up to 82.5 per cent on tomato and okra. The estimated losses worth of 91.46 and 27 per cent in okra, tomato and brinjal, respectively have been reported due to root-knot nematode.

At present, about 140 species of nematodes belonging to 45 genera are reported to be associated with the potato crop throughout the World causing about 12 per cent potential yield loss. About 90 species of nematode belonging to 38 genera have been reported to be associated with potato crop alone in India. Among these, the root-knot and potato cyst nematodes have been recognized as the major pests.

Considering the current price structure of some of the important vegetables, specially in urban areas where most of the vegetables are consumed. The share of the farmer is not proportionate to profit. However, the vegetables to be grown by improved technology can be highly remunerative to the farmers, specially in those areas where the communication and transportation are not serious problems and this fact is now being increasingly realized by scientists, administrators as well as cultivators. Some specific research programmes have been initiated to improve the genetic base of vegetables so that high yielding varieties are made available for commercial cultivation. The cultivators on their part are acquiring the knowhow to grow better varieties not only during the main season but also cultivate early or late season varieties which can fetch higher prices. Thus it can be claimed that the future of vegetable cultivation is bright and with the successes already achieved in other crops there is no doubt that attempts to improve the vegetable production by developing the new technology will meet with similar success.

It is an established fact that any attempt at improvement of agricultural production by using new cultivars with the recommended package of agronomic practices has invariably resulted in the increase in activity and damage by insect fests, diseases, mites and nematodes. The new high yielding hybrids and varieties generally proved to be susceptible to already known pests but in addition some minor pests became the major perts. Some of the outstanding examples are those of brown plant-hopper on rice, shootfly and midge on sorghum and white grubs in many rainy season (*kharif*) crops. Emergence of *Helicoverpa armigera* as a foremost pest of national importance can be included in this category. It is therefore considered necessary to focus the attention of scientists, extension workers and enlightened cultivators to the pest problems are causing economic losses even under the present conditions. Similarly the pests that are of minor importance today tomorrow are likely to become serious, once the efforts are intensified to produce more vegetables by using improved technology. Mention of such perts has therefore become necessary in this book. The pest and disease problems on vegetables can be more serious because of the favorable conditions which are provided for their multiplication by the present methods of cultivation. The monocropping, dense cropping and overlapping of crops create highly suitable microclimate and continuous availability of preferred host plant for pests infestation.

The strategy for the control of insect pest, mites, disease and nematodes in vegetable crop is necessarily to be different from the other crops because of the nature of utilization of vegetables. This calls for extra attention specially in the case of chemical control pesticides which are highly toxic and are known to leave hazardous residues cannot be recommended off hand. Proper care and precautions are to be taken while recommending the use of pesticides. Thus preference has to be given to other known effective method of control by which population build-up of insects, mites, diseases and nematodes can be avoided. These observations led to seriousness on integrated pest management including use of resistant varieties, manipulation of agronomic practices and biological control measures to minimize the use of pesticides. For this it is of paramount importance to know the habits, habitat, distribution, nature of damage, life cycle and seasonal occurrence of various pests. It is also essential to work out the economic injury levels so that control operations are undertaken at suitable time. Varieties which are less damaged by pests may have to be identified and developed for commercial cultivation. It can be claimed that relevance of integrated control and pest management is require. More in the vegetable crops than in other crops.

2

COLE CROPS

SEA CABBAGE (colewrat) : *Brassica oleracea* Linnaeus, that is still growing wild in Northern Europe, is native of coastal regions of England as well as the Western and Southern Europe. It is of great antiquity and has given rise to several varieties that are now used as vegetables and cultivated all over the world, from Arctic to subtropical regions and at higher altitudes in the Tropics. The different varieties are collectively called cole crops. The plants though grown as herbaceous - annuals for vegetables and biennials for seed production - are agriculturally a diverse group and differ in stem, leaf and inflorescence characters. The varieties grown in India include cabbage, cauliflower, knolkhol, broccoli and Brussels sprouts, the first two being comparatively more common.

CABBAGE

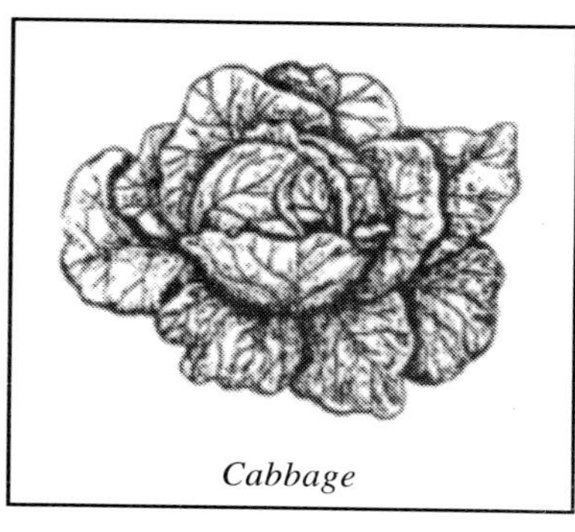
Cabbage

Brassica oleracea var. *capitata* Linnaeus has been grown for at least 4000 to 5000 years (*roman, 1980*). In India, it is cultivated on the hills and during the cold season in the plains. The flowers are bisexual and the edible portion known as head, is made up of numerous thick, smooth leaves overlapping each other and covering the terminal bud. The head may be round, conical or flat. Round-headed varieties mature earlier followed by conical varieties whereas flat-headed varieties are usually late (*Nath, 1976*). It is a rich source of vitamin A (2000 I.U.), B1 (50 IU) and C (124 mg/100g) and also contains various minerals like calcium, phosphorus, potassium, sodium and iron.

CAULIFLOWER

Brassica oleracea var. *botrytis* Linnaeus is grown mainly in Punjab, Uttar Pradesh, Bengal and Karnataka (Singh *et al.,* 1983). Introduced in India during the Moghul period, it requires cold and moist climate and is less hardy than cabbage. Its IARI-hybrid H-44 is highest yielder that gives more than 500 quintals per hectare. The edible portion is the white curd-like mass composed of a close aggregation of flowers, developed on thick branches of the inflorescences. It is eaten as vegetable and is also pickled. The flowers are bisexual. It is rich in minerals, namely, iron, magnesium, phosphorus, potassium and sodium and also a good source of vitamin A (38 IU) and B_1 (110 IU). 'Pusa Kathi' is an early variety whereas Snowball types are late maturing high-yielding varieties.

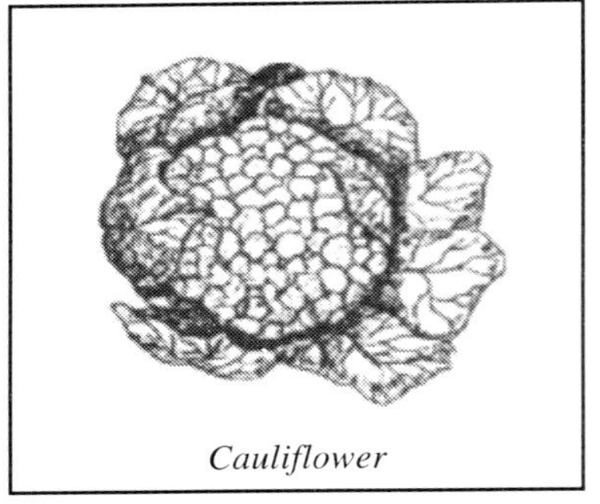

Cauliflower

KNOLKHOL

Brassica oleracea var. *gongylodes* Linnaeus, is grown in Maharashtra, South Gujarat, Uttar Pradesh and Punjab (Singh *et al.,* 1983). The plant has a short stem which is swollen just above the ground level into a spherical edible portion and has large leaf scars on its surface. It is eaten as salad or cooked as vegetable only when tender, later it becomes tough and stringy. It is rich in minerals like, calcium, magnesium, potassium, phosphorus, sodium and sulphur; it also contains vitamin A (36 IU), vitamin C (85 mg/100g) and iron. Purple Vienna and Kyote No. 3 are the commonly grown varieties

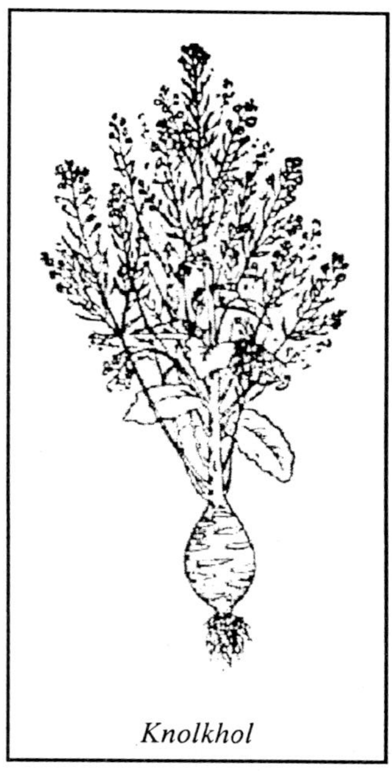
Knolkhol

whereas 'White Vienna' is an introduced early variety with dwarf growth habit.

BROCCOLI

Brassica oleracea var. *italica* Plenck is akin to cauliflower, but has large leaves and small green flowering heads. It is reported to be unsuitable for cultivation in the Tropics and is grown in India only as pot-herb.

BRUSSELS SPROUTS

Brassica oleracea var. *gemmifera* Zenk has been named after Brussels, the capital of Belgium, as it is reported to have been grown first in that area as early as 13th century. The stem is 60 cm to one metre long and bears a large number of delicately flavored small sprouts or heads crowded along its length in the upper portion. Grown chiefly in Maharashtra and South Gujarat, its young shoots, buds and leaves are edible. These are rich in vitamin A (210 IU) and vitamin C (72 mg/100g) besides calcium, phosphorus and iron.

INSECT PESTS

The major pests of cole crops are diamondback moth, cabbage butterfly, mustard sawfly, aphids, painted bugs, cabbage leaf webber and cabbage borer. Besides these, many other insects that appear regularly but do not cause any severe damage include brown cricket, cutworms, sap sucking bugs, Bihar hairy caterpillar, flea beetles, thrips and termites.

Diamondback Moth : *Plutella xylostella* Linnaeus (Plutellidae) is a major pest of almost all the cruciferous crops though its preferential hosts are cabbage and cauliflower. It is cosmopolitan in distribution (CIE map No. A-32). Eggs are laid singly, glued to ventral surface

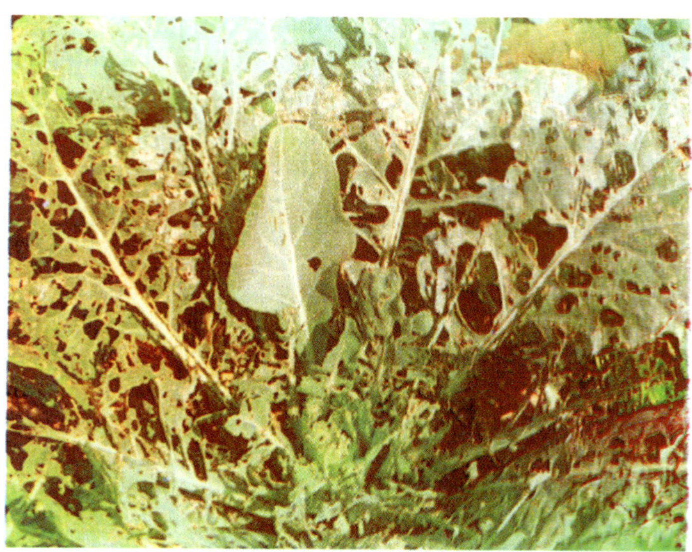

Cauliflower foliage damage by diamondback moth caterpillars

Cauliflower foliage damage by diamondback moth caterpillars

Caterpillars of diamondback moth

Cabbage leaves damages by caterpillars of diamondback moth

of leaves. A female lays on an average 40 to 60 eggs. On hatching, young caterpillars feed by scrapping epidermal leaf tissues thus producing typical whitish patches. Advance stage larvae bite holes in the leaves. The infestation is more severe during dry season when it causes retardation of growth resulting in undersized cabbage-heads and cauliflower curds. In plains the pest is active during winter while on the hills attack is more severe during April to August and less during November to March (Nayar *et al.,* 1976).

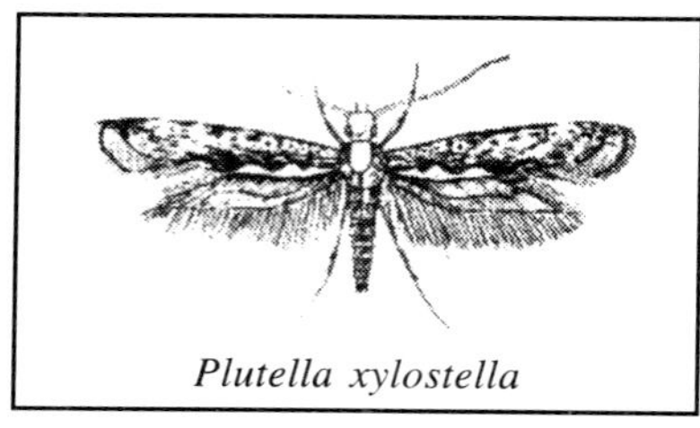
Plutella xylostella

Eggs are yellowish-white with greenish tinge. Caterpillars, when full-grown, measure 8 to 12 mm in length and are pale yellowish-green in colour with fine erect black hair scattered all over the body. Pupation takes place in beautiful transparent cocoons, fastened to ventral side of leaves. Each cocoon is about 20 mm long, the pupal cover is so thin and loosely spun that it hardly conceals the pupa within. Adults are 8 to 10 mm long, grayish-brown in colour having pale whitish narrow wings with inner margins yellow. There are 3 pale whitish triangular markings on hind margins of each forewing and when at rest, a dorsal median patch of three diamond-shaped yellowish-white spots is clearly visible by joining both the forewings-hence the common name 'diamond back'. Hind wings have a fringe of long fine hair. Wing expanse is about 15 mm. Incubation period is 3 to 8 days, larval 10 to 30, pupal 7 to 14 and total life-cycle is completed in about 4 to 5 weeks. Longevity of females is 16 to 18 days. There are 8 to 10 overlapping generations in a year. After the harvest of the crop, the moths hide under the remnants of plants during the rest of the Winter.

To prevent the infestation intercropping of cabbage with tomato (Srinivasan and Veeresh, 1986) or carrot (Varela and Guharay, 1988) has been suggested. Intercropping or mixed cropping makes microclimatic conditions less favourable for the pest. To prevent the carryover of the pest according to Gera and Bhatnagar (1992) three fortnightly sprayings with 0.02% cypermathrin or 0.05% monocrotophos or 0.04% phosphamidon are most effective. Mane and Chiandele (1995) recommended spraying 0.12% diafenthium or 0.05% quinalphos. Remove and destroy all the debris remnants and stubbles, after the harvest of the crop and plough the fields. Spraying 0.05% malathion or carbophenthion is effective in checking the pest population. Synthetic pyrethroids like cypermethrin, deltamethrin, fenvalerate, permethrin etc. have also been found very efficacious. Spraying the pathogenic bacteria. *Bacillus thuringiensis* Berliner has been reported to give complete check of this pest within two days (Oka, 1957). Recently, Ranganathan and Govindan (1996) found infenuron @ 400 or 600 ml per hectare to be highly effective in reducing the larval population and increasing the yield.

1. The pest has developed higher levels of resistance to a wide range of commonly used insecticide. Sannaveerappanavar (1995) recorded 100 fold resistance to acephate in Bangalore. The low degree of resistance to malathion was reported by Kinura (1989), Mehrotra and Phokela (1995) and Sannaveerappanavar (1995) ovserved resistance to cartap in Varanasi and Bangalore population of DBM. Nirmal and Singh (2004) also reported resistance to endosulfan, monocrotophos, malathion, acephate, carbaryl and cartap hydrochloride in different location around Hyderabad.

   Emamectin 5 sg reported ideal for control DBM, particularly where resistance has developed and it is also comparatively safe for natural enemy (Suganya Kanna *et al.,* 2005)
2. A synthetic fusion gene of *Bacillus thuringiensis* encoding a translational fusion product of Cry 1 B and Cry 1 Ab 8-endotoxins was transferred to a tropical cabbage breeding line by Agrobacterium-mediated transformation. Selection of transformants was carried out on media containing kanamycin. Polymerase chain reaction (PCR) analysis revealed that twelve of the putative transformants contained the transgene. Insect bioassays carried out with the leaves of PCR-positive plants and neonate larvae of Diamondback moth (DBM) showed that one of the transgenic plants was completely resistant to repeated infestation by the larvae. Southern hybridization confirmed

gene integration in the DBM-resistant plant. Double-antibody sandwich Enzyme-Linked Immunosorbant Assay (ELISA) analysis revealed accumulation of fusion protein up to 0.16% of total soluble protein in the leaves of the transgenic plants. Progeny (T1 generation) of the selfed transgenic plants were analyzed for the transgene segregation and insect protection. These studies clearly demonstrated the efficacy of Cry 1 B-Cry 1 Ab fusion protein to confer protection to cabbage against DBM infestation. The transgenic cabbage plants will serve as a good systm to study the role of gene pyramiding in resistance management strategies intended to prevent evolution of resistance in DBM. (Paul *et al.,* 2005).

In nature, the caterpillars are parasitised by *Diadegma fenestralis, D. varuna, Thyraeclla collaris*, *Apanteles reficrus*, *A. plutellae*, *A. sicarius, Chelonus* species, *Brachymeria* species, *Tetrastichus sokolowskii* Kardj and Voria *ruralis* Fall., whereas the pupae are parasitised by *Brachymeria excrinata* Gahan, *B. pultellophaga* and *Horogenus* species (Nayar *et al*., 1976). But none of these natural enemies have been useful in checking the pest population effectively.

Leaf Webber : *Crocidolomia binotalis* Zeller (Pyraustidae) - commonly called cabbage cluster-caterpillars-is a serious pest of almost all almost all *Brassica* crops especially cabbage. It is widely distributed in the Indian sub-continent, South-east Asia. Australia and Africa. Eggs are laid in batches of 40 to 100 each on the ventral surface of leaves. On hatching, the young caterpillars feed gregariously on leaves; later, they web together the leaves and feed within-hence the common name. These webbed leaves are further spoiled by the accumulation of excreta of the caterpillars. Pupation takes place in silken cocoons within the webbed leaves but occasionally in the soil underneath the infested plants.

Full-grown caterpillars are 18 to 20 mm long and pale brown in colour with a paler dorsal stripe and two lateral and one sub-lateral series of black spots and specks. Pupae are red in colour and adults are pale ochreous moths with forewings suffused in parts with ferruginous and fuscous having distinct and indistinct wavy lines and prominent white spots. Hind wings are semi-hyaline

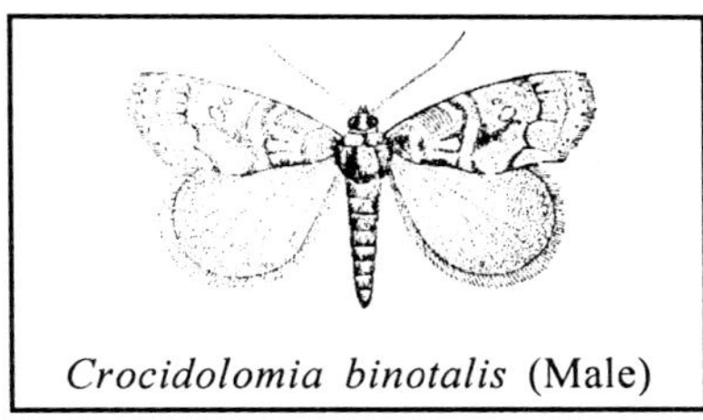
Crocidolomia binotalis (Male)

ochreous white, apical area suffused with fuscous. Wing spread is 24 to 28 mm. Eggs hatch in 5 to 15 days; larval development takes 24 to 27 days during summer and extends upto 51 days during winter whereas pupal period ranges between 14 and 40 days depending upon the climatic conditions.

To control the leaf webber, remove and destroy promptly the webbed leaves with caterpillars within. In case of severe infestation, spray with 0.2% carbaryl or 0.15% malathion. Two sprayings at fortnightly interval will be sufficient to check the pest effectively. In nature, the caterpillars are parasitised by *Microbracon mellus* Ram and *Apanteles crocidolomiae* Ahmed.

Cabbage Butterflies : *Pieris brassicae* (Linnaeus), *P. canidia (S. parrman)*, *P. napi* (Linnaeus) and *P. rapae* (Linnaeus) are all oligophagous pests having a limited host range of cruciferous crops. The different species look alike and have same habits and habitat. These have been reported from North America, Europe, Turkestan to Myanmar along the Himalayan range extending to China and Malaysia (Talbot, 1939). *P. brassicae* is comparatively more common and destructive. It causes severe damage to cabbage, cauliflower, radish, turnip as also to mustard and rapeseed. In India, it is widely distributed along the entire Himalayan region. The pest passes winter in the plains and migrates to hilly regions during Summer and from September to April, it breeds on mustard and rapeseed.

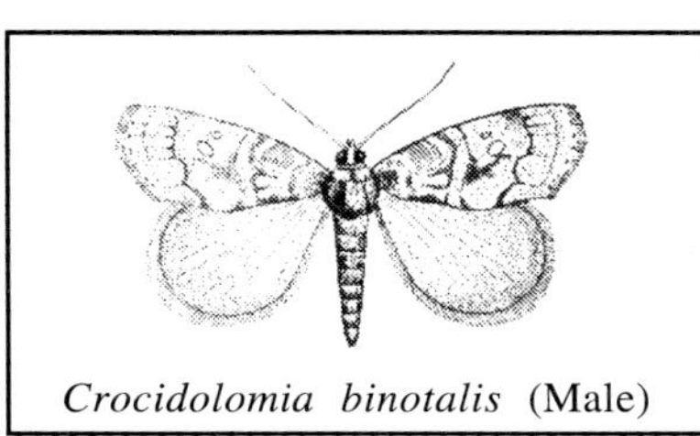
Crocidolomia binotalis (Male)

Eggs are laid in clusters on ventral surface of leaves. Each female lays 2 to 3 egg-masses of 50 to 80 eggs each. On hatching, young caterpillars feed gregariously on leaves for a couple of days, then disperse, spreading infestation to the adjacent plants and fields. As a result of their feeding, the leaves are skeletonised, sometimes the caterpillars bore into the heads of cabbage and cauliflower. Pupation takes place on leaves and stems of the host plants.

Eggs are flask-shaped, about one mm long and yellowish in colour. Full-grown caterpillars are 38 to 44 mm long. Velvety bluish-green in colour with black dots and yellow dorsal and lateral stripes covered with white filaments. Pupae are yellowish-green with black spots and dots. Adult butterflies have snow-white forewings with black distal margins more developed in females than in males; hind Wings are also pure white with black apical spots. Wing spread is 60 to 70 mm. Moths emerging in Summer are larger in size than those of winter. Incubation, caterpillar and pupal periods are on an average 3.2, 5.6 and 7.3 days respectively during May extending upto 17.6, 40.7 and 28.8 days respectively in January (Nair, 1975). There are two generations during Winter (plains) and 4 to 5 in Summer (hilly region).

Population build-up of this pest can be checked by hand-picking and mechanical destruction of caterpillars during early stage of attack when these are feeding gregariously. In case of widespread infestation spray, 0.05% dichlorvos or 0.1% malathion (Butani *et al.*, 1977). Vankova (1962) suggested spraying *Bacillus thuringiensis* Berliner. In nature, caterpillars are parasitised by *Apanteles glomeratus* Linnaeus.

Mustard sawfly : *Athalia lugens proxima* (Klug) (Tenthredinidae) is a cold weather pest found all over Indian sub-continent. It is a major pest of almost all cruciferous crops. The maximum activity is during September to December after which the activity declines and is practically nil during March to July. The pest appears on radish by the end of July.

Eggs are laid singly during day time inserted into leaf tissues near the periphery of leaves. A female lays on an average 35 eggs (20 to 150). On hatching, the grubs nibble margins of tender leaves and later, bite holes in the leaves. Grubs are diurnal and feed generally during dawn and dusk. With slight disturbance they fall on the ground and feign death. For pupation, in case of young crop, thin white silken cocoons are spun between two leaves of the plant but later on, before aestivation, the full-grown grubs enter into the soil, construct elongated oval cocoons within which they remain quiescent for 3 to 4 days before pupating. These cocoons are waterproof structures made of silken threads secreted by grubs and thickly inter-woven with soil particles. Since a number of grubs live and feed on one plant and enter the soil to pupate, there is always a cluster of cocoons, adhered to each other, found lying in the soil.

Newly hatched grubs are 2 to 3 mm long, smooth, cylindrical and greenish-gray in colour; full-grown ones are 16 to 20 mm long and greenish-black in colour; look and behave like caterpillars but have 8 pairs of prolegs. Adults are 8 to 12 mm long, having dark head and thorax, orange coloured abdomen and transluscent smoky wings with black veins. Females have a strong saw-like ovipositor-hence the name sawfly. They generally do not fly long distances but hop from leaf to leaf or fly from one plant to another plant. Their activity is pronounced during day while the insects remain practically motionless at night. Egg period is 6 to 8 days, grub development takes 21 to 31 days, prepupal and pupal periods last for 3 to 4 and 7 to 10 days respectively (Tripathi, 1963). Severe Winter is passed in pupal stage which may last 14 weeks. Average longevity is 10 days (maximum 20 days). According to Lefroy and Ghosh (1908) eggs laid in October. November and December become adults after 26 to 32, 48 to 51 and 68 to 85 days respectively. In South India where there is no severe Winter, as against only three in Northern India, the pest undergoes as many as 10 overlapping generations in a year (Nair, 1975).

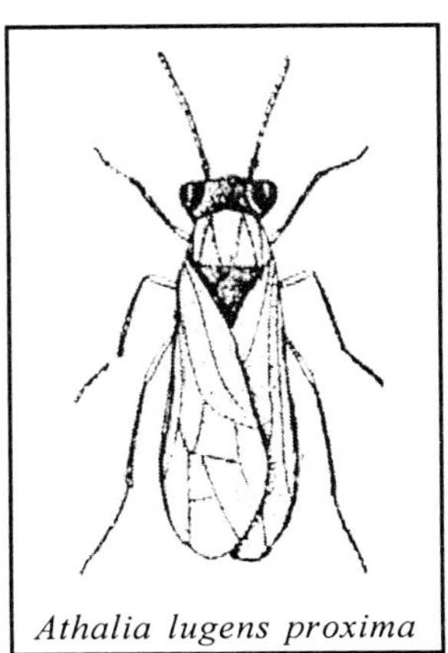

Athalia lugens proxima

Hand-picking of grubs is suggested if the area under crop is limited. In large areas chemical control is the only feasible method. Sohi (1958) suggested spraying 0.2% HCH: Jagtap and Kadam (1981) observed that spraying 0.02% parathion. 0.05% malathion, 0.02% diazinon and parathion being highly toxic their use on vegetable crops must be avoided. Butani *et al.* (1977) also suggested spraying 0.05% dichlorvos or 0.1% malathion. Dusting 5% malathion or carbaryl is also effective. Narayanan (1953b) reported that in nature grubs were parasitised by *Exacrodus populans* which helped in keeping down the pest population.

Aphids (plant-lice) : These are polyphagous pests and feed on a very wide range of host plants including all cole crops. Colonies of these insects are ofter found on tender shoots and as a result of their sucking vital sap from the tissues, the plants remain stunted in growth and head formation is poor. In case of severe infestation plants may completely dry up and die away. When seedlings are infested, they lose their vigour, get distorted and become unfit for transplanting. The aphids also produce copious quantity of honeydew, which makes the plants sticky and favours the growth of sooty mould. As a result, a black coating is formed on affected plant parts hindering the photosynthesis and adversely affected the plant's growth. The main period of activity of aphids generally coincides with the growing period of cruciferous crops, extending from October to March. Thus the aphids have often proved to be a limiting factor in obtaining high yield potential and quality of vegetables.

The aphids are soft bodied pear-shaped insects. The outstanding feature which makes these insects the most formidable pests is their tremendous rate of reproduction which is mainly parthenogenetic and viviparous. The high rate of multiplication often causes overcrowding, which means less food per individual. This, coupled with high temperature and low humidity, results in appearance of winged forms which migrate from plant to plant and field to field, thus spreading the infestation. It is a common site to see the winged aphids flying about during the sultry evenings. Low wind velocity and high relative humidity favour the migration of alate aphids (Kareem and Basheer, 1965).

The species most destructive to cole crops in South-east Asia include *Brevicoryne brassicae* (Linnaeus) and *Lipaphis erysimi*

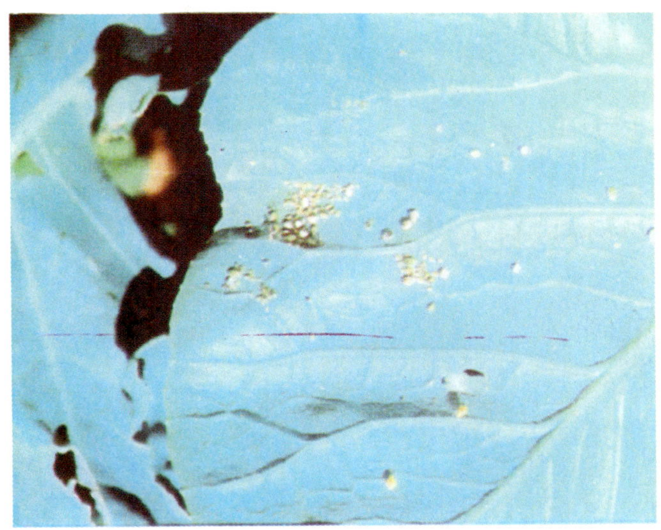

Brevicoryne brassicae on cauliflower leaf

Brevicoryne brassicae on cauliflower leaf

Brevicoryne brassicae on cabbage head

Myzus persicae (pink) and *brevicoryne brassicae*

(Kaltenbach). Besides, *Aphis gossypii* Glover, *Myzus persicae* (Sulzer) and *Siphocoryne indobrassicae* Das have also been reported damaging the cole crops.

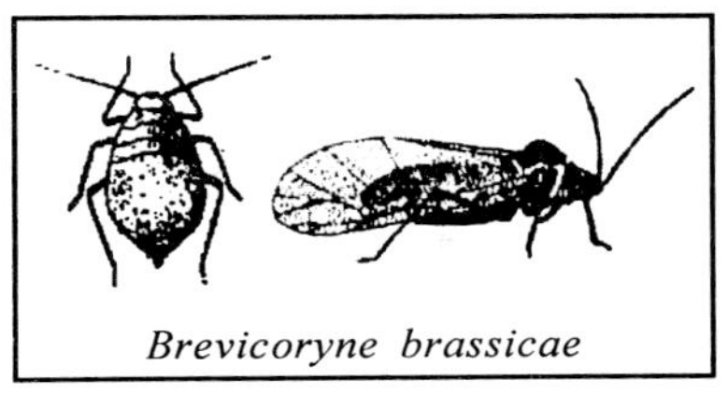
Brevicoryne brassicae

Cabbage aphid : *Brevicoryne brassicae* was originally confined to Palaearctic or Holarctic regions but at present it has a very wide range of distribution (CIE map No. A-37). Good rains followed by dry season are the most favourable conditions for rapid multiplication of this aphid. Reproduction is mostly viviparous parthenogenetic during Summer and mild Winter. However, during severe Winter sexual reproduction may also occur (Batra, 1960). Eggs are pale-yellow with greenish tinge. Nymphs are 1.0 to 1.5 mm long and yellowish-green in colour, while adults are 1.8 to 2.0 mm long and darker in colour than the nymphs. Eggs are laid during November-December and hatch in 20 to 22 days. The nymphs mature in 10 to 15 days and immediately start producing young ones without mating. A single female may produce 40 to 45 young ones during her life time. The life-cycle is completed in 11 to 45 days and as many as 21 generations have been recorded during a year under favourable conditions. The longevity of winged forms is 8 to 12 days during which each winged female produces 8 to 14 young ones.

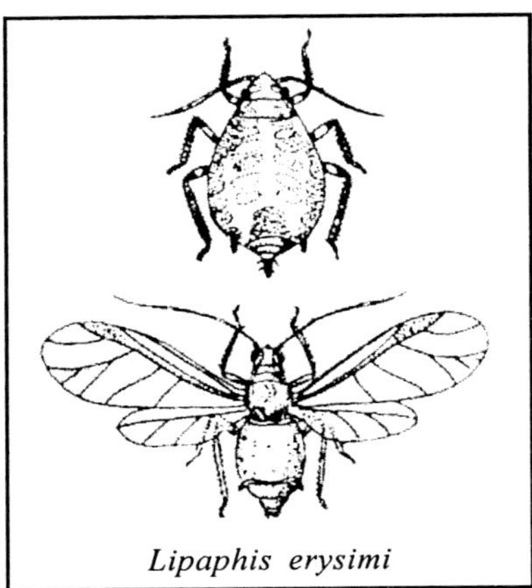
Lipaphis erysimi

Mustard aphid : *Lipaphis erysimi* is distributed throughout the country. These aphids appear in November and remain active till April. Cloudy and moist weather favours the rapid multiplication of this pest (Narayanan, 1954). Application of fertilizers containing nitrogen, phosphorus or potash do not have any effect on reproduction or development period of this aphid (Kundu and Pant, 1967). Both nymphs and adults are similar in appearance to those of *Brevicoryne brassicae* except that these are lighter in colour and slightly longer in size, the nymphs being 1.5 to 2.0 mm long and adults 2.0 to 3.0 mm.

Males of this species are rare and have been reported by Verma and Mathur (1966) from Jammu while Menon *et al.* (1968) recorded oviparous females at Delhi. A single female produces 5 to 10 young ones per day during February to May and 2 to 5 per day during July to January, but practically none during June when it is extremely hot. The life-cycle occupies 11 to 18 days during April to September and 23 to 60 days during October to March. Thus there can be 11 to 20 generations in a year under favourable conditions. Pre-reproduction, reproduction and post-reproduction periods during October to March last for 8 to 18 13 to 30 and 5 to 18 days respectively, whereas during April to September these stages occupy only 6 to 8, 4 to 9 and 4 to 5 days respectively. A female produces on an average 67, 13 and 26 young ones during October to December, January to March and April to September respectively (Sidhu and Singh, 1964).

Regular surveillance of the crop is of utmost importance to check the intensity and dispersal of aphids. As soon as aphid infestation appears, cut and destroy the infested shoots mechanically. This will prevent the population build-up of the pest. However, as soon as 5% Plants are infested, spray with 0.025% phosphamidon or methyl demeton or 0.05% endosulfan. Repeat the spraying after a fortnight if still 5% infestation is there. The spraying should be done only during afternoons or evenings when the activity of pollinators is low and do not spray if the population of parasites and predators is high. The predators can often check the aphids, if the infestation is not widely spread.

Aphid, *Myzus persicae* (Sulzer) is a key pest of cabbage, *Brassica oleracea* var. *capitata* (Pride of India) causing substantial damage. Bijaya *et al.*, 2005 evaluated efficacy of five plant extracts *viz; Artemisia vulgaris*, *Acorus calamus*, *Ageratum conzoides*, Neemall and *Achook* against the aphid. The overall results indicated that neem product *Achook* irrespective of concentration was found to be the most effective. The mortality rate of all the bio insecticides

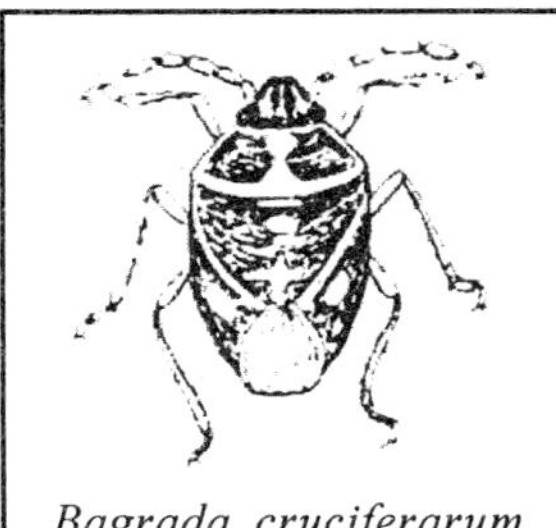
Bagrada cruciferarum

tested was found to increase with an increase in duration of exposure as well as increase the concentration.

In nature, *Coccinella septempunctata* (Linnaeus), *Chilomenus sexmaculata* (Fabricius), *Adonia variegate*, *Syrphus balteatus*, *S. catractus*, *Chrysoperla scelestes* (Banks) and Agromyzid flies have been found to prey upon *Brevicoryne brassicae* (Puttarudriah and Channa Basavanna, 1958; Wesley, 1957). *Lipaphis erysimi* is parasitised by *Diaeretiella rapa* (Curtis) Kundu *et al.*, 1965) and *Lipolexis gracitis* Forester (Sharma and Subba Rao, 1964). The predators include. *Ischiodan scuteliralis* (Fabricius) (Cherian, 1934). *Syrphus issaci* B. and S., *S. serarius* (Wiedmann) (Bhatia and Shaffi, 1932), *Erisatalis qinquelineatus* Fabricius, *Lasiopticus selenticus* Brunetti, *Xanthogramma* species (Anand *et al.,* 1967), *Chrysoperla scelestes* (Banks) (Nasir, 1947), *Coccinella septempunctata* (Linnaeus) (Atwal and Sethi, 1963) and *Menochilus sexmaculatus* (Fabricius) (Modawal, 1941). Some mortality is also caused by fungal infection, *Entomophthora coronata* (Ramaseshiah and Dharmadhikari, 1968) have been recorded to infect *Brevicoryne brassicae*, *Lipaphis erysimi* and *Myzus persicae*.

Sap Sucking Bugs : Painted bugs, *Bagrada cruciferarm* Kirkaldy and *B. hilaris* (Brumester) Pentatomidae have been recorded as major pests of various cruciferous crops and weeds while *B. cruciferarum* has been reported from East Africa, Afghanistan, Pakistan, Sri Lanka, India and South-east Asia (Hill, 1983). *B. hilaris* is found in Africa, Italy, Iran, Iraq, Pakistan, India, Sri Lanka and Russia.

The adults appear in field around October and their activity decreases with the onset of Summer but is again accelerated in Autumn. Both nymphs and adults suck cell sap from tender plant parts causing yellowing of leaves which gradually dry up and ultimately fall down exposing the plants to secondary invasion of bacteria and fungi. The plants wilt and wither affecting adversely the yield both quantitatively and qualitatively. Eggs are laid singly or in batches of 2 to 12 on leaves, stems and flower buds. These are oval in shape, about one mm long, pale-yellow when freshly laid gradually becoming pinkish-orange. Nymphs are beautifully patterned with a mixture of black, white and orange colour and are 1.5 to 4.5 mm long depending on their age. Adults are also black and orange coloured bugs, similar in colour pattern as nymphs. Males are 6 to 7 mm long and females 7 to 8 mm. The mating takes place 2 to 6 days after the final nymphal moult and the oviposition commences a week after first mating and may continue intermittently throughout the life span of the female. A female lays on an average 230 eggs @ 15 to 20 eggs per day (Batra and Sarup, 1962). Egg and nymphal duration is 2 to 5 and 18 to 20 days respectively and a single life-cycle is completed in 20 to 30 days. Adults live for 16 to 18 days with 6 to 8 generations in a year.

As these bugs can breed on a number of weeds, clean cultivation is imperative for avoiding infestation of these bugs. In case of heavy infestation spray with 0.05% dichlorvos or 0.1% malathion or lindane (Butani *et al.*, 1977). Care should be taken not to spray the crop at the time when the crop is ready to be harvested for marketing. At least 7 to 10 days waiting period should be there between treatment and harvest.

Other sucking bugs causing minor damage include, *Creontiades pallicifer* Walker (Miridae), *Spilostethus pandurus* (Scopoli) (Lygacidae) and *Tricentrus bicolor* Distant (Membracidae). These are all polyphagous and cause severe damage to a number of crops. On vegetables, *C. pallicifer* is recorded as a major pest of potato, while *T. bicolor* is more common on tomato and *S. pandurus* on cabbage and cauliflower. These bugs suck the sap from the infested tissues and devitalize

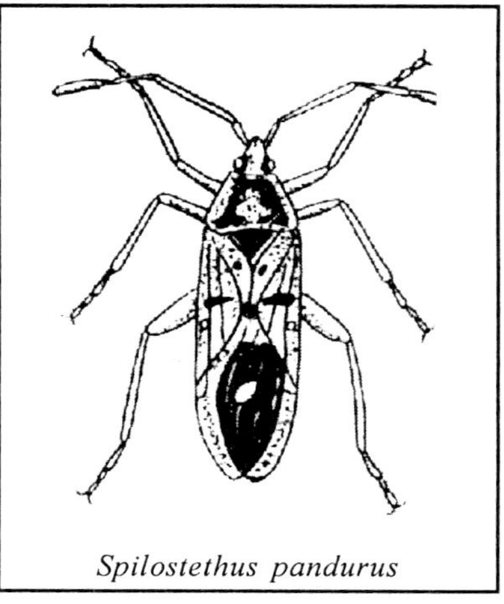
Spilostethus pandurus

the plants. A severe infestation, which is rare in case of cole crops, can affect adversely the yield and quality of the produce. In case of severe infestation control measures recommended for *Bagrada cruciferarum* may be adopted against these bugs as well.

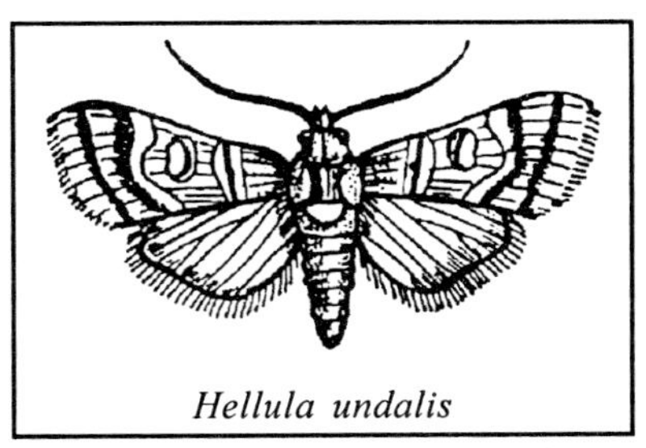

Hellula undalis

Cabbage borer : *Hellula undalis* Fabricius (Pyraustidae). Cabbage borer is cosmopolitan in distribution. It attacks almost all the cole crops and some root crops as well. Eggs are laid on ventral side of the leaves. Young caterpillars mine the leaves along the side veins and the leaves become white papery skeletons filled with excreta; third instar onwards, the caterpillars feed on chlorophyll of leaves and leaf petioles protected by a silken covering and finally these caterpillars bore into the stems or cabbage and cauliflower heads. The infested plants lose vigour and bear deformed heads.

Eggs are oval in shape and pinkish in colour. Full-grown caterpillars are 15 to 18 mm long, pale whitish-brown in colour with 4 to 5 purplish-brown longitudinal stripes. Moths are pale grayish-brown, suffused with fuscous; forewings have gray wavy lines, a pale apical spot and pale-edged dark lunule; hind wings are pale dusky with slight fuscous suffusion on apical area. Wing expanse is 18 to 22 mm. Incubation, caterpillar, pupal and total life-cycle durations occupy 2 to 4, 7 to 12, 6 to 10 and 15 to 25 days respectively.

Mechanical destruction of caterpillars in the early stage of attack helps to check the infestation. In case of severe infestation, spray 0.05% quinalphos or endosulfan or 0.1% malathion. Dusting 5% carabary1 or malathion is also effective in checking the pest population In nature, *Habrobracon hebetor* (Say) has been found as larval parasite.

Soil Insects : Soil, or subterranean, insects are those that live underground during one or both of their active stages and their immature stages, feed mostly on the roots. Some soil inhibiting insects may come out at night to feed on seedlings as in case of cutworms. All these insects are polyphagous and their incidence is more generallly in the sandy or sandy-loam soils and drier regions. These are conspicuously absent in areas with clayey soils or in low-lying and water-logged areas.

Termites : *Microtermes anandi* Homlgren (Termitidae) is a polyphagous pest, sugarcane being its main host. These social insects live in colonies known as termitaria. Each termitorium consists of a queen, king, soldiers and workers. The damage is done by the worker-caste. The winged termites, or sexual forms, come out of the termitaria with the onset of rains; they shed their wings, pair, mate and reenter the soil to start new colonies. Seven to ten days after swarming, the mated female, now known as queen, lays first batch of eggs numbering 100 to 130. the oviposition continues for several years and the number of eggs laid per day also goes on increasing. One queen can lay upto 30,000 eggs per day (Nair, 1975). This shows the phenomenal rapidity of multiplication of this pest. The damage to cole crops is usually not serious, but occasionally cabbage crop suffers some loss especially when there is drought. The outer leaves of the infested plants begin to dry. Such plants when pulled out show their roots riddled with holes caused by termites that are found feeding inside the roots.

Brown Ant : *Dorylus orientalis* Westwood (Formicidae) is another subterranean enemy of a large number of economic crops. Among the vegetables, its preferred host is potato and it causes only minor damage to cole crops. These insects live in colonies and build their nests underground, from where the workers tunnel through the soil to reach the roots of their host plants for feeding thereupon.

Cetonid Beetle : *Chiloloba acuta* (Wiedemann) is another minor polyphagous pest. Eggs are laid on roots of the seedlings and in case of severe infestation the affected seedlings wither and die away. Adults come out of the soil and feed on foliage, but the loss caused is negligible. Adults are 14 to 18 mm long, metallic green in colour, clothed with long and dense yellow hair. Head is declivous and finely setose on either side; pronotum is closely punctured and setose; scutellum and elytra are thinly setose; pygidium is clothed with long hair.

Cutworms : These noctuid borers cause severe damage to a number of economic crops all over the World. Among the vegetables, they prefer potato and are minor pests of cole crops. The species known for reported damaging cole crops include, *Agrotis ipsilon* (Hfnagel), *A. segetum* (Dems and Schiffer-muller) and *Xestia c-nigrum* (Linnaeus). These are commonly found in nurseries or at the time of transplanting. Eggs are laid in soil; the caterpillars are nocturnal in habit; come out of the soil at night; cut the seedlings at ground level and feed only on tender parts. Thus they destroy many more seedlings than what they actually eat.

To keep these subterranean pests at bay, avoid growing cole crops in a termite-infested field, irrigate the field copiously and regularly. If and when an infestation is observed mix thoroughly with the soil 5% HCH dust @ 20 to 22 kg per hectare, or add lindane emulsion @ 1kg a. i. per hectare in irrigation water. These treatments will check most of the soil-borne insects.

Earwigs : These are generally considered to be beneficial insects because most of these have been reported to be predators of insect pests, especially of the young caterpillars. A few of the species do cause serious damage especially in nurseries and to tender transplanted seedlings. These insets are abundant in the tropics of Old and New World and live mostly concealed in the soil, under clods, stones or even bark of the trees. Eggs are laid in clusters in the soil. A female lays 1 to 4 clusters, each of 25 to 40 eggs. Parental care of the eggs and young ones is characteristic feature of earwigs; the female guards its eggs till these hatch and young ones can move about freely. The species found feeding on cole crops, specially cabbage, include *Euborellia annulipes* (Lucas) and *E. stali* (Dohrn.) (Labiduridae).

Groundnut Earwig : *Euborellia annulipes* is a minor pest of cabbage and onion. It is a medium-sized, slender, black shining insect. Basal two segments of antennae are reddish in colour, apical two segments whitish and the remaining intermediate ones grayish-brown. Elytra are absent. Telson is smaller than pygidium; ovipositor in females is wanting while males have paired aedeagus. Branches of forceps in males are sub-contiguous at base, stout, strongly incurved, right branch crossing above the left at apex; in females the branches are straight, conical and sub-contiguous. Body length of males and females is 10 to 11 and 12 to 14 mm respectively, forceps being 2 to 3.5 mm long.

Euborellia stali, another groundnut earwig, is similar in colour and shape as *E. annulipes* except that its intermediate segments of the antennae are black and elytra are present as small ovate leathery flaps on either side of mesonotum. Forceps are robust with branches not contiguous at base, trigonal in basal halves straight at first, bent, tapering and cylindrical in apical half. Body measures 8 to 9 and 9 to 10 mm in length with forceps 1 to 2 mm in males and slightly longer in females.

Mating takes place 7 to 10 days after emergence. Preoviposition, oviposition and incubation duration last for 10, 13 and 7 to 11 days respectively while nymphal stage is 50 to 54 days . adult longevity is reported to be 106 to 252 days (Cherian and Basheer, 1940).

Large Brown Cricket : *Brachytrypes protentosus* (Lichtenstein) (Gryllidae) is a major pest of jute and a minor pest of cabbage and cauliflower, occasionally causing severe damage in certain areas. Eggs are laid at the end of rainy season when the females make burrows in moist soil and deposit 40 to 50 eggs in each burrow. These eggs hatch in September and the nymphs make fresh burrows in the soil. It is the appearance of fresh excavated soil near the holes that indicates the presence of this pest. Nymphs and adults feed on seedlings of cabbage and cauliflower at night and hide during day.

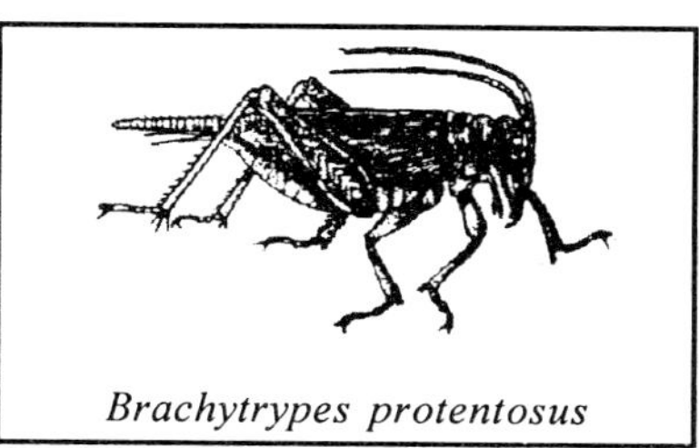

Brachytrypes protentosus

Eggs are cylindrical in shape and white in colour. Nymphs and adults are grayish-brown dorsally and pale yellow ventrally. They are more or less similar in shape and appearance except that last 2 instars of nymphs have wing-pads and adults have fully developed wings. Forewings are rather hard, their anterior halves lie horizontally on the dorsum of the insect while the posterior halves remain vertical against the sides of abdomen. Hind wings are large and membranous and lie folded in fan-like fashion below the forewings. Adults are 40 to 60 mm long; males have a pair of anal styles and females a long and pointed ovipositor. Incubation and nymphal periods last for 2 to 4 and 28 to 32 weeks respectively while adult longevity is reported to be 16 to 20 weeks (Sen, 1921).

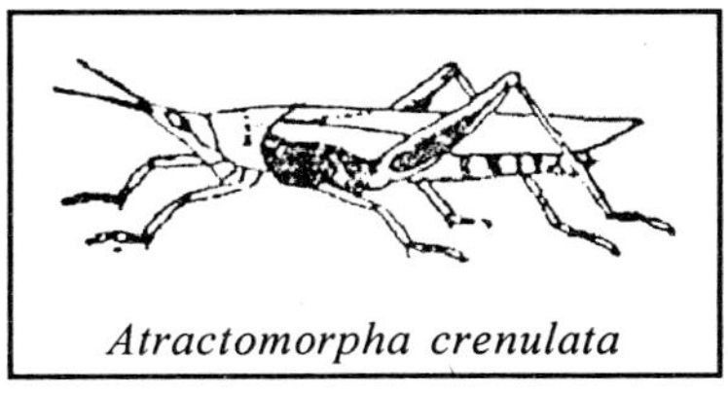

Atractomorpha crenulata

Grasshopper : *Atractomorpha crenulata* Fabricius (Acrididae) is a polyphagous pest and has countrywide distribution. It is a serious pest of crucifers and a minor one of brinjal, leafy vegetables. Under suitable conditions the pest breeds profusely and is most active from July to September (rainy season); with fall in temperature the activity decreases and is at its lowest during December to February. Eggs are laid in clusters of 20 to 30 eggs in the soil, 75 to 100 mm deep. A female lays 65 to 135 eggs. Both nymphs and adults feed on green leaves and after the plants are completely defoliated. Nymphs and adults are more or less of the same colour and shape but differ in size, and while the latter have wings, the former have only wing-pads. Adults are 28 to 38 mm long and green in colour.

The adults mate 4 to 13 days after emergence; preoviposition and oviposition periods are 1 to 5 and 11 to 17 days respectively. Incubation period is 15 days during April to August and extends upto 2 months in Winter. Nymphal development takes 31 to 39 days during July-August and 69 days in November. Adult males and females live for 25 to 36 and 30 to 68 days respectively (Agarwal, 1955).

Acrotylus inficita Walker, a surface grasshopper, is also a minor pest that feeds on cabbage and cauliflower leaves.

**Bombay locust : ** *Patanga succincta* (Linnaeus) (Acrididae) is widely distributed in Western and Southern India (Gujarat to Tamil Nadu). Sri Lanka and South-east Asia. It is a minor pest of cabbage and cauliflower. Though it is called 'locust', its hoppers usually do not congregate like other locust species, but are found scattered among the various crops and wild grasses. The outbreaks in gregarious swarming form in 1883 and 1904 were exceptional phenomena. These hoppers prefer low vegetation and unlike *Atractomarpha crenulata*, have no climbing habit.

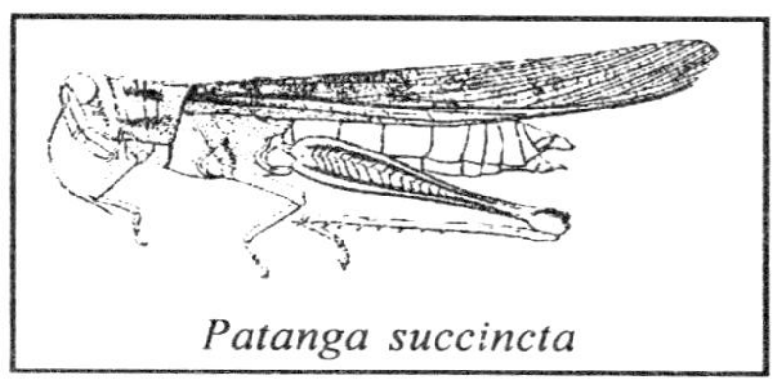

Patanga succincta

Eggs are like rice grains, cylindrical in shape, 3 to 3.5 mm long and yellowish in colour. Full-grown hoppers are 25 to 35 mm long, light green in colour with dark brown speckles and a black dorsal line on abdomen. There are two dark spots on each wing-pad. The adults are 35 to 42 mm long and keep growing another 5 to 8 mm. General colour is light green with brownish tinge; this gradually changes to dry-grass colour. Head and thorax become brownish-green speckled with dark brown, abdomen is light green with a black dorsal stripe. Wings are hyaline with dark veins. In advanced stage, a red suffusion gradually develops and the veins become less distinct and the wings appear red. During mating period the bright red colouration gives way to brownish-yellow except on wings which remain crimson. Such colour changes are not unusual in migratory locusts whereas grasshoppers do not undergo any such changes. A female lays only one egg-mass, that is more or less cylindrical in shape 25 to 40 mm long and may contain 90 to 120 eggs. The females die soon after oviposition. Incubation period is 7 to 8 weeks and hopper development in captivity takes 61 to 71 days (Lefroy, 1906). The most striking feature is the habit of living about 8 months as full-grown insect, waiting till the onset of rains before the reproductive system develops. Thus there is only one generation in a year.

In case of serious infestation, which is rather rare, dust the crop with 5% malathion or carbaryl or 4% endosulfan.

Thrips : Thrips are tiny, 0.5 to 1.0 mm long, fragile, slender insect found all over the World; mostly polyphagous, infest a large number of cultivated and wild plants. Nymphs and adults lacerate the tender tissues of leaves, stems or flowers and rasp the oozing sap. As a result of their attack and due to physical damage as well as drainage of sap, the infested leaves become spotted showing white patches and subsequently the leaf tips fade and gradually the lower portion of the leaves also becomes blighted. This adversely effects the yield.

Nymphs and adults look alike except for size and wings. Adult females have distinct ovipositor and terminal abdominal segment is conical while in males the terminal abdominal segment is bluntly rounded. Wings have microtrichia and forewings possess marginal veins. Thrips have rapid rate of multiplication. Sexual reproduction is more common. Females are bigger in size than males and more abundant in number. A life-cycle is generally completed in 10 to 35 days. Females live longer than males and there are several overlapping generations in a year.

Three species of terebranian thrips *Caliothrips indicus* (Bagnall). *Thrips tabaci* Lindemann and *Frankliniella schultzei* Trybom) (*F. dampfi* Priesner) have been found on cole crops in India. The first two feed on leaves whereas the last one infests flowers. Forewings of *F. schultzei* are broad and round at apex with two longitudinal veins and front margin is without fringe of long hair. Females have ovipositor curved upward. *C. indicus* and *T. tabaci* look more or less alike. Forewings are pointed at apex and fore margins have fringe of hair; ovipositor is curved downwards. *C. indicus* differs from *T. tabaci* in having terminal antennal segment long, thin and needle-like and dorsum of body deeply reticulate with polygonal areas.

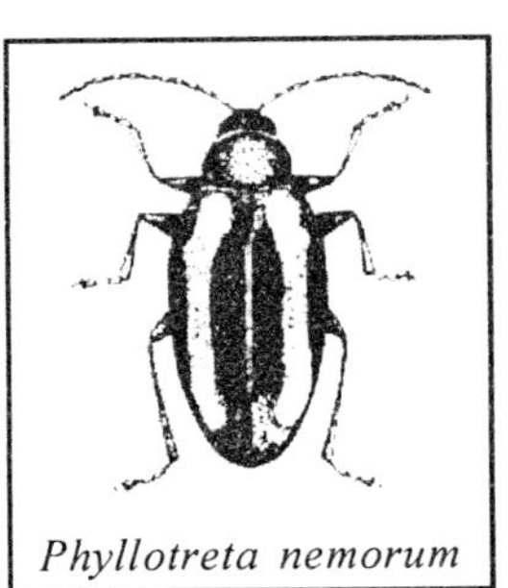

Phyllotreta nemorum

To control thrips, if and when necessary, spray 0.05% monocrotophos, endosulfan or fenitrothion.

Leaf-eating Beetles : Flea beetles are common and widely distributed polyphagous pests. In spite of their regular occurrence, the loss caused by these insects to cole crops is not of an economic importance except when heavy infestation occurs, especially in the nurseries. The females lay eggs in soil near host plants. On hatching, the grubs feed on roots but generally do not cause much damage. Pupation takes place in soil. The adults feed on foliage of cabbage, cauliflower, radish etc. and make typical small shot-holes. Adults have very stout femora which help them to jump like fleas - hence the common name flea beetles. The species commonly found on cole crops in India, include *Altica caerulescens* (Baly), *Chaetocnema basalis* Baly, *Phyllotreta cruciferae* (Goeze), *P. chotanica* Duvivier, *P. nemorum* (linnaeus) *P. vittata* Fabricius and *Psylliodes tenebrosus* Jacoby 4 nemorum. Of these, *P. cruciferae* is comparatively more common and occasionally attains status of major seedling pest.

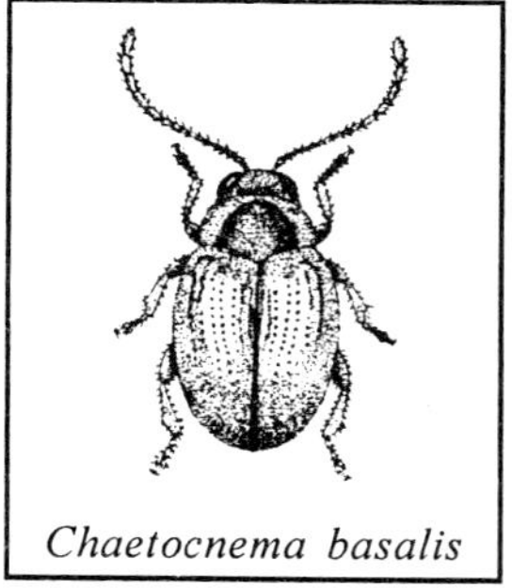
Chaetocnema basalis

Altica caerulescens adults are slightly bigger in size, 4 to 5 mm long, body convex and elongate-oval in shape, shining metallic blue above and blue-black underside. Head with vertex smooth and impunctate; antennae black. Elytra are oblong, broader than prothorax and distinctly punctuate.

Chaetocnema basalis are tiny beetles, 1.5 to 2.0 mm long, ovate in shape and shiny black in colour. Head has impunctate vertex and front is very finely granulate; antennae have basal 4 segments brown and remaining ones pitch-black. Scutellum is small, triangular with apex broadly rounded and impunctate. Elytra are convex, broader than thorax at the base, attenuated towards apex and regularly punctuate-striate, each elytron having 11 rows.

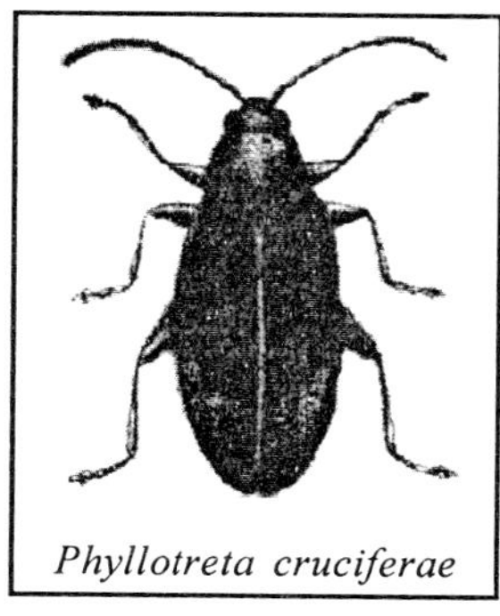
Phyllotreta cruciferae

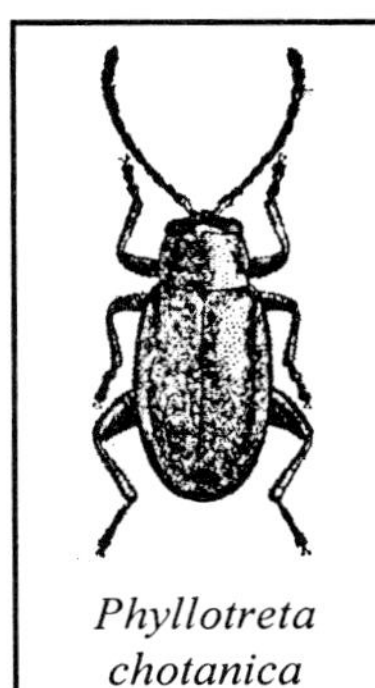
Phyllotreta chotanica

Phyllotreta cruciferae has been reported from Europe, Egypt, Middle East, Asia, Russia and North America. In India, its peak occurrence has been recorded during March-April after which the pest migrates to root-crops like radish, turnip etc. Adult beetles are 1.5 to 2.0 mm long, elongate-oval in shape and metallic bluish-green in colour. Head has impunctate vertex and black antennae. Prothorax is more broad than long and scutellum is small. Elytra are closely covered with punctures. A single female lays 50 to 80 eggs. Incubation, grub, prepupal and pupal periods last for 5 to 10, 9 to 15, 2 to 4 and 8 to 14 days respectively, with 7 to 8 generations in a year (Varma, 1961).

Phyllotreta chotanica adults are slightly bigger than *P. cruciferae* adults, being 2 to 2.5 mm long; narrow and oblong in shape and metallic bronze in colour. Head has front impunctate, bearing short whitish hair and black antennae. Scutellum is small, triangular with apex rounded and shining having impunctate surface. Porthorax and elytra are more or less similar to those of P.cruciferae.

Monolepta signata Olivier – the white spotted beetle is another chrysometid beetle recorded on cole crops. It is highly polyphagous. Adults 3 to 4 mm long, with reddish-brown body and pale brown elytra having 2 big white spots on each elytron, are very conspicuous.

Generally no control measures are adopted against these beetles. However, dusting 5% HCH at seedling stage is quite effective in warding off the pest attack. If attack is noticed at a later stage, the crop may be sprayed with 0.2% carbaryl W.P. or dusted with 10% carbaryl (Butani *et al.,* 1977). Narayanan *et al.* (1960) reported *Microctonus indicus* Narayanan as new parasite.

Leaf-eating Weevils : *Tanymecus circumdatus* (Wiedemann) (Curculionidae) is commonly met with in nurseries of cole crops. The grubs live in soil feeding on roots of various grasses and often attack the freshly sown seeds of the host plants. Adults hide under big clods of loose soil both during day and night. They come out only during dawn and dusk and may be seen nibbling the leaves of seedlings. The adults are grayish-fawn weevils with metallic coppery reflection; elytra are strongly acuminate behind, more specially in males, apices shortly mucronate in both sexes with fine distinctly punctuate striae (Marshall, 1916). Eggs hatch in 2 to 3 weeks, grub development takes 4 to 5 months and pupal period occupies 8 to 10 weeks. Only one generation has been observed in a year : pre-oviposition period may extend upto 5 months.

Myllocerus blandus Faust (Curculionidae) is another minor pest of cole crops and has feeding habits similar to those of *Tanymecus circumdatus* except that these weevils are also found on transplanted plants in the fields. Adults are 2 to 3 mm long, black weevils with dense grayish scaling. Elytra usually with 2 irregular whitish patches, one before and other behind the middle and also with some small darker spots (Marshall, 1916).

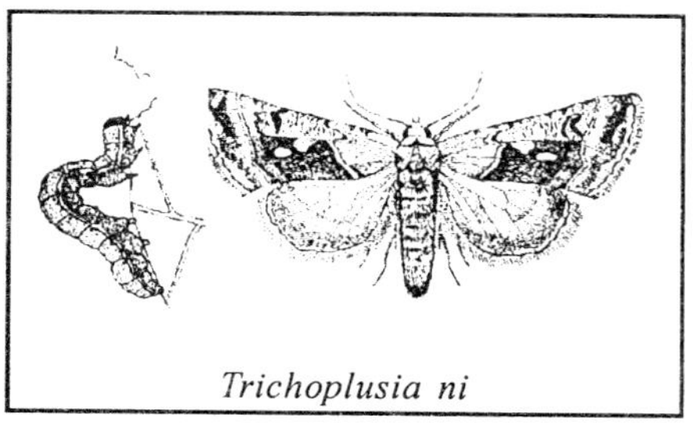

Trichoplusia ni

Since these are minor pests, no separate control measures against them are required. Dusting the seedlings with 5% carbaryl or 4% endosulfan is quite effective in checking the adult population. In heavily infested areas, soil treatment with 5% HCH, chlorpyrifos is suggested for the control of the grubs.

Leaf-eating Caterpillars : *Trichoplusia ni* (Hubner) (Noctuidae), a cabbage green semilooper, is a formidable pest of cabbage in USA. It is also widely distributed in Indian sub-continent and often inflicts damage to cabbage in South India and Sri Lanka. It is a polyphagous pest and among the vegetables, besides cole crops, it also attacks tomato and other cruciferous vegetables. Eggs are laid singly on ventral surface of leaves. On hatching, the caterpillars start scrapping and feeding on the leaves. In case of severe infestation, the entire plant may be defoliated, leaving behind midribs and main veins. The pest is active from September to April and causes comparatively more damage in the nurseries than in the fields.

Eggs are greenish-white, spherical and sculptured. Full-grown caterpillars are 35 to 40 mm long, slender, attenuated anteriorly and green in colour with light wavy white lines and a broad lateral stripe on either side. Pupation takes place in thin transparent cocoons on ventral surface of leaves. Adults are stout moths; head and thorax ferruginous-gray in colour while abdomen is ochreous-white with basal tufts ferruginous. Forewings are gray; sub-basal, ante-and post-medial lines are ferruginous and more wavy and a slender Y-mark is present. Hind wings are pale-fuscous. Wing expanse is 35 to 40 mm. Life-cycle occupies, on an average, one month.

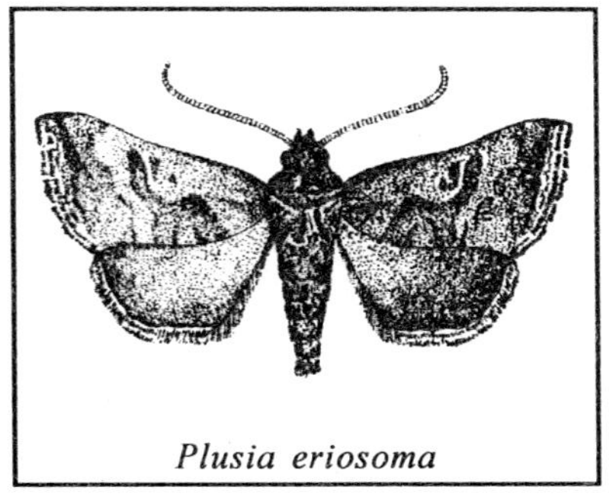

Plusia eriosoma

Other semiloopers reported damaging cole crops include, *Plusia chalcites* Esper, *P. eriosoma* Doubleday, *P. nigrisigna* Walker, *P. signata* Fabricius and *Diachrisia orichalea* (Fabricius). All these are minor pests, noctural in habit and are attracted to light. Feeding habits are same as those of *Trichoplusia ni.* The immature stages also look more or less alike but the adults can be easily distinguished from one another (Hampson, 1894).

Plusia eriosoma moths have head and thorax reddish in colour, forewings also reddish with medial line oblique and post-medial line more oblique and sinuous; the Y-mark is large, prominent and golden in colour. Wing expanse is 40 to 44 mm. *P. nigrisigna* moths also look like those of *P. eriosoma* but are slightly smaller in size, wing span being 38 to 42 mm. The Y-mark on forewing has its lower arm nearly straight and there is a small bright chestnut patch beyond the end of its tail; a sinuous golden mark on outer edge of reniform, with 3 black specks in its undulations.

P. signata moths have head and thorax pale reddish-brown and abdomen paler. Forewings are also pale reddish-brown but with cupreous thinge; sub-basal, ante-and post-medial lines are more distinct and less wavy; a prominent silver Y-mark below the cell-the tail of Y detached from the arms. Hind wings are uniformly dark fuscous. Wing expanse 30 to 34 mm.

Diachrisia orichalcea moths have head, collar and vertex of thorax reddish-orange in colour. Forewings are pale reddish-brown, sub-basal, ante-and post-medial lines are fine, very indistinct and white. Hind wings are pale at the base, outer area being fuscous. In size these are the biggest moths having wing expanse of 45 to 48 mm.

The other lepidopterous larvae found feeding on leaves of cole crops are an arctiid *Spilarctia obliqua* (Walker) and a few noctuids *Spodoptera littoralis*' (Fabricius), *S. downsei* Baly and *S. exigua* (Hubner). All these are polyphagous pests. Among the vegetables the preferred host of *Spilarctia obliqua* is potato and that of *Spodoptera* spp, is tomato. These caterpillars feed gregariously, scrapping and skeletonising the leaves; later, the caterpillars segregate and feed voraciously on leaf lamina.

To control these leaf defoliators, if and when necessary, spray 0.05% endosulfan or 0.2% carbaryl (Butani and Jotwani, 1984).

Leaf Miner : *Chromatomyia (Phytomyza) horticola* (Goureau) (Agromyzidae) is a polyphagous pest, causing minor damage to cole crops, especially cabbage. Adult flies appear in January. Eggs are thrust into leaf tissues that causes some minor injury, whereas the maggots mine the leaves, forming whitish zigzag galleries which often cause drying of infested leaves. On the whole, the damage caused is negligible, hence no control measures are warranted against this pest on cole crops.

3

ROOT CROPS

Root crops grown commercially in India include radish, carrot, turnip and beet-root. The other less known vegetables belonging to this group are parsnip *Pastinaca sativa,* rutabaga *Brassica napus* var. *napobrassica*, Salsify *Tragopogon porrifolius*, Skirret *Slium sisarum* and celeriac *Apium graveolens* var. *rapaceum*. These are cold season crops, grown mostly in temperate regions in small areas and very little information is available on the pest problems of these crops.

RADISH

Raphanus sativus Linnaeus, a native of China, is a common crop grown practically all over the World. It is one of the most ancient crops and its cultivation and use is reported as early as 2780 B.C. Radish and garlic were supplied in the rations to the labourers who built the great pyramids (Burkill, 1953). Radish is an annual or biennial herbaceous crop having short hairy leaves with large terminal lobe and several pairs of smaller side lobes. Tender leaves as well as enlarged fusiform roots are eaten raw as salad or sometimes used as cooked vegetable. It can also be boiled and turned into soft pulp to be eaten with salt and vinegar or with sauce. Radish greens are rich in vitamins A, B and C with little iron and proteins. The roots have cooling effect and also act as appetizer. Fruits are thin, 30 to 90 mm long pods with tapering beak-like terminal part, each containing 2 to 8 globose yellow or

Radish

brown seeds; these are also eaten raw or as cooked vegetable. The roots and leaves also serve as an excellent fodder and are fed to a number of animals including milch cattle. The characteristic smell and the peculiar or pungent taste of the root is not imparted to milk.

Asiatic varieties (biennial) produce edible roots in the first year and seed in the second year, whereas exotic (European) types are annual, and produce edible roots under conditions of Tropical and subtropical climate (plains) and seed under temperate climate (hills). The edible roots of Asiatic type are more pungent than European types. In India, the commonly grown European cultivars are 'White Icicle' and 'Rapid Red'. The roots of the former are white and tapering, 200 mm or more in length and slightly more bitter in taste than the red variety, whose roots are roundish, having a diameter of 20 to 30 mm. The Asian varieties include, 'Japanese White' and 'Pusa Hemani'. Both are 250 to 350 mm long, pure white, crisp and mildy pungent.

INSECT PESTS

Radish, though a short duration and quick growing crop, is severely attacked by a number of insect pests, of which those causing major loss to crop are - aphids, mustard sawfly and diamondback moth. The sporadic pests of minor importance are grasshopper, whitefly, painted bugs, thrips, leaf-feeding caterpillars, flea beetles and weevils. Most of these major and minor pests are also the pests of cole crops.

Aphids : The species that feed on radish foliage comprise, *Brevicoryne brassicae* (Linnaeus), *Lipaphis erysimi* (Kaltenbach), *Myzus persicae* (Sulzer) and *Toxoptera aurantii* (Boyer de Fonscolombe). The first one prefers cabbage and cauliflower; the second one (mustard aphid) attacks almost all crucifers. *M. persicae* (peach green aphid) and *T. aurantii* (citrus black aphid) are highly polyphagous pests having a very wide range of host plants; the former has been recorded on almost all temperate fruit trees and winter vegetables while the latter is a pest of citrus, litchi, tamarind, tea, coffee, bread-fruit and jack-fruit.

Brevicoryne brassicae and *Myzus persicae* on turnip stems

Brevicoryne brassicae and *Myzus persicae* on radish leaf

Cutworm caterpillar boring into radish

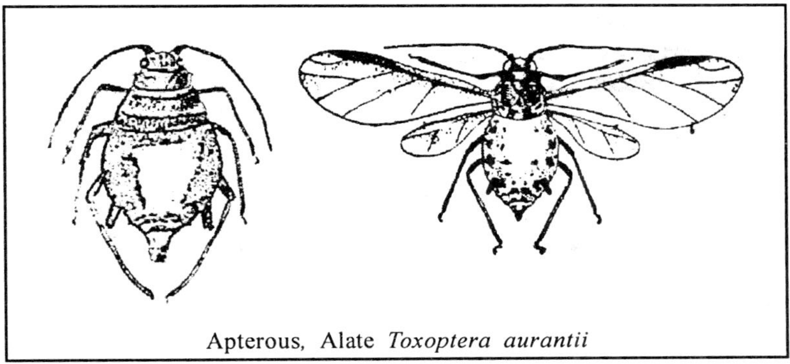

Apterous, Alate *Toxoptera aurantii*

Colonies of the aphids consisting of various stages of nymphs and adults are seen on tender stems and underside of leaves sucking the cell sap. The affected plant parts fade, curl and dry up. Besides the direct damage by feeding, these insects also secrete honeydew which favours the growth of sooty mould fungus; as a result, growth of the plants is retarded and quality as well as quantity of the edible fleshy root is adversely affected. The damage is reported to be relatively more on crops grown for seed output; it is caused by the sucking of the sap from pods which adversely affects the seed quality.

To control aphids, remove and destroy the affected plant parts along with the aphids thereon, as soon as infestation is observed. Only in case of severe infestation, spray 0.05% acephate, dimethoate or monocrotophos. Appropriate waiting period should be observed before harvesting the crop.

Mustard Sawfly : *Athalia lugens proxima* (Klug) and diamondback moth *Plutella xylostella* (Linnaeus) are the only other major pests of radish. Both these are polyphagous pests, causing severe damage to almost all the cruciferous vegetables and are cosmopolitan in distribution. These have been described under cole crops.

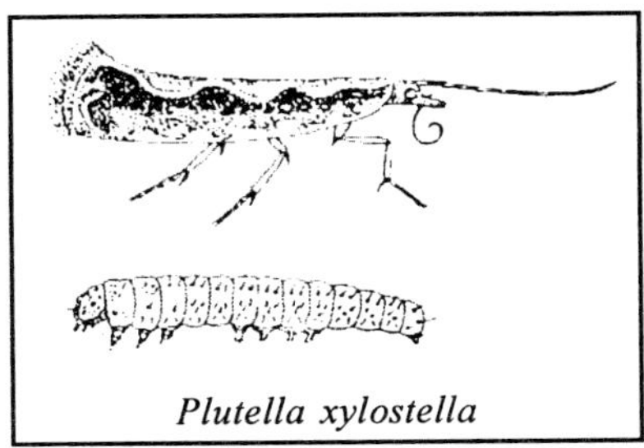

Plutella xylostella

Minor Pests : Grasshopper

Atractomorpha crenulata Fabricius is a polyphagous pest. Cosmopolitan in distribution, it sometimes causes severe damage to the seedlings. If and when necessary, dust with 4% carbaryl or 5% endosulfan.

Bemisia tabaci (Gennadius), (Aleyrodidae), the cotton whitefly is a polyphagous pest feeding on numerous food plants belonging to different families. Among the vegetables, it has been recorded on eggplant, bean, chilli, cowpea, cururbit, okra, potato, radish and tomato; the last one being the preferred host. No separate chemical control measures are generally necessary for this pest on radish. The chemical control adopted for aphids will check whitefly also.

Bagrada cruciferarum Kirkaldy and *Eurydema pulchrum* Westwood are the two pentatomid bugs recorded damaging radish crop; the former prefers cabbage and cauliflower while the latter is known as radish bug and has been recorded damaging only the second year crop, maintained for seed production. Besides radish, it is found on other cruciferous crops as well as on wheat, millets and pulses. The nymphs and adults feed on leaves and pods sucking the vital sap and devitalizing the pods and thereby adversely affecting the seed formation.

Thrips tabaci Lindemann (Thripidae) is a major pest of onion but is commonly found damaging almost all the cruciferous vegetables including radish. Its damage is noticed only when large population develops on the plant. Under such conditions, spraying with contact insecticides like 0.03% endosulfan, 0.1% carbaryl or systemic insecticides, recommended for aphid control, can be effectively used.

Lepidopterous larvae feeding on foliage include, *Crocidolomia binotalis* Zeller, *Hellula undalis* Fabricius, *Spodoptera littoralis* Fabricius and *Spilarctia obliqua* (Walker). All these are minor pests of radish. Out of these four species, the first two have been recorded on most of the cruciferous plants whereas S. *litura* and S. *obliqua* are highly polyphagous and among the vegetables crops, prefer tomato and sweet potato respectively.

Flea beetles, *Chaetocnema basalis* Baly, *Phyllotreta chotanica* Duvivier, *P. downesi* Baly and *P. vittata* Fabricius (Alticidae) have been reported feeding on radish leaves, especially of the young plants. Of these, *P. chotanica* is more common. It is a small flea beetle, about 2 mm long and oblong-narrow in shape; its upper side is metallic bronze in colour with greenish-blue reflections, while underside and antennae are black. Prothorax is broader than long and slightly narrow in front. Elytra are rounded at apices and are closely covered with punctures (Maulik, 1926). An allied species, *P. brassicae* Alam, has been reported from Bangladesh but not from India.

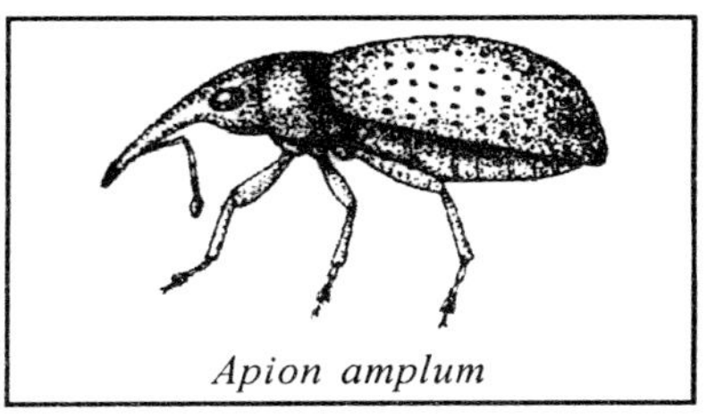
Apion amplum

Besides these flea beetles, *Apion amplum* Faust (Curculionidae), a tiny dark shining weevil having slender long snout has been reported nibbling tender leaves of radish. This weevil also feeds on black gram, green gram and cashewnut.

CARROT

Daucus carota Satova de Condolle, is the cultivated type believed to have originated from wild carrot, *D. carota* Linnaeus which is widely distributed all over Europe, Africa and Asia; it occurs as a troublesome weed in temperate regions of these countries. It is a biennial herb with twice or thrice pinnate leaves and flattish umbel of small white flowers. Fruits are oblong and about 30 to 40 mm long. The cultivated type is an annual herb with an erect much-branched stem, half to one metre in length; leaves are pinnate and decompound; flowers are white or yellowish, small, borne in compound terminal more or less globose umbels; fruits are oblong with bristly hair along the ribs. The varieties grown in South-east Asia were originally imported from Europe. In India, Singh (1963) evolved a new variety 'Pusa kesar' which has good qualities of both Asian and European types. Carrot root, the edible part, is exceptionally rich in iron and is used as vegetable, for preparing soup, stew, curries and pies; grated roots are used as salad, in puddings

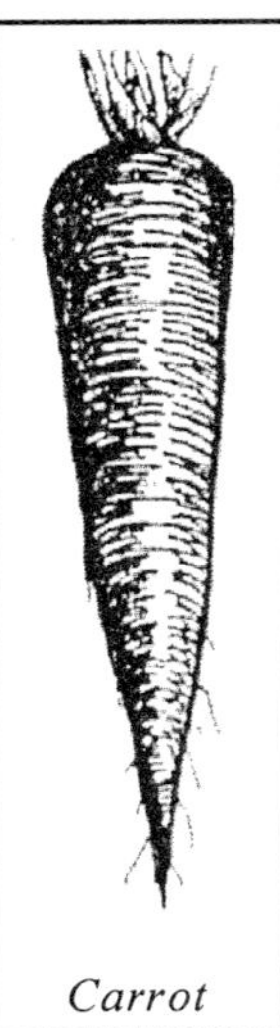
Carrot

and in sweetmeats; tender roots are picked or made into jams and jellies. Carrot juice is a rich source of carotone which is used for colouring butter and other food articles. Carrot leaves are generally fed to cattle but are also sometimes cooked and eaten by the poor people.

INSECT PESTS

Carrot is one of the few crops free from the depredations by any major insect pests. However, the minor pests recorded on this crop include leaf-hopper, pea leaf-miner, cutworm, red ants and a few species of beetles. Some sporadic soil borne may also occasionally damage the fleshy roots.

Leafhopper : *Empoasca punjabensis* Pruthi (Cicadelldae) is a major pest of eggplant, but occasionally it also infests carrot crop and causes some damage by sucking the cell sap from tender leaves.

Pea-leaf Miner : *Chromatomyia (Phytomyza) horticola* (Goureau), as the name suggests, is a pest of pea that sometimes invades carrot crop and mines the leaf stalks and stems; the damage caused is however negligible.

Cutworm : *Agrotis ipsilon* (Hufnagel) is a plyphagous pest having a wide range of host plants including eggplant, carrot, cole crops, okra and potato. Its serious infestation causes severe damage to any of these crops, often killing the seedlings outright. The caterpillars cut the seedlings at ground level and eat away the tender parts.

Red Ant : *Dorylus orientalis* Westwood is another polyphagous pest. Its preferred hosts being cole crops and potato.

Flea Beetles : *Chaetocnema basalis* Baly and *Chlorophorus* species (Cerembicidae) have been reported feeding on carrot leaves. *Anthrenus* spp. (Dermestidae) which generally feed and breed on dried animal and vegetable matter, as well as stored products, are also reported as minor sporadic pests of carrot. The grubs cause the damage by making small holes in carrot roots. The initial damage may result in secondary infection of other micro-organisms and may lead to rotting of the roots. Eggs are elongated in shape having spine-like projection at the end with which they remain attached to the surface where they have been laid. Grubs are elongate-oval in shape and have tufts of long hair on their bodies. Adults are small hemispherical beetles clothed with fine scales or hair and characterized by very short antennae. The species known for damaging carrot in India, are, *A. coloratus* Rottenburg, *A. jardanicus* Picard and *A. oceanicus* Favral.

As all the pests listed above are not significantly harmful, usually no control measures are adopted against these insect pests on carrot crop. However, if or when serious infestation occurs, these pests can be controlled by dusting the crip with 4% endosulfan or 5% carbaryl for foliage pests and raking into the soil 5% HCH, chlorpyrifos or heptachlor dust @ 26 kg/hectare for the root feeding grubs.

Carot Weevil : *Listronous oregonensis* (Le Conte) (Curculionidae), native of Northern USA and Southern Canada, has been reported as a potential pest of carrot, celery and parseley in USA (Simonet, 1981), but has not yet been reported in India.

TURNIP

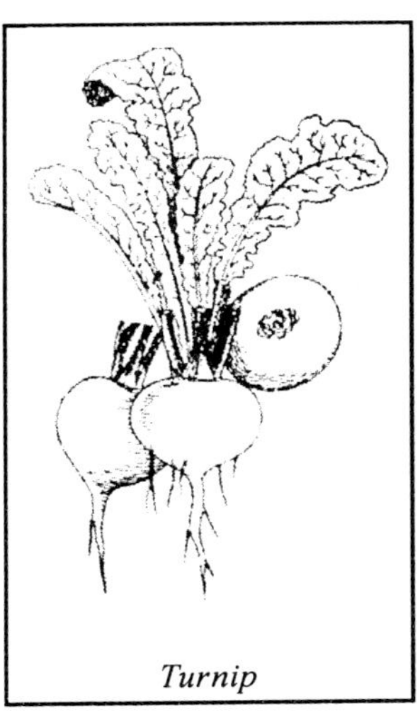

Turnip

Brassica rapa Linnaeus, is of Russian origin, as it is known to be growing wild in Russia and Siberia (Thompson and Kelly, 1957) extending from Baltica to Caucasus. Though it is a biennial crop producing swollen fleshy tap-root and a rosette of hairy leaves during first year and a half to one metre long stem bearing leaves and flowers next year; it is normally grown as an annual. Its enlarged root is napiform, consisting of the hypocotyl which swells and becomes spherical. Leaves are long, narrow, soft and hairy with a large terminal lobe and several side-lobes. Flowers are small, yellow in colour and hermaphrodite. Turnip greens are excellent source of vitamins A, B and C and also contain calcium and are sometimes used as vegetable but mostly as fodder. The roots are edible and used in curries, pickles and candy. Turnip is a cold season crop. In Europe the seed-crop is grown in plains but in India, it is grown only in hilly regions. European types are relatively sweeter and more palatable and are eaten even as salad, while Asian types are good for pickles and candy. The common varieties grown in India are, 'Pusa Sweti' (Asian type), 'Pusa Chandrima' (European type) and 'Pusa Kanchan' - a selection from cross between 'Red Round' (Asian) and 'Golden Ball' (European), Early 'Milan' top, 'Punjab safed-4', 'Purple top' white globe and 'snow ball'.

INSECTS PESTS

The major pests reported to cause economic losses to turnip crop are mustard sawflies and cabbage butterflies. Besides these, other insect pests which are sporadic in nature and cause minor damage are aphids, red ant, pea-leaf miner, cutworm, cabbage borer, diamondback moth and flea beetles.

Mustard Sawfly : (Klug) is an oligophagous pest attacking various cruciferous vegetables. According to Srivastava (1957), cabbage, cauliflower, knol khol are attacked by this pest in nurseries only; most of other food plants are also damaged in the early growth stage of the plants but turnip is the only crop of which leaves are infested even when plants are fairly grown up, showing that the insect has a high gustatory preference for turnip crop. In addition to this sawfly, Rishi (1967) reported an allied species *Athalia colibori* Fabricius as a major pest of turnip in Kashmir valley. The author stated that the pest is so destructive in Kashmir that damage caused by it may range from 60 to 70 per cent of the crop. The pest appears in May and remains active upto middle of December, passing through 4 to 5 generations. The egg, grub, pupal and adult

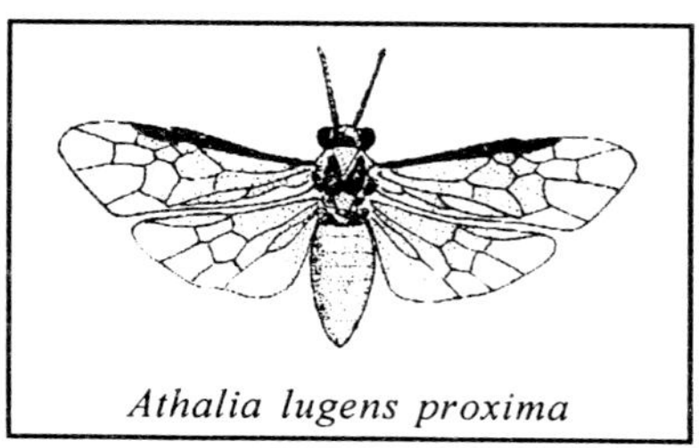

Athalia lugens proxima

longevity has been found to be 3 to 4, 12 to 13, 10 to 14 and 4 to 5 days respectively; the grub and pupal periods of third and fourth generations prolong for 4 to 5 days each. Hibernation takes place under the soil in pupal stage - from middle of December to the end of March.

Hand-picking and mechanical destruction of grubs in the early stage of attack can keep the incidence of pest attack low. If the attack is severe, spray 0.2% carbaryl or 0.1% acephate or malathion or 0.05% dichlorvos.

Cabbage Butterfly : A group of several different species, is the most common pest of cruciferous vegetables. *Pieris rapae* (Linnaeus), *P. napi* (Linnaeus), *P. brassicae* (Linnaeus), *P. oleracea* Harris and *P. protodice* have been reported from Kashmir (Rishi, 1967); of these the first one is comparatively more common and destructive.

Pieris (Artogeia) rapae is a palaearctic species, native of Europe. In India it is found throughout the Kashmir valley on almost all cruciferous vegetables. It has been reported to cause, on an average, 25 to 30% loss in Kashmir valley. It is a major pest of turnip. Eggs are tiny, spindle shaped with bright greenish-yellow ridges running lengthwise and crosswise. Full grown caterpillars are 30 to 35 mm long, greenish in colour with yellow dorsal stripe and covered all over with black dots, each bearing pale hair. Pupae are angulated, 24 to 30 mm in length and variable in colour, chiefly gray freckled with black but may be even dull green. Adult butterflies look like *P. brassicae* but are smaller in size having wing span of 50 to 58 mm. The head, thorax and abdomen are black in colour checkered with white scales. Wings are creamy-white. Forewings with upper half of cell and costal margin above it sparsely irrorated with black scales, have a round black spot in the middle. Hind wings are uniformly cream coloured with only a very short transverse laterally compressed and generally diffused black spot in interspace 7 (Bingham, 1907). Though adults appear in the fields from February to mid-March, they complete 5 generations by the end of September, when the pest enters into hibernation in pupal

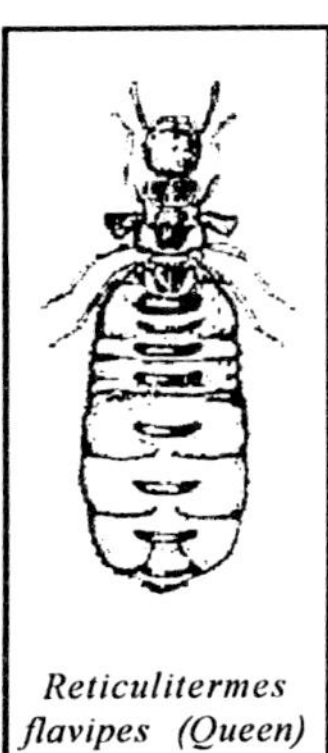

Reticulitermes flavipes (Queen)

stage. The egg, larval and pupal stages last for 8 to 12, 25 to 32 and 9 to 13 days respectively (Rishi, 1967).

Minor pests : Some of the minor pests attacking turnip include the mustard aphid *Lipaphis erysimi* (Kaltenbach), the red ant, *Dorylus horticola* Westwood, pea-leaf miner, *Chromatomyia (Phytomyza) horticola* (Goureau), Cutworm *Agrotis ipsilon* (Hufnagel), cabbage borer *Hellula undalis* (Fabricius), diamondback moth *Plutella xystella* (Linnaeus) and flea beetles *Chaetocnema basalis* Baly and *Phyllotreta vittata* favricius.

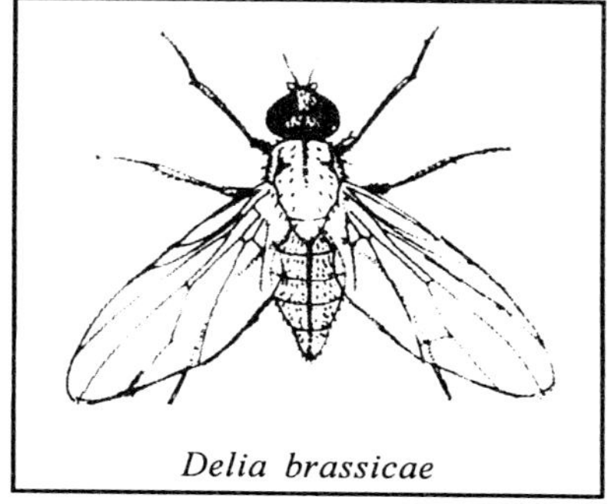

Delia brassicae

Termite *Reticulitermes flavipes* Kollar has been reported causing severe damage to turnip crop in Florida (Osburn, 1937); cabbage maggot *Delia brassicae* Bouche (Anthomyiidae) has been reported as a major pest of turnip in South-east Asia (Stilt, 1953; Doane and Chapman, 1962) and blue beetle *Attica cyanea* (Weber) (Alticidae) has been found feeding and breeding on turnip leaves in Bangladesh (Alam, 1970) but none of these pests has yet been recorde as a pest of turnip in India.

BEET-ROOT

Beta vulgaris Linnaeus (garden-beet) is known as a crop of economic importance since ancient times. It owes its origin from the wild sea-beet, *Beta martima* Linnaeus, a native of coastal area of Mediterranean region and now cultivated all over the world. It is said that when Greeks paid homage to Apollo, they served him beet-root on a silver platter (Roe,1974). In India, it is usually grown during winter in Northern states and throughout the year in Maharashtra, Goa and coastal area of Kerala. *B. uvlgaris bengalensis* Hortorm is grown in North India as a pot-herb and is used purely as salad. Another cultvar *B. vulgaris sapa* Dumorier, commonly called sugar-beet, is extensively cultivated in Temperate regions of the world and is used for extracting sugar while its molasses are used for manufacturing alcohol. It is smaller in size than the garden-beet and light in colour with high sugar contents (over 20 per cent).

Beet-root is a biennial herb, producing a large swollen fleshy root and a rosette of leaves in the first year while in the second year, 8 to 12 tall seed stalks are produced bearing numerous flowers in an open panicle. The so called seed is a corky fruit containing several seeds. The root is generally deep red in colour, occasionally pink or even sometimes white. The root are eaten raw or boiled and eaten as salad and sometimes even pickled; the smooth young leaves are also eaten as greens. Beet-root contains vitamin B_1 (70 IU), vitamin C (88 mg/100g) and iron (1.0 mg) whereas beet-greens contain more iron (5.1 mg) and are also richer in vitamin A (2100 IU), vitamin B_1 (110 IU) and vitamin C (50 mg/100g).

INSECT PESTS

Butani (1958b) has reported *Plusia nigrisina* Walker, *Diacrisia orichalcea* (Fabricius), *Chalciope hyppasia* (Cramer), *Diacrisia todara* Moore, *Marasmia trapezalis* (Guenee). *Diaphania indica* (Saunders). *Tryporyza (Schoenobius) incertulus* (Walker). *Spodoptera mauritia* (Boisduval) and *Hymenia facialis* (Cramer) as pests of sugar beet in Bihar. In addition to these leaf eating caterpillars, Chhiber (1975) has reported greasy cutworm, grasshopper, green peach aphid, thrips, whitefly, pentatomid bugs as minor pests.

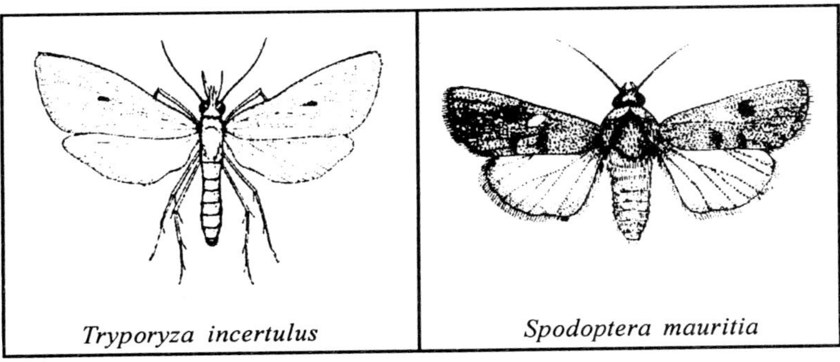

Tryporyza incertulus | *Spodoptera mauritia*

Besides these, a blue beetle and a leaf eating weevil have also been reported feeding on beet-root leaves. The hairy caterpillar, *Estigmene acrea* (Drury), a major pest of beet-root in Latin America has not yet been reported from India.

Hymenia facialis (Pyraustidae) which is the most common pest in Northern India appears around March; its activity increases till July and then gradually decreases. Incubation, larval and pupal stages last for 2, 9 to 11 and 5 to 8 days respectively (Butani, 1958 b) and adult longevity is 2 to 5 days. *Spilarctia* species (Noctuidae) and *Spodoptera* species (Noctuidae) are the most destructive pests in Tarai area of Uttarakhand. The damage by these leaf-eating caterpillars can be easily checked in the early stage of their attack by hand-picking the prominent egg-masses and gregarious early stage caterpillars, then destroying the same mechanically.

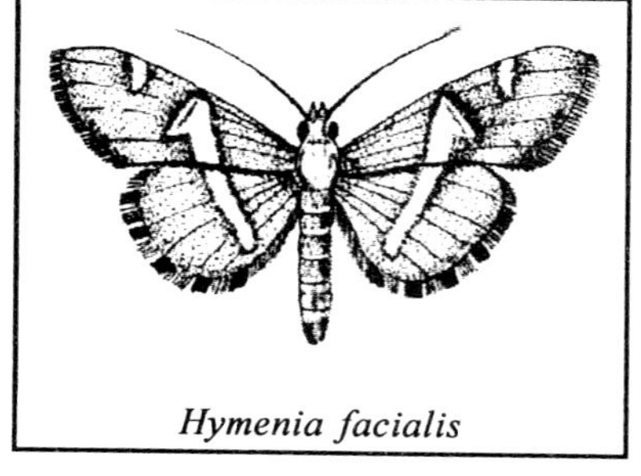

Hymenia facialis

Greasy cutworm has been reported to be active in sugar-beet fields during October-November and again in February-March. It can be checked by mixing thoroughly 5% HCH dust with the soil just before sowing the crop (Chhibber, 1975). This soil treatment will take care of the white grub as well.

Green peach aphid infestation starts in December and continues till the end of February. Thrips appear during September-October and if not checked in time, may reappear in March-April. To control aphids, thrips and other sap sucking pests, spray 0.05% acephate, dimethoate or oxydemeton methyl.

Occasionally, the grasshopper appears during October-November, but the damage caused by this pest is negligible and does not call for any control measures, except hand-picking of nymphs and adults and their mechanical destruction.

Blue Beetle : *Altica cyanea* (Weber) (Alticidae) was reported by Lefroy and Howlett (1909) as one of the perfectly harmless insect. Later, Fletcher (1920) recorded this beetle from different parts of

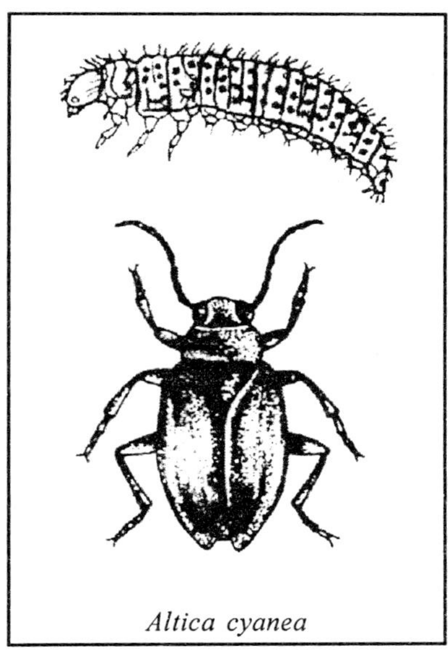
Altica cyanea

the Indian sub-continent feeding on a number of weeds as also on cabbage, cauliflower and sugar-beet (flower-heads); occasionally, it has also been found in large numbers on berseem, barley, wheat, rice etc. but has not been observed to cause any substantial damage. According to Batra (1961), the only economically important crop damaged by this beetle is waternut *(Trapanatans* var. *bispinosa* Roxburgh). The grubs feed on epidermis of leaves and when abundant they leave behind only the network of veins. Adults nibble and feed on the leaf margins. Pupation takes place in the soil in earthen cocoons. Fully grown grubs are 8 to 10 mm long and dark brown in colour. Pupae are 12 to 16 mm long, orange-yellow when freshly formed and gradually darken, finally become light to dark brown. Adults are shining steel-dark blue beetles about 5 mm long with black antennae (Maulik, 1926).

Leaf-eating Weevil : *Ptochus ovulum* Faust (Curculioidae) has been recorded as a minor pest of beet-root. Hand-picking and mechanical destruction of eggs, larvae and pupae of these pests can be effective in checking their pest population.

To control these coleopterous pests, dust the crop with 5% carbaryl or 4% endosulfan dust @ 20 kg. per hectare.

4

CUCURBITS

CUCURBITS, a common name given to a group of crops having trailing habit, are extensively grown all over the Tropical and subtropical countries and include the largest number of Summer and Rainy season vegetables. Of these only gourds and pumpkins are cooked and eaten as vegetables, fruits of cucumbers are usually taken as salads while musk-melon and water-melon are consumed mainly as dessert. Cucurbits in general are a good source of vitamins A and C and various vital minerals. Over ripened and dried pumpkin and bottle gourd fruits have been used since ages, for making various musical instruments; these as well as ash gourd and pointed gourd fruits are also used in making ketchups, sweets and candies.

The various types of gourds commonly cultivated in India, include, ash gourd *Benincasa hispida* (Thunberg) (=*cerifira* Savi), bitter gourd *Momordica charantia* Linnaeus, bottle gourd *Lagenaria siceraria* Standley, Ivy gourd *Coccina grandis* (Linnaeus) (=*indica* Wight and Arnott), pointed gourd *Trichosonthes dioica* Roxberg, snake gourd *T. anguina* Linnaeus, ridge gourd *Luffa acutangula* Roxberg, sponge gourd *L. aegyptiaca* Milliere (=*cylindrical* Roemer), squash gourd *Citrullus lanatus fistulosus* Duthie and Fuller, pumpkin *Cucurbita pepo* Linnaeus, and red pumpkin *C. maxima* Duchesne.

Ash gourd : A native to Malaysia, it is grown all over the Indian sub-continent and South-east Asia. It is an extensively trailing or climbing herb having 5-angled leaves. The fruits are broadly cylindrical (45 to 50 mm long) or spherical (25 to 30 cm in diameter) in shape and have white pulp. Seeds are numerous, much compressed and marginal. The juice of ripe fruits is beneficial in curing phthisis; it is also useful in haemoptysis and other internal discharges. The seeds possess anthelmintic porperies and the oil extracted from seeds is equally efficacious.

Bitter gourd : A monoecious climber with sub-orbicular leaves 5 to 7 lobed; flowers are yellow and solitary. Fruits are 5 to 25 cm long, pendulous, fusiform beaked, ribbed with numerous tubercles.

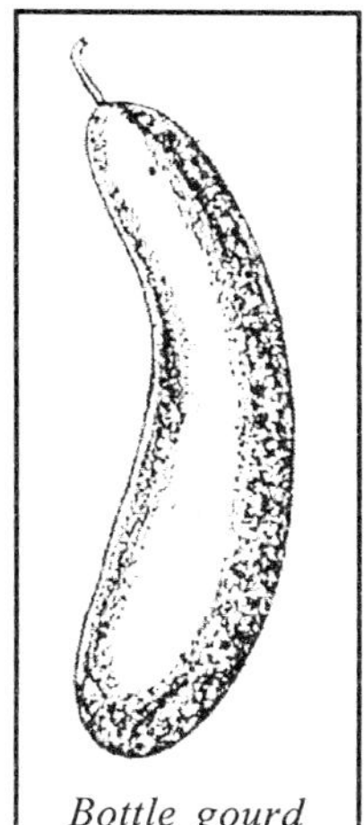

Bottle gourd

Bottle gourd : a climbing or trailing herb, is native to Northern coast of Peru, where it grew almost 5000 years ago. Cultivated all over India; the vines have heart-shaped leaves, white flowers; long and cylindrical (bottle-shaped) fruits. The fruits are used as cooked vegetable as also stuffed with rice and mince-meat then boiled. Fruit pulp is a good source of vitamin B and ascorbic acid. Seeds are also edible.

Pointed gourd : A dioecious climber with perennial root stock; leaves are cordate or ovate-oblong and flowers dioecious. Fruits are globose, oblong, smooth, 5 to 12 cm long, with light green stripes on young fruits and red on ripe ones.

Snake gourd : An annual creeper having very long and slender stems that are furrowed and sub-glabrous. The leaves are 5 to 8 cm long and of different shapes. Flowers are monoecious. Fruits are half to one metre long (maximum 4 metres) and often twisted; the variety White Long has pale green fruits with irregular white stripes whereas Green Long variety has dark green fruits with pale green stripes.

Ridge gourd : A large monoecius climber with palmately 5 to 7 angled or lobed leaves; male flowers have 3 stamens whereas female flowers are solitary. Fruits are 15 to 30 cm long (maximum one meter), cylindrical or club-shaped with 10 prominent longitudinal ribs or ridges.

Sponge gourd : Vines are more or less same as those of ridge gourd except that the flowers have 5 stamens and the fruits are smooth, cylindrical about 20 to 50 cm long (maximum 250 cm). Tender fruits are used as vegetable. On maturing the fibrovascular bundles harden and pulp becomes bitter and inedible. Tender fruits are also considered to be diuretic and lactagogue. Ripe fruits after burning and pulverizing, are used in China, as carminative and anthelmintic; juice of fruits acts as purgative.

Squash gourd : One of the most popular summer vegetables; its fruits are generally cooked, and occasionally pickled as well. This is diffuse annual creeper or climbing herb with stout stem. Leaves are cordate at the base and pinnate into 3-4 pairs of lobes. Fruits are pale to dark green in colour, rounded in shape and 2 to 7 in diameter.

Ivy gourd (little gourd) : A perennial herb, climbing or prostrate with angular or lobed leaves. Fruits are smooth, ovoid or elliptical, 25 to 50 cm in diameter with bright green stripes when immature which turn to bright scarlet in ripe fruits. Various preparations of roots, stems and leaves have been mentioned as being efficacious against skin disease, bronchial catarrh, bronchitis and diabetes.

Pumpkins : Native to New World, these are annual trailing herbs. *Cucurbita maxima* Duchesne (Winter Squash) has long running prickly or hairy stems, bearing large flaccid leaves, orbicular or renifom in outline with rounded lobes. Fruits are brownish-yellow in colour; when ripe, round or oval in shape and variable in size with soft spongy peduncle. Seeds are white or brown to bronze in colour with margin of the same colour. All plant parts are edible. Tender shoots and leaves are used as salad; flowers and fruits are cooked as vegetable. Seeds too are edible; these are anthelmintic and are used as taeniacide, diuretic and tonic. *C. pepo* Linnaeus, also

Winter squash

known as red gourd, has 5 sided stems devoid of prickly or hairy covering. Its leaves are limpid, velvety often with mottled or whitish blotches. Fruits vary in shape and colour. Seeds are flattened, grayish-white to tan in colour with dark marginal bands (CSIR, 1950)

Cho-Cho : *Schium edule* (Jacq) and 'Mitha karela' *Cyclanthera pedata* (Linnaeus) are recently introduced cucurbits from West Indies. These are grown in limited areas mostly on hills of Northern India, though cultivation of cho-cho is also being attempted in and around Pune (Maharashtra) and Mysore (Karnataka).

INSECT PESTS

Cucurbits are attracted by a number of insect pests, mites and nematodes but fortunately, in India, only fruit-flies and a few species of beetles are of economic importance; aphids and blister beetles though of regular occurrence, seldom cause severe damage, rest of the insect pests are of minor importance.

Fruit flies : One of the most important groups of pests of cucurbits. The species reported damaging the cucurbit fruits include, *Bactrocera (Dacus) cucurbitae* (Coquillett), *B. ciliatus* (Loew). *B. diversus* (Conquillett), *B. latifrons* (Hendel), *B. parvulus* (Hendel), *B. tau* (Walker), *B. zonatus* (Saunders) and *Myriopardalis pardalina* (Bigot) (Kapoor, 1970). Of these the first one is comparatively more common and destructive. Most of the fruit flies are polyphagous pests having a wide range of host plants. The female flies puncture the soft and tender fruits with their stout and hard ovipositor and lay eggs below the epidermis @ 4 to 10 eggs per fruit each time. A single female can lay about 200 eggs in her life span of 8 to 10 weeks. A puncture made by one female is often used by others also for ovipositing and a single fruit may have more than one puncture made by one or more females. On hatching the maggots feed inside on the pulp of fruits and the infested fruits can be identified by the presence of brown resinous juice which oozes out of the punctures made by the flies for oviposition. These punctures also serve as an entry for various bacteria and fungi - as a result the infested fruits start rotting, get distorted and malformed in shape and fall off from the plants prematurely. These fallen fruits are unfit for human consumption and have no market value. The fully fed maggots come out of the fallen fruits and pupate 10 to 15 cm deep in the soil. Where fruits do not fall, the maggots either pupate inside the fruits (which is not common) or come out, drop down and pupate in the soil. Late maturing varieties are generally more prone to the attack of these flies than the early maturing ones. The exact loss caused by these flies has not been systematically worked out though it is estimated that more than 50 per cent of fruits are either partially or fully damaged by one or the other species of the fruits flies (Singh, 1966).

Eggs, maggots and pupae of the different species look alike and are very difficult to distinguish, whereas adult flies have some specific diagnostic characters (Kapoor, 1972). Eggs are 1.0 to 1.5 mm long, whitish in colour, elongate-cylindrical in shape, slightly curved and tapering at both ends. Full grown maggots are 5 to 10 mm long, cylindrical in shape tapering anteriorly, blunt at posterior end and pale-white in colour. Pupae are 5 to 8 mm long, barrel-shaped and brown to ochraceous in colour. *Bactrocera* species belong to sub-family Dacinae. These are medium to large sized flies, ferruginous-brown in colour with hyaline wings. *B. cucurbitae* adults are 4 to 5 mm long having a wing expanse of 11 to 13 ad 14 to 16 mm in males and females respectively. The hyaline wings have costal band broad and prominent, anal stripes well developed and hind cross veins thickly margined with brown and gray spots at the apex. Adults of *B. tau* are more of less similar in shape and colour except that the hind cross veins are not margined. *B. diversus* adults have greenish-black thorax with 3 longitudinal lemon-yellow stripes or vittae on mesonotum. Costal and anal bands are narrow and weakly developed. Males are 5 to 6 mm long with a wing expanse of 9 to 11 mm while females are 8 to 9 mm long with a wing span of 12 to 14 mm. *B. zonatus* flies have yellowish-red body with pale yellow band on 3rd tergite and a wing expanse of 10 to 12 mm; costal band incomplete and anal band wanting. Females have red ovipositor with black tip. *B. ciliatus* are similar to *B. zonatus* in size and wing pattern but their body colour is ferruginous-brown; tergite are fused together and instead of pale yellow band there is a prominent dark brown oval spot on either side of 3rd tergite. *Myrioparadalis pardalina* belongs to the subfamily Trypetinae. The chaetoraxy of these flies is complete; median longitudinal line on thorax is present and there is only one medio-posterior spot with semi-circular posterior margin present on scuttellum. Wings are not reticulate and have dark cross bands.

Bactrocera cucurbitae, commonly called melon fruit fly, is widely distributed and has been recorded from East Africa, some parts oa USA, Northern Australia, Taiwan, Okinawa in Japan, South China, South-east Asia and the Indian sub-continent (CIE map No. A-64). It is highly plyphagous and though its preferred hosts are musk melon, snap melon, bitter gourd and snake gourd (Butani, 1975b); as many as 70 host plants have been listed by Batra (1953). In nature the population is generally low during dry weather and increases rapidly with adequate rainfall. Pre-oviposition, egg, maggot and pupal periods last for 9 to 21, 1 to 1½, 3 to 9 and 6 to 8 days respectively. During Winter the grub and pupal stages are extended upto 3 and 4 weeks respectively. A single life-cycle is completed in 10 to 18 days but it takes 12 to 13 weeks to complete a single life-cycle in Winter. Adult longevity is 2 to 5 months; females live longer than males. Generally, males die soon after fertilizing the females whereas the females die after completing the egg-laying.

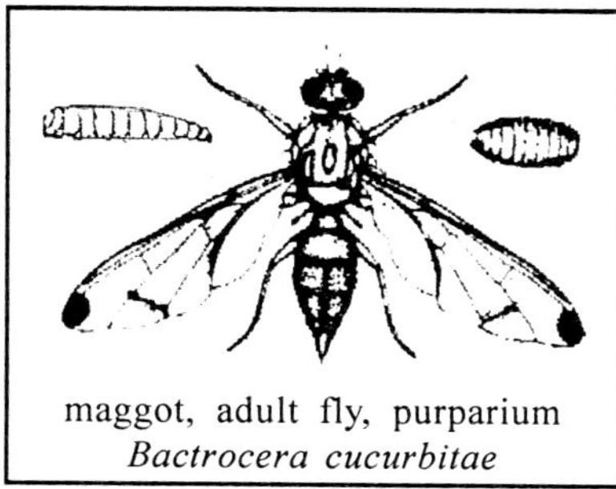

maggot, adult fly, purparium
Bactrocera cucurbitae

Bactrocera tau is widely distributed in South-east Asia, Philippines and Indonesia and is highly polyphagous. It has been found feeding on *Citrus* (specially pomelo), mango, mulberry, sapota, cucumber, bitter gourd, bottle gourd, ribbed gourd, snake gourd, sponge gourd, squash gourd, melons, pumpkins, tomato etc. This species closely resembles *B. cucurbitae* and there is every possibility that one has been confused for the other by various workers during their studies on bionomics and control.

Bactrocera diversus, guava fruit fly, is also widely distributed in South-east Asia. Besides guava and loquat, its preferred hosts,

Bitter gourd damaged by *Bactrocera* species

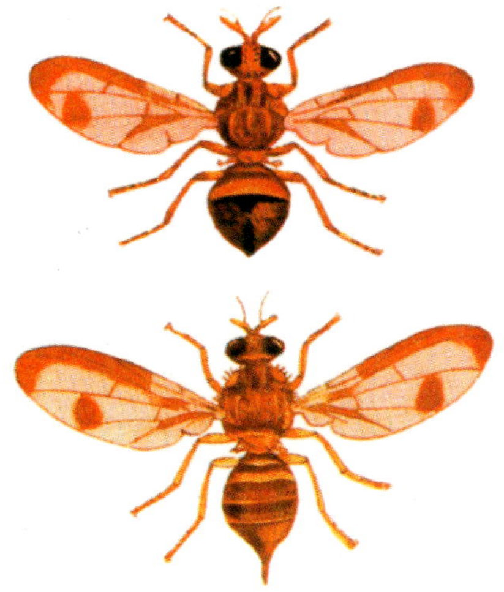

Bactrocera cucurbitae

Whitefly puparia on bottle gourd leaf

Adult whiteflies on bottle gourd leaf

it also attcks banana, *Citrus*, java plum (*jamun*), mango, papaya and some gourds especially bitter gourd and bottle gourd.

The pest is active all the year round except during severe Winter when it over Winters in adult stage. In Summer, it breeds excusively in lower flower buds of cucubitaceous plants and migrates to orchards in Winter. The adult flies feed on honeydew secreted by aphids but are often deterred by the presence of black ants,

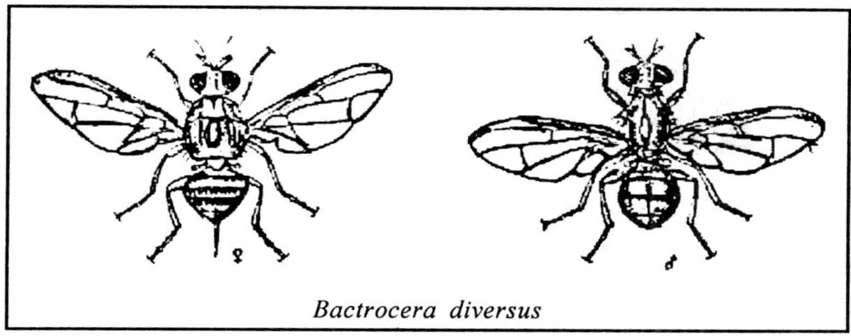

Bactrocera diversus

Camponotus compressus Fabricius which also feed on the honeydew. Eggs hatch in 1 to 4 days during July; maggot development takes 4 to 5 days during August while the pupal stage lasts for 7 days in August and extends upto 13 days in November (Batra, 1953).

Peach fruit fly, *Bactracera zonatus*, as the name suggests, is a major pest of peach. It is widely distributed in South-east Asia and has been recorded on apple, *ber,* eggplant, cherry, *Citrus*, pomegranate, sapota etc. Among cucurbits, it is more commonly found on melons, bottle gourd and ridge gourd. The pest is active all the year round except from the end of December to February, when it overwinters in pupal stage. Incubation period is 3 to 4 days; maggot and pupal stages last for about a week each during Summer and little longer in Winter. Adult longevity is one to four months (Butani, 1979).

Ethiopian melon-fly *Bactrocera ciliatus* is another fruit fly found in association with *Bactrocera cucurbitae*. It is a native of Africa and widely distributed in different countries of Europe, Africa. Middle East and the Indian sub-continent. In Bangladesh, it is a major pest of melons causing severe losses during February to April (Alam *et al.,* 1964). Besides cucurbits, it occasionally attacks apple and *Citrus* fruits. The characteristic feature of this pest is the cavities the female makes on the fruits by piercing the surface with its ovipositor and lays 3 to 8 eggs in each cavity; pupation often takes place in the affected fruits even if the fruits are drying and rotting; overwintering takes place in pupal stage (Pruthi and Batra, 1960). Egg, maggot and pupal stages last for 1 to 2, 4 to 6 and 8 to 10 days respectively (Cherian and Sundaram, 1939). Total life-cycle occupies 15 to 17 days in October and there are six generations in a year in South India. In North India, where there is a distinct winter season, there are only 4 to 5 generations in a year.

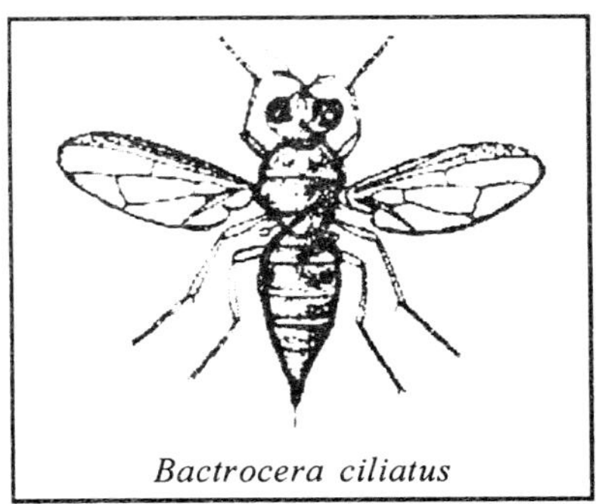

Bactrocera ciliatus

Myriopardalis paradalina has been reported from East Africa, Israel, Afghanistan, Pakistan and Northern India. It attacks almost all varieties of cucurbits and has been recorded as a major pest in Western countries though it is only a minor pest in India. Egg, maggot and pupal stages last on an average for 3, 15 and 13 days respectively while the adult longevity is 3 to 4 weeks (Cleghorn, 1914). The pest overwinters for about 4 months in pupal stage and has 2 to 3 generations in a year. A single female lays as many as 120 to 150 eggs, of which hardly 6 to 8 reach maturity.

Satisfactory control of fruit flies is still a far cry. To avoid infestation by fruit flies, growing of resistant or early maturing varieties has been recommended. 'Arka-tinda', a variety of round gourd, and 'Arka-Saryamakhia' variety of pumpkin are reported to be resistant to *Bactrocera cucurbitae* (Nath, 1971). To check the damage by these flies, fruits should be harvested before they start ripening. In areas where damage is observed every year, change in sowing dates is suggested. All the fallen and infested fruits should be collected and destroyed to prevent the carry over of the pest. The infested fruits should be buried more than 60 cm deep (Agarwal *et al.,* 1987). Frequent raking of the soil under the vines or ploughing the infested fields after the crop in harvested can help in killing the pupae. In case of severe infestation, poison-baiting or bait-spray is recommended. For preparing the bait, mix 20g malathion 50% WP with 500 g molasses or gur + 20g yeast hydrolysate (if available): to this two litres of water are added for poison-baiting or 20 litres for bait-spray. For obtaining good results, the control measures suggested should be undertaken as a cooperative campaign in all the fields in large areas; otherwise the adults being good fliers will keep on migrating and invading from the neighbouring fields inspite of one's best individual efforts.

In nature, *Bactrocera cucurbitae* is parasitised by *Opius fletcheri* Silvestri, *O. compensans* Silvestri, *O. incisus* Silvestri, *O. watersi* Fullaway, *Pachycrepoideus dubius* Ashmead, *Dirhinus giffardi* Silvestri, *D. luzonensis* Rohwer, *Spalangia philippinensis* Fullaway, *Syntomosphyrum indicum* Silvestri, *Ipobracon* spp. etc., whereas *B. ciliatus* is parasitised by *O. compensans, D. luzonensis, P. dubius*, *Spalangia afra* Silvestri, *S. grotivsi* Girault, *S. philippinensis*, *S. stomyxyshine* Girault and *Galesus* species (Narayanan and Batra, 1960). But none of these parasites bring about effective check of the pest population.

Survey in the mid hill region (1200-1500m) of Himachal Pradesh revealed an infertation of fruit fly *B.tau* on tomato, summery squash, pumpkin and Chow-Ehow, *Sechium edule* (Jack) Sw. *B.scutellaris* was also recorded on bottle gourd and pumpkin flowers (Sunandita and Gupta, 2007).

Pumpkin Beetles : Red pumpkin beetle, *Rhaphidopalpa (Aulacophora) foveicollis* (Lucas), blue pumpkin beetle, *A. lewisii* (Baly) and gray beetle, *A. cincta* (Fabricius) have been reported damaging cucurbits. The first one is most destructive and is widely distributed all over the South-east Asia as also Mediterranean region towards West and Australia in the East. In India, it is found in almost all the states though it is more abundant in northern states, generally in association with *A. lewisii;* in South India *A. cincta* is more common. These beetles are polyphagous pests and though they prefer cucurbitaceous vegetables and melons, some leguminous crops are their main alternate hosts. One female can lay 150 to 300 eggs. Eggs are laid in the moist soil usually around the host plant. Dry and waterlogged soils are not preferred for egg-laying. On hatching, the grubs feed on the roots and underground portion of host plants as also fruits touching the soil. The damaged roots and infested underground portion of host plants as also fruits touching the soil. The damaged roots and infested underground portion of stems start rotting due to secondary infection by saprophytic fungi and the unripe fruits of such vines dry up. Infested fruits become unfit for human consumption. Adult beetles feed voraciously on leaf lamina making irregular holes. They prefer young seedlings and tender leaves and damage at the stage may even kill the seedlings. These beetles are active from March to October, though the peak period of activity is April to June (Butani, 1975b).

Eggs are spherical in shape and yellowish-pink in colour, becoming orange after a couple of days. Freshly hatched grubs are dirty white in colour whereas full grown ones are creamy-yellow and about 22 mm long. Pupae are pale-white and are found in earthen cells 15 to 25 mm deep in the soil. Adult beetles are 6 to 8 mm long, having glistering yellowish-red to yellowish-brown elytra that are uniformly covered with fine punctures. Adults of *A. cincta* are similar in size and appearance except that the colour of the elytra is grayish-yellow to brownish-gray with distinct paler margin all round. Adults of *A. lewisii* are slightly smaller (5 to 6 mm) and the colour of elytra is blackish-blue. The taxonomic identities based on male and female genitalia of these and other five *Aulacophora* species have been studied in detail by Menon *et al.*, (1972).

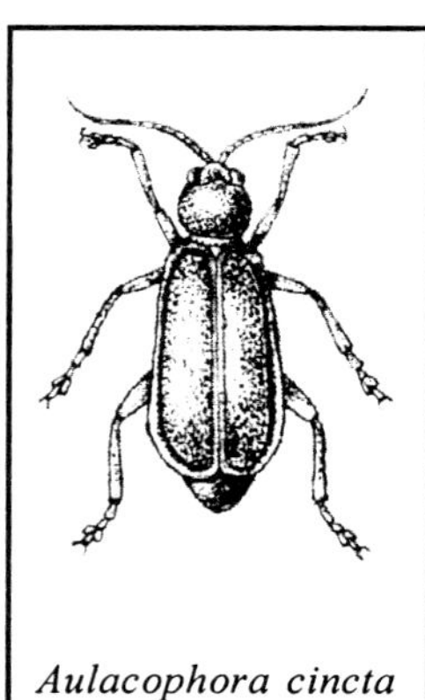
Aulacophora cincta

Eggs of *R. foveicollis* hatch in 5 to 8 days (maximum 15 days in Winter). Grub, pre-pupal and pupal stages last for 13 to 25, 2 to 5 and 7 to 17 days respectively and total life-cycle occupies 32 to 65 days (Narayanan, 1953). Adults live for abut a month. Overwintering takes place in adult stage (mid-November to end February) and 75% of the adults, mostly males, perish during this. period. The pest population in February-March may comprise six females to one male.

Other chrysomelids reported feeding on cucurbit leaves include, *Galerucida bicolor* (Hope) from Himachal Pradesh and *Lema praeusta* (Fabricius), *L. semiregularis* Jacoby and *L. signatipennis* Jacoby from Orissa. Beside cucurbits, these beetles also feed on leaes of eggplant.

To prevent the damage by these beetles, cultural practices, like clean cultivation, early sowing and use of resistant varieties have been recommended. After harvesting the infested fields, these must be immediately deep ploughed to kill the grubs present in the soil. As regards chemical control, since a number of organo-synthetic insecticides including HCH, endrin and parathion have been reported phytotoxic to various cucurbits (Banerjee and Chatterjee, 1955; Mookerjee and Wadhi, 1956; Nagaraja Rao, 1959), great care need be exercised in selecting proper insecticides. Dusting with 5% carbaryl or 4% endosulfan or spraying with 0.2% carbaryl or 0.05% endosulfan, as suggested by Butani and Verma (1977a) are effective and safe. Some of the recently developed synthetic pyrethroids can also be recommended.

Hadda Beetles : *Epilachna* species are major pests of various vegetable crops including eggplant, cucurbits, potato, tomato and peas. Among cucurbitaceous vegetables, bitter gourd is their preferred host. Both grubs and adults feed voraciously by scrapping the chlorophyll of leaves, causing characteristic skeletonisation of leaf lamina; the affected leaves gradually dry and droop down. A severe infestation kills the young plants outright; while the older vines show stunted growth and poor yield. In early stage of attack, and picking and mechanical destruction of egg-masses and adults, which can be easily located due to prominent colouration and damage symptoms, can keep the population under check. Chemical control suggested against pumpkin beetles will control these beetles as well.

Blister Beetle : *Mylabris pustulatus* Thunberg is a polyphagous pest widely distributed in Africa and South-east Asia (Hill, 1975). Its main hosts are various cucurbits including melons, though it is also found feeding on cotton, groundnut, millets, okra, rose etc.

Eggs are laid in soil of cultivated fields. On hatching the grubs feed on egg-pods of various grasshoppers and locusts, found in the soil - thus this stage can be considered as highly beneficial (Singh, 1970). Full grown grubs are coarctate and form pseudopupae, which become pupae later. Pseudopupae are devoid of functional appendages and hibernation takes place in this stage in the soil. The adults emerge out of the soil around August and are active till early December. They feed on pollens and petals of flowers and flower buds, as a result, fruit setting is adversely affected.

The beetles are 22 to 26 mm or more in length having 3 black and 3 yellowish orange bands running vertically and alternately on elytra. When handled or disturbed these beetles exude an acrid yellow fluid which contains cantharidine, which is irritant to touch and causes blisters on human skin -hence the common name.

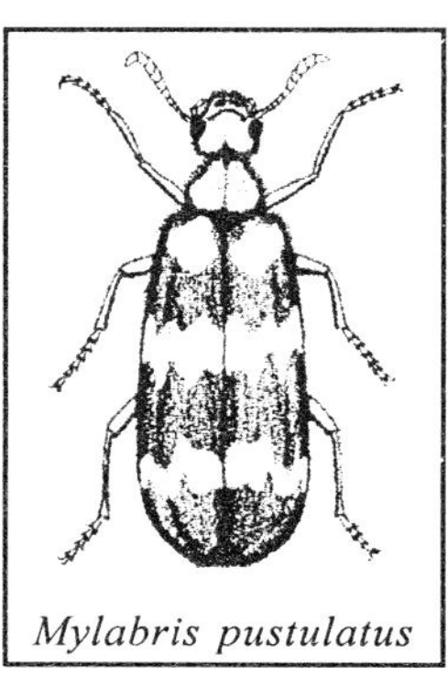
Mylabris pustulatus

Mylabris phalerata Pallas has also been occasionally found in association with *M. pustulatus*, specially on melons. This beetles is slightly smaller in size (20 to 24 mm long) and has its yellowish-red bands narrower than black bands.

To control these beetles is rather difficult because of their rapid mobility. Hand collection and prompt destruction of beetles can keep the population under check. This may be done during early morning hours, when the beetles are comparatively less active. Chemical control measures to be adopted exclusively against this pest may not be economical: dusting or spraying carbaryl or endosulfan as suggested for pumpkin beetles will control these beetles as well.

Flea Beetle : *Phyllotreta crucifera* (Goeze) - a major pest of brassicas is sometimes found feeding on leaves of various cucurbitaceous vegetables making typical shot-holes. The seedlings may be completely destroyed while the grown up vines do not suffer any significant damage. Being a minor pest, generally no control measures are warranted.

Aphids : *Aphis gossypii* Glover - the notorious cotton or melon aphid - is a major pest of melons and a number of cucurbitaceous vegetables as also various malvaceous and solanceous plants. It has a very wide cosmopolitan distribution, absent only from extremely cold regions of Asia and Canada. Large colonies of nymphs and adults are found on tender twigs and shoots as also on ventral leaf surface sucking the vital sap from the tissues; the affected parts turn yellow, get curled, wrinkled and deformed in shape and ultimately dry and die away. Fruit size and quality is also reduced. The aphids also exude copious quantity of honeydew on which sooty mould develops which in turn hinders the photosynthetic activity of the vines - resulting in stunted growth. The fruits covered by sooty mould look unattractive and lose their market value. Besides the damage caused by desapping the vines, this aphid transmits various virus diseases - the loss on this account is collosal and irrepairable. The affected plants have to be uprooted and destroyed to prevent the spread of the disease. The pest is active throughout the year, but maximum activity on cucurbits has been observed during April to June. High humidity and cloudy weather with a little rainfall is favourable for rapid multiplication of this pest (Reddy, 1968). Both nymphs and adults are soft bodied. pear-shaped insects. There are alate and apterous forms, the latter being more common and abundant. Appearance of apterous forms is generally associated with over crowding, and these individuals are slender and fragile while the alate ones are comparatively robust.

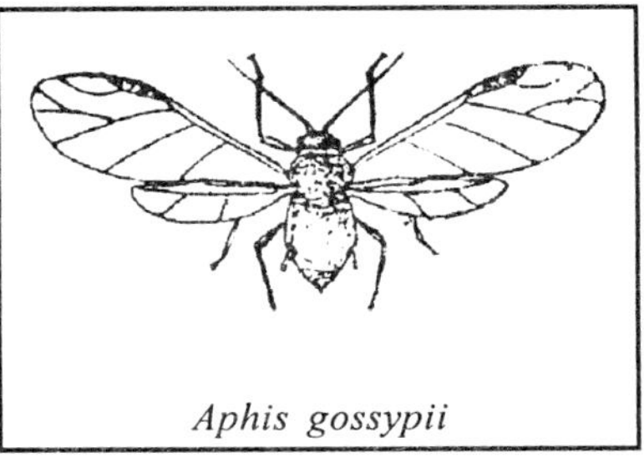

Aphis gossypii

Aphids have a complex life-cycle. Reproduction is generally viviparous and parthenogenetic. A single apterous female produces 8 to 22 nymphs per day. These mature in 3 to 4 days and total life-cycle lasts for 7 to 9 days. The winged forms appear after several generations, specially when there is over crowding. Overwintering is in egg stage; which are laid just before Winter starts and the adults normally perish during winter.

Myzus persicae (Sulzer) peach green aphid, is another species found on cucurbits. It is also cosmopolitan in distribution and highly polyphagous; peach and other temperate fruit trees being its main hosts. This aphid cannot live long on cucurbits and it keeps on moving from plant to plant, sampling the same for food and incidentally spreading mosaic virus disease of cucurbits (Dickson *et al.,* 1940). Both alate and apterous forms appear at different limes of the year depending upon environmental conditions. The alate forms have a characteristic black sclerotic patch on the mid-abdominal dorsum,

To prevent population build-up and damage by aphids, clip-off and destroy immediately the affected shoots and twigs along with the crowded pest population thereon, in the initial stage of attack. In case of severe infestation spray the crop with 0.05% dimethoate, monocrotophos or endosulfan. Repeat the spraying after 10 to 12 days to effectively check the pest population.

Stink Bugs : *Apspongopus brunneus* Thunberg. *A.janus* (Fabricius) and *A. observus* Fabricius have been often found on cucurbits, specially on pumpkins and various gourds. *A.janus* is widely distributed all over India while *A. observus* is found mainly in North-east India and *A. brunneus* in South India. Clusters of these bugs may be found clinging to the leaves and tender shoots. Both nymphs and adults suck the cell sap and thereby devitalise the plants and retard their growth. The bugs also emit characteristic buggy smell – hence the common name stink bug.

Adults are flat, medium-sized insects. *A. janus* is about 30 mm long; pronotum and base of elytra are bright red while head and wing membrane are black. *A. brunneus* is pale brown in colour and slightly smaller in size. Eggs are deposited in long rows on leaves. Egg and nymphal durations last for 9 to 10 and 24 to 28 days respectively.

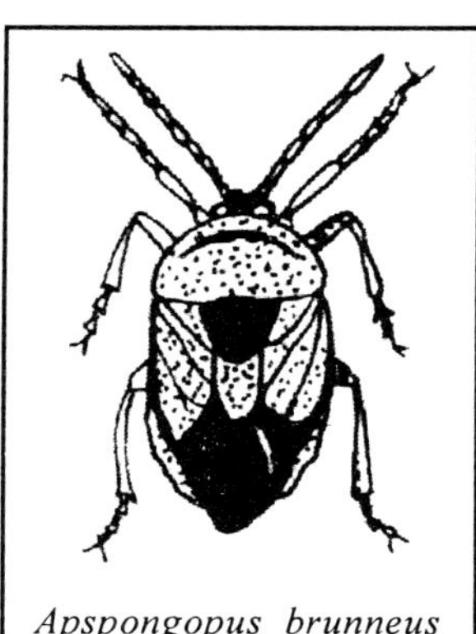

Apspongopus brunneus

In early stage of attack collect and destroy the leaves and twigs bearing congregating bugs. To control, spray 0.2% carbaryl or 0.05% endosulfan or dust 5% malathion or carbaryl dust.

Leaf-footed Plant Bug : *Leptoglossus australis* (Fabricius) (Coreidae) is another pest, commonly found in South-east Asia. It has also been recorded from Western, Central and Eastern Africa, South China, Philippines, Indonesia, New Guinea, Pacific Islands and North Australia (CIE map No. A-243). This is polyphagous pest and though cucurbits are its main hosts, bitter gourd is the most preferred one (Fernando, 1957; Jadhav *et al.,* 1979). Besides it has also been found feeding on cotton and orange in Malaysia (Pagden, 1928), *Citrus* spp. and tomato in Sri Lanka (Hutson, 1936), passion fruit in Kenya (Hill, 1975), banana, citrus, cotton and pomegranate in India (Jadhav *et al.,* 1976) as also cocoa, coffee, groundnut, oil palm, sweet potato, rice, yam and various legumes (Butani and Jotwani, 1984). The bugs congregate on terminal shoots or puncture the young fruits and suck the juice therefrom. The terminal shoots wither and die off beyond the point of attack whereas affected fruits develop dark spots at points where the punctures are made and gradually these fruits shrivel and fall prematurely. Though the pest is quite common and occurs widely, the damage is seldom severe. Eggs are cylindrical in shape and light brown in colour. Full grown nymphs are 12 to 16 mm long and blackish in colour. The head is characterised by two black spines on the upper side in between the eyes; thoracic plate carries a spine laterally on either side and abdominal segments are also protruded laterally into spines. Wing–pads appear after third instar, when hind tibia also flatten out as leaf-like structures. Adults are about 20 mm long, black coloured bugs. Antennae are 4-segmented, black and yellow alternately; mesothorax shield is produced laterally into a spine on either side; a minute yellow spot is present in the centre of each forewing and on hind flat tibia. Despite the common occurrence of this pest, control measures are seldom required against it. Spray 0.2% carbaryl or 0.05% endosulfan to check the pest population.

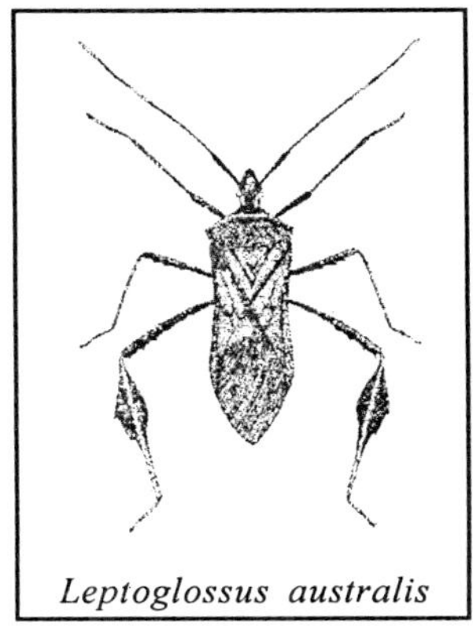

Leptoglossus australis

Leafhoppers (jassids) : *Eutettix phycitis* Distant and *Empoosca binotata* Pruthi are found feeding on leaves of cucurbit vines; the former prefers bitter gourd while the latter is a pest of potato. Damage caused to cucurbits is usually minor and needs no control measures.

Mirid Bugs: *Creontiades palidifer* Walker and *Cyrtopeltis tennis* (Reuter) have been found feeding on cucurbitaceous plants but not as pests of economic importance. These have been reported as pests to potato and tomato respectively. *Metocenthus pulchellus* Dall (Miridae) is another sap sucking bug found feeding on bottle gourd and pumpkin leaves causing negligible damage; its main host being tobacco.

Snake gourd Semilooper: *Anadevidia peponis* (Febricius) as the name suggests, is a serious pest of snake gourd. Eggs are laid singly on ventral surface of leaves. Caterpillars cut edges of leaf lamina, fold it over the leaf and feed within the leaf roll; Pupation takes place in silken cocoons inside these folded leaves. The infestation thus becomes quite conspicuous.

Eggs are spherical in shape, beautifully sculptured and greenish-white in colour. Caterpillars are 30 to 40 mm long with whitish-green elongated body having black warts, off-white longitudinal stripes and a hump on its anal segment. Adults are stout, dark-brown moths with wing expanse of 30 to 35 mm. Egg, larval and pupal periods last for 4 to 5, 24 to 30 and 7 to 8 days respectively and a life-cycle is completed in about 6 weeks.

To cheek the damage by this semilooper, collect the folded leaves and destroy all the caterpillars and pupae. Dust 5% carbaryl or 4% endosulfan or spray 0.2% carbaryl or 0.05% endosulfan. In nature, caterpillars are parasitised by, *Apanteles plusiae* Viereck, *A. taragamae* Viereck and *Mesochoris plusiaephlus* Viereck.

Plume Moth : *Sphenarches caffer* (Zeller) Pterophoridae is widely distributed in West, South and East Africa, Maldive Islands, Indian sub-continent, Philippines, Japan, Indonesia, Australia, New Hebrides and Tonga Islands in Oceania and West Indies (Fletcher, 1920a). It is a polyphagous pest and among cucurbits, it has been reported damaging bottle gourd and *Luffa* spp. (dish-cloth gourds). The pest is found breeding throughout the year. Eggs are laid singly on leaves. On hatching the caterpillars start feeding on leaf lamina making small holes.

Eggs are oval (0.5 x 0.3 mm) in shape, bluish-green in colour with reticulate designs. The caterpillars are small, cylindrical, 7 to 9 mm long when full grown, green in colour, having a lateral brown stripe on either side and clothed with dense pubescence of short spines and long capitate hair. Pupae are 6 to 8 mm long, spiny and greenish-brown in colour. Moths are slender, having both pairs of wings divided into narow lobes fringed with scales. Wing expanse varies between 13 and 16 mm. Incubation, larval and pupal durations last for 2 to 6. 17 to 30 and 5 to 22 days respectively (Fletcher, 1920a). Total life-cycle occupies 24 to 27 days during September but during Winter, it may take as long as 58 days.

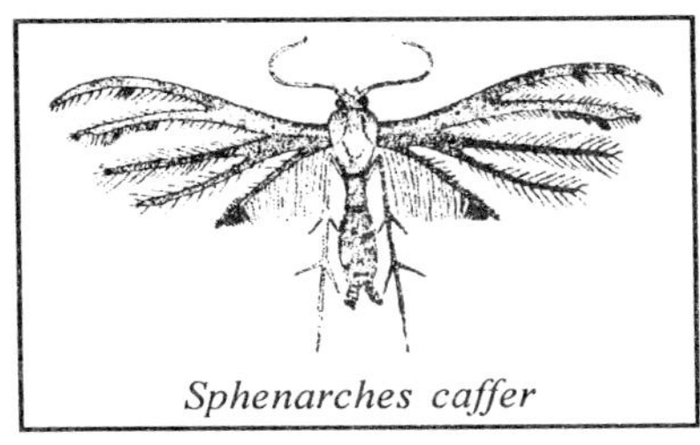

Sphenarches caffer

To check the damage by this pest, collect and destroy larvae and pupae in the early stage of damage. These can be easily spotted by typical damage symptom on leaves. In case of severe infestation, dust 5% carbaryl or spray 0.05% dichlorvos or endosulfan. In nature, the caterpillars are parasitised by *Apanteles paludicolae* Cameron, *A. pusaensis* Lal and *A. ruidus* Walker (Narayanan, 1941, Pruthi, 1946).

Pumpkin Caterpillar : *Diaphania indica* (Saunders) (Pyraustidae) is widely distributed all over the world. It is a minor pest, found on almost all cucurbitaceous plants. The eggs are laid on leaves. On hatching, the young caterpillars lacerate and feed on chlorophyll of foliage. Later, they fold and web together the leaves and feed within. Pupation takes place inside these folded leaves in flimsy silken cocoons. The caterpillars have also been observed damaging the ovaries of flowers and boring into young developing fruits (May, 1946). Affected flowers bear no fruits and infested fruits become unfit for human consumption.

The caterpillars are elongated in shape and bright green in colour, having two narrow longitudinal stripes dorsally. Adults are medium-sized moths, quite conspicuous, as all the four wings are white and transparent with big brown marginal patches. Females have tuft of orange coloured hair at their anal end. Incubation, caterpillar and pupal periods last for 4 to 6, 11 to 14 and 5 to 15 days respectively (Patel and Kulkarny, 1956).

In early stages of infestation, hand-picking of caterpillars and their mechanical destruction helps in keeping the pest population under check. In case of severe infestation, dust 5% carbaryl or spray 0.05% endosulfan or 0.2% carbaryl.

Stem Gall Fly : Dipterous fly, *Lasioptera falcata* Felt (Cecidomyiidae) is a minor pest of bitter gourd reported from South India. Eggs are laid on tender shoots. Maggots bore inside the distal shoots and feed within. The infested vines develop galls. These galls are slender, attennate-fusiform, solid indehiscent angulated and costate but generally cylindrical swelling of branches, finely pubescent and otherwise smooth (Mani, 1973).

The only control measure recommended is to remove and destroy the affected shoots in early stage of infestation.

Lasioptera falcata

Stem Boring Beetles: Number of longicorn beetles, *Apomecyna* spp. (Lamiidae) are widely distributed all over the world. The better known large species are mainly the pests of forest trees as their grubs can bore in hard woody tissues. A smaller number of less robust forms recorded on cucurbits include *A. saltator* Fabricius (=*pertigera* Thomson), *A. alboguttata* (=*quadrifasciata* Thomson), *A. histrio* Fabricius and *A. perotetti* Fabricius. Of these *A. saltator* is widely distributed all over India whereas *A. perotetti* is more common in South India while *A. histrio* has been recorded only from Bihar. All these beetles are polyphagous. *A. alboguttata* has been reported on ridge gourd and smooth or sponge gourd, while *A. sallator* has been reported damaging ivy gourd, bottle gourd, ridge gourd, snake gourd, sponge gourd, pumpkin etc. (Beeson, 1941; Nair, 1975). Eggs are laid singly in the epidermis of the stems of cucurbit vines. On hatching, grubs bore into the long trailing stems usually at or near a node and tunnel inside along the central pith eating away the surrounding tissues. Pupation takes place inside these tunnels. The beetles emerge by biting their way out of the stems and feed by gnawing the leaf petioles and softer parts of the stems, occasionally the leaves are also eaten making irregular holes in leaf lamina. Though the beetles occur quite often, the damage caused is seldom severe.

Eggs of *Apomecyna saltator* are elongated-oval in shape, about 1.5 mm long and creamy-white in colour with reddish tinge. Grubs are brownish in colour having flattened head and thorax, soft and distinctly segmented abdomen and when full grown these measure 18 to 22 mm in length. Pupae are exrate type, pale yellowish in colour, 10 to 13 mm long and abdomen having lateral projections and hair. Adults are 12 to 18 mm long, brown coloured beetles, densely pubescent above, less closely below. The head is bent inwards; thorax and elytra are marked with a number of big and small irregular white spots. These spots run into each other, to appear as a pair of large white irregular patches on each elytron; one is little below the base and the other between the apex and the middle region. Adults of *A. alboguttata* are comparatively smaller in size being 8 to 10 mm long, dark brown in colour with round white spots on thorax and elytra. *A. histrio* beetles have a pair of white spots which are as big as in A. saltator but differ in design of elytra. In these beetles at the base of each elytron there is an oblique row of 4 white spots and just behind these another 4 spots between the middle and apex. Adults are able to make cocophony (sharp squeaking noise) by rubbing the inner surface of hind margin of pronotum on a small oval shaped

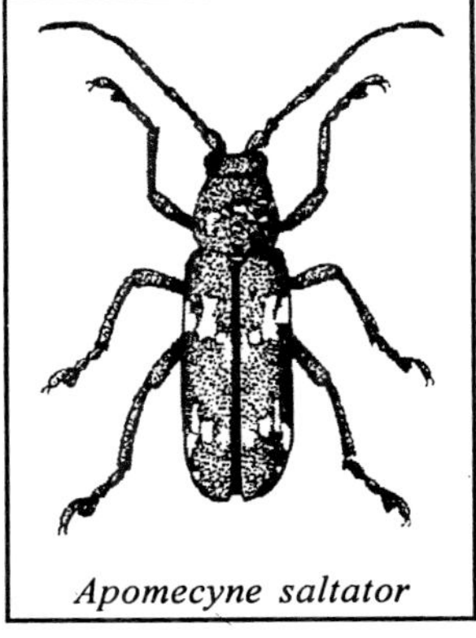

Apomecyne saltator

smooth and highly polished raised surface on mid-dorsal region of meso-thorax. Incubation, grub and pupal periods last for 5 to 6, 22 to 33 and 6 to 8 days respectively; a life-cycle is completed in 35 to 46 days and adult longevity is 33 to 39 days (Lefroy, 1910). There are 3 to 4 generations in a year; 2 to 3 short ones during Summer and monsoon seasons and a large one during Winter. The pest overwinters as grub inside the stems of host plants from October to end of February. Emergence from dried stems of vines takes place usually during May. Mating takes place soon after emergence. During mating the females move about freely and keep on feeding, carrying the copulating males on their back.

As these are minor pests, no control measures are normally adopted against these beetles. Jotwani and Butani (1980) suggested cutting and destroying the affected stems. This will prevent the population build-up as also spread of the pest infestation.

Weevils : Small dark weevil *Acythopius citrulli* Marshall called water-melon weevil, commonly bores into ripening fruits; though its preferred host is water-melon, it occasionally bores into other cucurbit fruits as well. Gray weevil. *Myllocerus blandus* Faust has been reported nibbling the leaves of pumpkins. This is a polyphagous pest and has been found feeding on leaves of eggplant, crucifers, okra, etc. A small dark weevil, *Bans trichosanthis* Subramanium has also been reported feeding on tender shoots of snake gourd in South India.

Generally, no control measures are undertaken against these weevils. However, when serious, dusting 5% carbaryl or 4% endosulfan or spraying 0.2% carbaryl or 0.05% endosulfan will check these weevils effectively.

MITES

Tetranychus neocatedonicus Andre (Tetranichidae), is one of the most destructive pests of cucurbits. Though various cucurbits are its main hosts, it is also found on a number of other crops including beans, eggplant, cole crops, okra, onion, peas, potato, sweet potato, tamarind, tomato and some ornamental plants.

Colonies of these mites can be seen on ventral surface of thick leaves of the infested plants sucking the plant sap, protected under fine silken webs. The damage caused by desapping and covering the leaves with thick webs on which soil particles collect during windy weather results in leaf drop which in turn affects the growth as also the flowering and fruit formation adversely. The maximum activity of the pest has been observed during post-monsoon period and -the activity declines with drop in temperature. All the stages except eggs, are destroyed in rains and there is rapid multiplication after rainy season that continues till November.

Eggs are laid at random on the webbings on ventral surface of leaves. Pre-oviposition period is 1 to 2 days whereas oviposition lasts for 8 to 12 days during March to September and 20 days during October to December. There is generally no egg laying in January and February. Fertilised female lays 61 to 93 eggs and unfertilised one 39 to 59 eggs. The duration of various immature stages varies considerably depending upon the prevailing climatic conditions. The eggs hatch in 2 to 6 days during April to October but take as long as 30 days in Winter. Larval and nymphal development takes 1 to 6 and 2 to 9 days during May to October in case of males and females respectively and as long as 20 and 27 days during January-February. The total life-cycle is completed in 4 to 8 and 6 to 10 days during April to October in case of males and females respectively and 50 and 57 days in winter. There may be as many ay 32 overlapping generations in a year. Parthenogenesis is common but the progeny obtained comprises only males.

Eggs are spherical, minute (about 0.1 mm in diameter) and transluscent. Freshly hatched larvae are almost spherical in shape, 0.1 to 0.2 mm in diameter and light amber coloured; later these become elongated in shape and greenish in colour; they have three pairs of legs and 2 small dark specks dorsolaterally. one on either side. Protonymphs are slightly bigger than larvae, being 0.2 to 0.3 mm in diameter, with longer bristle on dorsum and 4 pairs of legs. They are deep green in colour; males are elongate in shape while the females are ovate and the dark specks on dorsum are also slightly bigger. Deutonymphal stage occurs only in case of females; these are slightly bigger than protonymphs and the genitalia are visible. Adults are ovate in shape - males being 0.3 to 0.4 mm and females 0.4 to 0.5 mm. These are greenish-gray to light brown in colour with deep red eyes; abdominal segmentation is quite distinct and there are 12 pairs of hair on the dorsum. the two black spots of the dorsum increase in 'size with age and finally cover the entire dorsum.

Besides *Tetranychus neocaledonicus*, Nassar and Ghai (1981) have reported, *T. angolensis* Meyer, *T. cinnabannus* (Boisduval), *T. urticae* Koch. *Bravipalpus californicus* (Banks), *B. phoenicis* (Geijskes). *Euretanychus anneckei* Meyer. *E. maximae*. Nassar and Ghai and *E. orientalis* (Klein) damaging different cucurbit vines. To this Butani and Jotwani (1984) added *T. Iudeni* Zacher, *T. macfarlanei* Baker and Pitchard.

To control the mites, if and when necessary, dust with fine sulphur or spray 0.1% wettable sulphur.

5

EGGPLANT

EGGPLANT, *Solanum melongena* Linnaeus (Solanaceae) is a native of Northern India and is extensively grown in China and South-east Asia. Seeds were carried to China about 1500 years ago and to Europe by the Arab traders during the Dark Ages; it was introduced into Brazil before the close of seventeenth century (Verma, 1980). It is a herbaceous, erect or semi-spreading plant. The fruit also called aubergine is a berry, borne singly or in clusters. Though a perennial plant, it is usually grown as seasonal crop but in almost all the seasons. It can withstand varied climatic conditions except frost. A wide genetic variation with regard to colour, shape and vegetative growth exists among the indigenous varieties. 'Pusa Kranti' is a high yielding variety that can be grown both during Spring and Autumn seasons in Northern India (Choudhury *et al.*, 1971). Eggplant is one of the most common vegetables and its fleshy violet coloured fruits are too prominent to be missed by anyone visiting a vegetable market. It contains vitamins A, B and C and has got *Ayurvedic* medicinal properties as well; white fruit is said to be good for diabetic patients (Choudhury, 1967). According to Tomar *et al.* (1996) eggplant seed-oil is rich in linoleic and linolenic acids (87.94%) as compared to groundnut oil (75.14%), soybean oil (87.24%) and sunflower oil (94.46%).

INSECT PESTS

The crop is attacked by a number of insect pest; those of major importance include shoot and fruit borer, stem borer, leaf roller, hadda beetles, jassid, lace-wing bug and aphids. The minor or sporadic insect pests are termites, grasshoppers, mealy bugs, thrips, leaf miners, bud worms, whitefly etc.

Shoot and Fruit Borer : *Leucinodes orbonalis* Guenee (Pyraustidae) also called eggplant caterpillar, is the most destructive pest of eggplant, widely distributed in the Indian sub-continent as also in Thailand, Laos, South Africa, Congo and Malaysia. Besides eggplant, which is its main host, it also damages potato and other solanaceous crops. The pest is active throughout the year at places having moderate climate but its activity is adversely affected by severe cold. The damage by this pest starts soon after transplanting of the seedlings and continues till harvest of fruits. Eggs are laid singly on ventral surface of leaves, shoots, flower-buds and occasionally on fruits. In young plants, the caterpillars bore into petioles and midribs of large leaves and young tender shoots, close the entry point with their excreta and feed within. As a result, the affected leaves dry and drop down while in case of shoots, the growing point is killed. Appearance of wilted drooping shoots is the typical symptom indicating damage by this pest; these affected shoots ultimately wither and die away. At later stage, the caterpillars bore into flower buds and fruits, entering from under the calyx, they leave no visible sign of infestation, while the caterpillars feed inside. The damaged flower buds are shed without blossoming whereas the fruits show circular exit holes. Such fruits, being partially unfit for human consumption, lose their market value considerably. Hami (1955) found that in affected fruits vitamin C (ascorbic acid) is reduced to the extent of 68 per cent. Peswani and Rattan Lal (1964), based on market surveys, reported that 20.7% fruits were damaged by this borer and further stated that even if only damaged portion of these fruits is discarded the loss in weight comes to 9.7 per cent. This is an underestimate, as damage caused at earlier stage by the boring of the leaves and shoots has not been taken into account. In Bangladesh, the pest is reported to cause 1 to 16% damage to shoots and 16 to 64% to fruits (Alam and Sana, 1964). Long and narrow fruits are usually less susceptible to the attack by this borer (Srinivasan and Basheer, 1961).

Eggplant

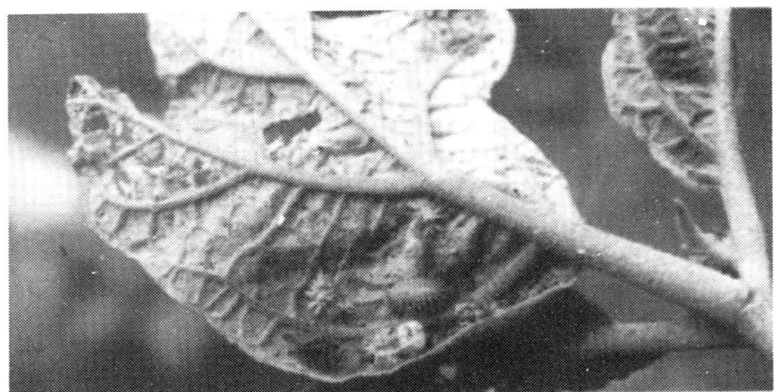

Epilachna on Eggplant leaves

Eggplant damaged by *Leucinodes orbonalis*

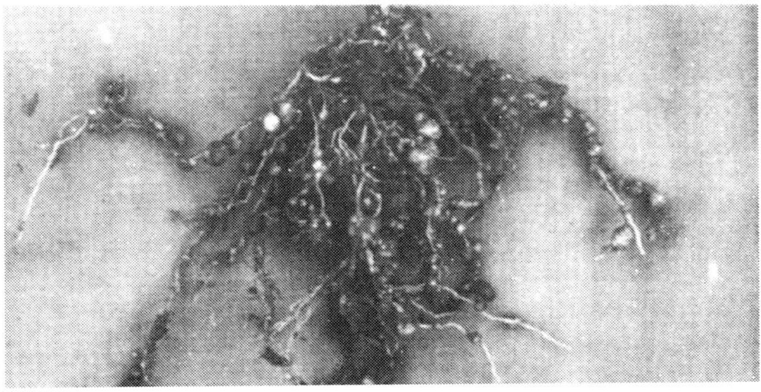

Eggplant roots affected by root-knot nematode

Eggplant fruits damaged by *Leucinodes orbonalis*

Eggplant fruits damaged by *Leucinodes orbonalis*

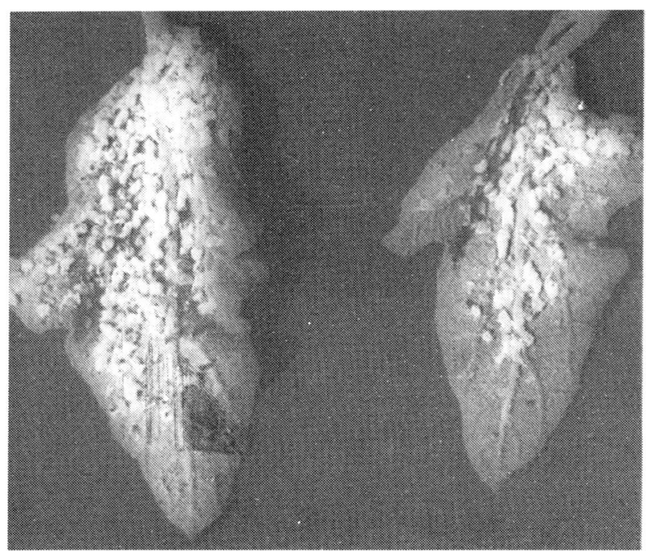

Coccidolystrix on eggplant leaves

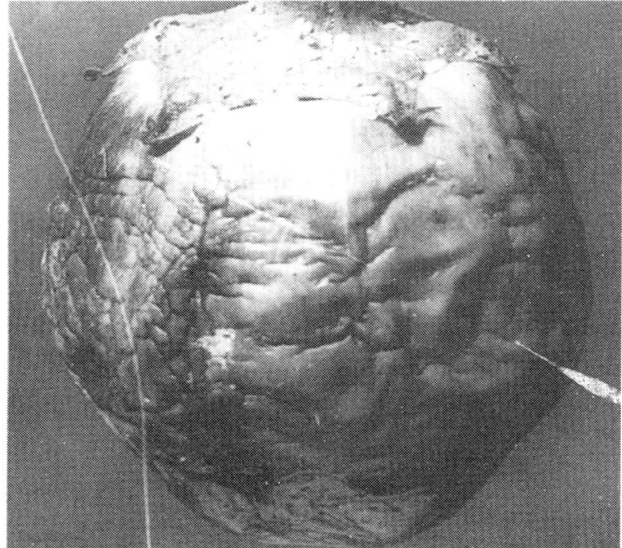

Coccidolystrix on eggplant fruit

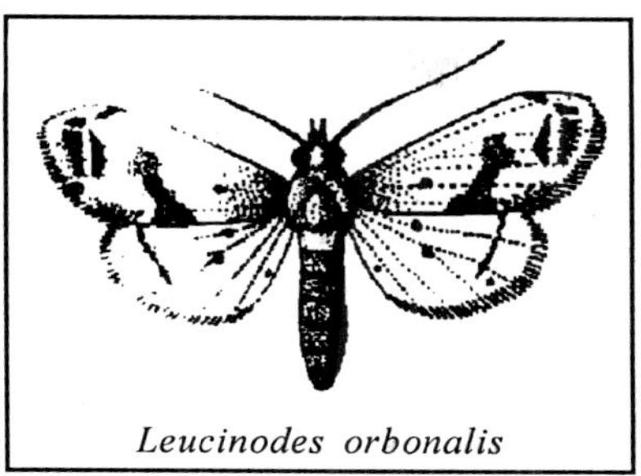
Leucinodes orbonalis

Eggs are flattened, elliptical in shape, 0.5 mm long and creamy-white in colour. Full-grown cater-pillars are 15 to 18 mm long and light pink in colour. Moths are medium sized with forewings having cons-picuous black and brown patches and dots; hind wings are opalescent with black dots along the margins. Wings span is 22 to 26 mm.

A female lays on an average 250 eggs. These hatch in 3 to 5 and 7 to 8 days during Summer and Winter respectively. Larval duration is 12 to 15 days during Summer extending upto 22 days in Winter. Pre-pupal period is 3 to 4 days whereas pupal period is 7 to 10 days and may last upto 15 days during Winter.

As continuous cropping of eggplant crop favours heavy infestation, the same should be avoided. Grow relatively less susceptible varieties like 'Arka Kusumakar', 'Pusa Purple Round' etc. (Kale, *et al.*, 1986; Bhatia *et al.*, 1995). Collect all attacked shoots and fruits and burry the same at least one metre deep (Prem Chand, 1995). Residue of DDT may persist on fruits (Jotwani and Sarup, 1963; David, 1963; Srinivasan and Gowder, 1959 a and b). application of HCH may impart off-odour and affect taste of fruits (Vevai, 1970). Dichlorvos or endosulfan being comparatively safe, may be sprayed 2 to 3 times at fortnightly intervals depending upon the infestation, the concentration recommended is 0.05% (Deshmukh and Udeaan, 1972). Jagan Mohan *et al.* (1980) suggested 5 fortnightly sprayings with synthetic pyrethroids - fenvalerate or permethrin - @ 0.1 kg a.i. per hectare, while Patil *et al.* (1996) found cypermethrin 0.25% dust @ 20 kg per hectare reduced bored infestation and increased the fruit yield.

In nature, caterpillars are parasitised by *Pristomerus testaceus* Morley, *Cremastus flavoobitalis* Cameron, *Shirakia schoenobii* Viereck, *Microbracon greeni* Ashmead, *Iphiaulax* species, *Phanerotoma* species and *Pseudoperichaeta* species but the parasitisation is generally low.

Eggplant Stem Borer : *Euzophera perticella* Ragonot (Phycitidae) is another destructive pest of eggplant which also attacks chilli, tomato and at times even potato. It is widely distributed all over the Indian sub-continent. Eggs are laid singly or in clusters on young leaves, petioles or even tender shoots. Soon after hatching, the caterpillar bores into stem and moves downwards. The attacked plants wither and wilt, the growth is stunted and fruit-bearing capacity is adversely affected.

Full-grown caterpillars are 16 to 18 mm long and light brown in colour. Pupae are dark brown. Moths are medium-sized; forewings are pale rufous with distinct dentate vertical black lines beyond middle of the wing and hind wings are whitish. Wing expanse is 26 and 32 mm in case of males and females respectively. Incubation, caterpillar and pupal stages last for 3 to 10, 26 to 58 and 9 to 16 days respectively with 3 to 4 generations in a year (Srivastava, 1961) extending upto 8 overlapping generations (Prem Chand, 1995).

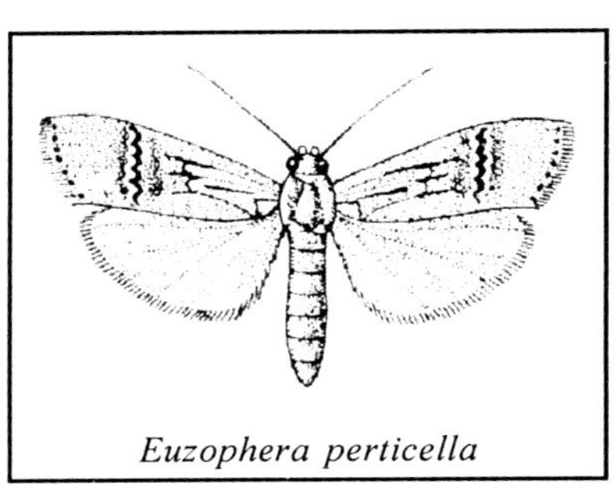
Euzophera perticella

The control measures suggested for shoot and fruit borer are effective against this borer as well. It is desirable to check the infestation in the initial stage by mechanical methods. Ratoon crop should be avoided.

Eggplant Leaf Rollers : *Antoba (Eublemma) olivacea* (Walker) (Noctuidae) is a common foliage pest of eggplant. Besides eggplant, it also attacks other wild solanaceous plants (Singh, 1970). Eggs are laid on leaves. On hatching, the caterpillars fold leaves from tip downwards and feed within by scrapping the green matter; as a result of this damage, the folded leaves wither and dry up (Tirumala Rao and Koteswara Rao, 1955). Pupation takes place within the folded leaves. The caterpillars may also bore in green shoots and feed on inner tissues resulting in withering of the entire plant.

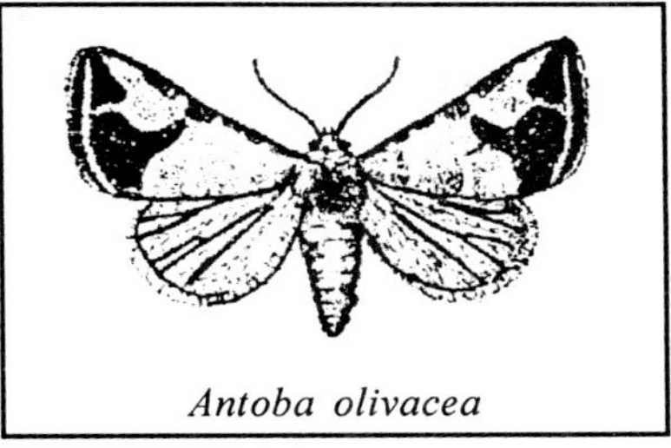
Antoba olivacea

Full-grown caterpillars are about 20 mm long, stout, purple-brown in colour and ornamented with yellow spots and hair. Moths are medium-sized and ochreous-white in colour; forewings are slightly suffused with brown tinge and a large triangular olive-green patch on the outer area; hind wings are white suffused with fuscous towards outer margin. Wing spread is 22 to 26 mm.

Pterophorus lienigianus Zeller is another leaf roller recorded as a minor pest of eggplant. It is found all over the Indian sub-continent. The caterpillars feed inside the rolled leaves, and are stout, pale yellowish with brown head and hairy body and are 8 to 10 mm long when full-grown.

To control these leaf rollers, collect and destroy the rolled leaves with caterpillars and pupae inside. If the infestation is severe, spray 0.05% endosulfan or 0.2% carbaryl.

Leaf Feeding Beetles : A group of coccinellids, commonly known as hadda beetles, have been reported as serious pests of various vegetable crops from different parts of the world. The species reported damaging eggplant are *Henosepilachna viginitioctopunctata* (Fabricius), *Epilachna dodecastigma* Mulsant, *E. implicata* Mulsant, *E. ocellata* Redtenbacher, *E. septima* Dieke and *E. demurili*. *E. varivestis* Mulsant - the Mexican bean beetle - is confined to Southern Canada, USA, Mexico and Guatemalaa (CIE map No. A-46) where it is a serious pest of beans (*Phaseolus* spp.) while *E. hirta* (Thunberg), *E. fulvosigmata* Reiche and *E. similes* (Thunberg) are common in Africa on cucurbits. These species have not yet been intercepted from South-east Asia. These phytophagous and harmful species are often confused with beneficial predatory conccinellid beetles popularly known as ladybird beetles. Of the species recorded in South-east Asia, *H. vigintioctopunctata* is the most common and destructive one, especially to eggplant, bitter gourd, bottle gourd, snap gourd, potato and tomato. The adults being strong fliers, infest wide areas during their peak activity period.

Eggs are laid in clusters on ventral surface of leaves. A female lays 120 to 180 eggs in 4 to 8 batches, each of 20 to 35 eggs. Both grubs and adults feed by scrapping chlorophyll from epidermal layers of leaves, the affected leaves get skeletonised and gradually dry away. Eggs are elongated, cigar-shaped and yellowish in colour. Grubs are also yellowish in colour, stout and have spines all over their bodies. Pupation takes place on leaves and the pupae are hemispherical in shape. Adults are spherical, pale brown and mottled with black spots; *H. vigintioctopunctata* has 14 spots on each elytron whereas *E. dodecastigma* has only 6 spots on each elytron.

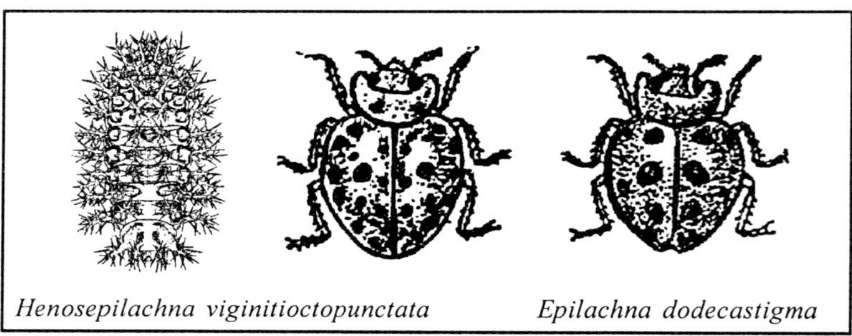

Henosepilachna viginitioctopunctata *Epilachna dodecastigma*

Incubation, grub and pupal periods last for 2 to 4, 12 to 18 and 3 to 6 days respectively (Pradhan, 1959). Entire life-cycle occupies 18 to 25 days during Summer and upto 50 days in Winter. There are 7 generations in a year, according to Krishnamurti (1932).

The infested leaves can be easily spotted in the field and in the initial stage of attack the infestation can be checked by collecting and mechanically destroying the affected leaves as also in the different stages of the pest. In case of severe attack, the pest can be controlled by spraying 0.05% dichlorvos or endosulfan (Butani and Verma, 1976a). According to Venkataramireddy *et al.* (1993), spraying with neem extract gives 95% mortality.

Amongst the other beetles, *Lema praeusta* (Fabricius), *L. semiregularis* Jacoby and *L. semiregularis* Jacoby have been recorded as minor pests in Orissa. Their main host is turmeric. Eggs are laid singly, thrust into leaf tissues. On hatching, the grubs and adults feed on leaf tissues. Affected leaves dry and wither.

Eggs of *Lema praeusta* are flate, ovoid (0.9 x 0.7 mm) and light brown in colour. Grubs are 6 to 8 mm long, pale yellowish-white in colour with black head; apex of anal segment is turned upward so that faecal matter falls on their dorsum and forms a sort of protective covering. The excrement is greenish-yellow which gradually hardens and becomes white. Pupation takes place between leaves under characteristic covering of excreta. Pupae are 5 to 6 mm long and orange-yellow in colour. Adults are small beetles 5 to 6 mm long having dark brown elytra with apical one-third blue. Grubs of *L. semiregularis* are smaller in size being 5 to 6 mm long. Pupae are 4 to 5 mm long with pale brown head and yellowish body. Adults ;are 4 to 5 mm long with bluish elytra.

Egg, grub and pupal periods of *Lema praeusta* are 8 to 10, 10 to 12 and 15 to 25 days respectively and total life cycle occupies 40 to 50 days (Sengupta, 1957), while that of *L. semiregularis* is 4 to 5, 15 and 19 days respectively and the total life-cycle is completed, on an average in 39 days (Sengupta and Behrua, 1956).

Leafhoppers : *Amrasca biguttula biguttula* (Ishida), *Empoasca binotata* Pruthi, *E. parathea* Pruthi, *E. punjabensis* Pruthi and *Cestius phycitis* (Distant) have been reported feeding on eggplant leaves - the first one being comparatively more common and destructive. *A. biguttula biguttula* commonly known as cotton leafhopper or jassid is and outstanding pest of cotton and okra. Nymphs and adults suck cell sap from ventral surface of leaves and inject toxic saliva into the plant tissues. Besides, these insects also transmit the viral disease, called little leaf.

Whenever necessary, spray 0.04% dimethoate or phosphamidon, or 0.01% fenvalerate or 0.005% permethrin. As synthetic pyrethroids have quick knock-down effect, low mammalian toxicity and comparatively over residues, these may be preferred. But again, lest the insects may develop resistance against these pyrethroids, these should not be used continuously. It is suggested to use a pyrethroid and and organophosphatic insecticide, alternately. Recently, Walunj and Dethe (1997) reported that carbamates - thiodicarb and methomyl @ 0.5 to 1.0 kg a.i./ha-effectively minimized the leaf hopper and witefly population. Carbosulfan was reported effective at 200, 250 and 300 g a.i. ha to control the leafhopper. Sheeba *et al.*, (2005).

Sap Sucking Bugs : Black bug, *Anoplocnemis phasiana* (Fabricius) (Coreidae), a cow bug *Tricentrust bicolor* Distant (Membracidae) and plant bug *Coptosoma nazirae* Atkinson (Corimaluenidae) have also been reported sucking the sap from leaves and tender shoots. *Halticus minutus* Reuter (Miridae) damages the seedlings; *Creontiades pallidifer* Walker (Miridae) and *Aspongopus janus* Fabricius (Pentatomidae), though pests of potato and pumpkin, respectively, are also occasionally seen on eggplant leaves.

If and when these bugs appear in large number dust 4% endosulfan or spray 0.05% dichlorvos or quinalphos to check their infestation (Butani and Jotwani, 1984).

Lace-wing Bugs : *Urentius hystricellus* (Richter.) (=*echinus* Distant) and *U. sentis* Distant (Tingidae) are specific pests of eggplant. They are more common in North-western India and adjoining region of Pakistan. Both nymphs and adults suck the cell sap from leaves. Affected leaves become yellowish and are found covered with nymphal exuviae and excreta. The bugs are active from April to October, maximum activity being during August-September; as such summer crop suffers comparatively more damage than the winter crop.

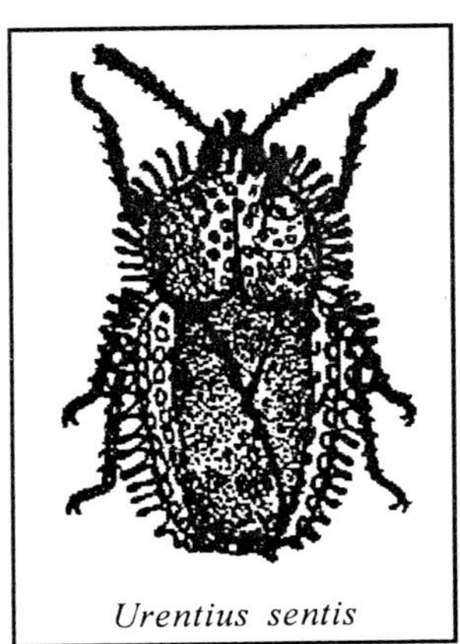

Urentius sentis

Nymphs of *Urentius sentis* are about 2 mm long, pale ochraceous in colour, stoutly built with prominent spines. Adults are about 3 mm long, straw coloured dorsally, and dark brown to blackish ventrally; females are oval in shape while males are more elongated. Pronotum and elytra are reticulated consisting of irregular thick lines forming a frame-work of cells, coastal area is hyaline with strong spines on the outer margin; hind wings are whitish and transparent.

A female lays 35 to 45 eggs singly in the leaf tissues. Eggs hatch in 3 to 12 days and nymphal period lasts for 10 to 23 days; longevity of male and female is 30 and 40 days respectively (Bhandari and Soni, 1962). There are 8 overlapping generations in a year.

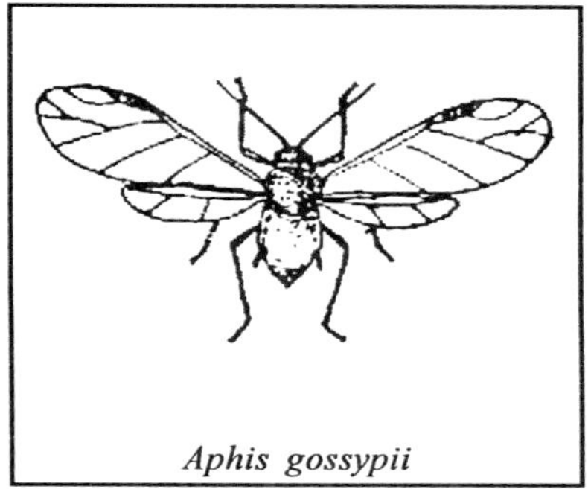
Aphis gossypii

Aphids : Aphids, or plant lice are polyphagous pests having a very wide range of host plants and are present in all the countries on cultivated crops and wild weeds. A number of crops often harbour more than one species and eggplant is one such crop. The two species commonly found on this crop are cotton aphid *Aphis gossypii* Glover and peach green aphid *Myzus persicae* (Sulzer). Both are cosmopolitan in distribution, absent only from colder parts of Asia and Canada (CIE map Nos. A-18 and A-45 respectively).

Both nymphs and adults, suck cell sap from leaves and tender apical shoots. The affected parts turn yellow, get deformed and dry away. Besides, the aphids also secrete copious quantity of honeydew on which sooty mould grows rapidly covering the affected parts with a thick black coating, which interferes with the photosynthetic activity of the plant. The infested plants become weak, pale and stunted in growth which consequently results in reduced fruit size.

Nymphs of *Aphis gossypii* are greenish-brown or yellowish in colour while adults are yellowish-green to dark green in colour, little over one mm in length and have a pair of siphunculi near the posterior side of abdomen. Wings, when present, are transparent with black veins. Adults of *Myzus persicae* are usually of green colour but may be pale-brown to pinkish, 1.5 to 2.5 mm long with long clavate siphunculi. *A. gossypii* breeds during winter on a number of vegetables, including eggplant, from where they migrate in the month of April to melons and by June-end return to cotton (Butani, 1975a). Reproduction in case of *A. gossypii* is parthenogenetic viviparous and rate of multiplication is often phenomenal. *M. persicae* reproduces by parthenogenetical viviparity during Summer, Monsoon and Autumn seasons but sexually in cooler regions during Winter. Adults normally perish due to severe cold and eggs overwinter lie in cracks and crevices of the bark of various temperate fruit trees. When the temperature rises, the eggs hatch and nymphs start feeding on blossoms. They mature in 3 to 4 days and reproduce partheno-genetically, young ones which develop into wingless adults. After 2 to 3 generations, when the temperature rises further, or when there is too much crowding of these aphids, the winged forms are produced and these migrate to other crops including eggplant; again, with fall in temperature, the aphids migrate back to temperate fruit trees.

Aphids being migratory in habit with a wide range of host plants, the chemical control can only alleviate the infestation for some time, making it imperative to give repeated applications. Two to three sprayings at 10-12 days interval with 40% nicotine sulphate (1:600) or 0.5% dimethoate, endosulfan or monocrotophos can effectively check these aphids. Thanki and Patel (1991) found the varieties 'GB-1', 'GB-6' 'Selection-1' and 'Selection-4' to be resistant (less susceptible) to aphids, leafhoppers, thrips and whiteflies.

In nature, these aphids are parasitised or predated upon by *Aphelinus semiflavus* Howard, *Chilomenes sexmaculatus* Fabricius, *Geocoris tricolor* (Fabricius). *Coccinella septempunctata* Linnaeus, *Coelophora bissellata* Mulsant, *Leucopis griseola* Fall, *L. nigricornis* Eggers, *Nephus regularis* Soc., *Pullus xeramphelinus* Mulsant, *Scymnus* species, *Amblynotus syrphidipagus* Narayanan, *Ischiodon scutellaris* Fabricius, *Sphaerophoria javana* Wiedemann, *Syrphus balteatus* de Geer, *S. confractor* Wiedemann, *S. serarius* Wiedemann, *Triphelps tantilus* Motschulsky, *Chrysoperla* species and *Hemorobius* species.

Some birds like the warbler *Phyllowcopus tritis* and rufous fantail *Cisticola cixsitans* also prey on aphids.

Termites : Termites or whiteants are cosmopolitan pests having a wide range of host plants. These abound in sandy or sandy-loam soils and cannot thrive under conditions of bad aeration and poor drainage. The ravages by termites may start from the time seed is placed in the soil and often continue till the harvesting of crop. The species reported damaging eggplant are *Trinervitermes biformis* Wasmann (Termitidae) and *Microtermes* species (Termitidae). These species being subterranean, gnaw the roots and stems below ground level and tunnel upwards through the stems, eating all the inner tissues. The affected plants become pale, wither and ultimately dry away. As the crop is usually grown under irrigated conditions, the damage by termites is usually not severe except during dry periods or when the crop is grown in light soils.

Soil application of 5% HCH, chlorpyrifos, cypermethrin @ 20 kg per hectare is effective in warding off the attack of termites. Avoid growing the crop in sandy soils or termite infested lands. In case of an attack, lindane emulsion may be applied along with irrigation @ 1 kg a.i. per hectare.

Grasshoppers : *Atractomorpha crenulata* Fabricius, *Orthacris simulans* Boliver *Oxya japonica japonica* (Thunberg), and *Poekilocerus pictus* (Fabricius) (Acrididae) are some of the species that often attack young eggplant plants. These are all polyphagous pests. Both nymphs and adults feed on leaves often eating away the leaf lamina completely, leaving behind only the network of major veins and midribs. The grasshoppers are found all the year round except during severe cold but are usually more active during rainy season.

Atractomorpha crenulata is found foraging on eggplant crop during August to September; thereafter its activity increases and the pest migrates and feeds on cabbage, cauliflower and radish grown during that season. *Orthacris simulans* - the wingless grasshopper, is occasionally found feeding on tender leaves. *Poekilocerus pictus*, a major pest of *ak* (*Calotropis* spp.), attacks cowpea, eggplant, okra, tomato etc., only when the main host is not available. *Oxya japonica japonica* is coparatively a small grasshopper (about 30 mm long) that feeds on tender leaves of eggplant.

If and when the grasshoppers appear in large numbers, the prefruiting crop may be dusted with 5 to 10% HCH depending upon the stage of growth of grasshoppers; young nymphs can be controlled with 5% HCH whereas advanced stage nymphs and adults require 10% HCH for satisfactory control.

Coccoids : Mealybug, *Coccidohystrix insolita* (Green) (Pseudococcidae) and scale insects, *Aonidiella orentalis* (Newstead), *A. aurantii* (Maskell), *Quadraspidiatus (Aspidiotus) destructor* (Signoret), *Cerococcus hibisci* Green, *Chionaspis manni* Green (Diaspidae) and *Orthezia insignis* Browne (Ortheziidae) have been found damaging eggplants. Of these onlyl the mealybug is common and comparatively more destructive pest, especially on older plants. Besides eggplant, this mealybug is also found on tomato and some wild plants. It is widely distributed all over the Indian sub-continent.

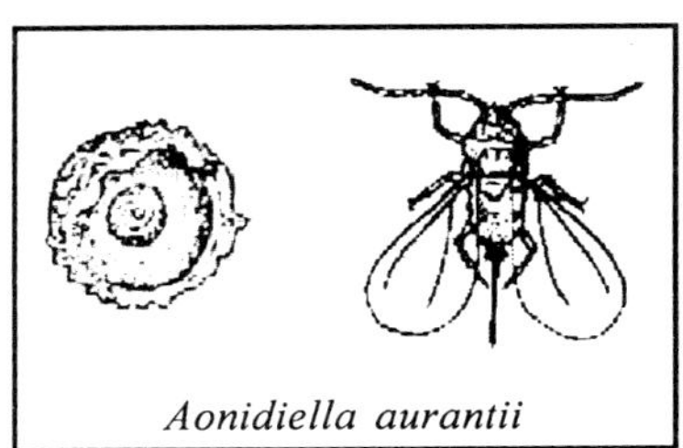

Aonidiella aurantii

Coccidohystrix insolita is a small, oval, soft-bodied insect, covered with white mealy wax. A cluster of these insects may be seen on ventral leaf surfaces, specially during the months of June ot August. The fertilized female lays 100 to 200 eggs in ovisac that protrudes out from the posterior end of the abdomen. Eggs hatch in 3 to 8 days. Males and females pass through complete and incomplete metamorphosis, respectively, and the post-embryonic development takes 10 to 20 days. No parthenogenesis has been recorded in this species.

Aonidiella aurantii, the citrus red scale, is a polyphagous pest found all over the World (CIE map No. A-2). It is found round the year but is comparatively more active from August to October. Adult females are flat and circular in shape, reddish in colour and apodous. Males are somewhat elongated and winged. Reproduction is ovo-viviparous; a female produces 2 to 3 young ones per day for over 2 months (Butani, 1979a). young ones remain protected under the body of the mother for 1 to 3 days, then crawl about a bit, find a succulent and suitable sport on leaves or twigs and settle there permanently. Once settled they soon secrete a waxy substance to cover their entire body.

Aonidiella orientalis, Oriental yellow scale, is a polyphagous pest. Eggs are laid under the protective armour of female body. On hatching, the crawlers move about for sometime, select a soft succulent spot and settle thereon permanently. During the process of moulting these scales lose their legs. They secrete a waxy substance with which they cover their entire dorsum. The crawlers either develop into adult females or pass through pre-pupal and pupal stages and become adult males. Males are winged whereas females are wingless and legless, about 2 mm in diameter and have yellowish-brown coloured body.

Quadraspidiatus destructor, coconut scale, is Pantropical in distribution (CIF map No. A-218). It is a polyphagous pest and besides coconut, it has been found feeding on a wide range of other economic crops including chillies, eggplant and yams. Its activity is maximum during hot months when hundreds of female scales may be seen crowding on ventral surface of leaves. These are circular in outline (1.5 to 2.0 mm in diameter), bright yellow in colour and covered with filmsy, semi-transparent, slightly convex scale. A female lays upto 90 eggs which are deposited under its shield-like scale. The crawlers move out in search of succulent spots and settle there. Males are winged. Life cycle takes 4 to 5 weeks, a little longer in case of females (Menon and Pandalai, 1958). Incubation period is 7 to 8 days and nymphal development takes 21 to 25 and 26 to 30 days in case of males and females, respectively.

Cerococcus hibisci are generally found in shallow pits on the mid-veins of the infested leaves. Adult females are small, elongate-oval insects; their bodies are covered with thick coating of pale yellow waxy mass.

All these coccids suck the cell sap causing yellowing, wilting and drying of affected leaves and thereby devitalise the plants. The coccids also excrete small amount of honeydew which favours the development of sooty mould fungus and attracts the ants, especially, *Camponotus compressus* Linnaeus.

Orthezia insingis, commonly known as lantana bug, has been reported from India, Sri Lanka, Malaysia, some African countries,

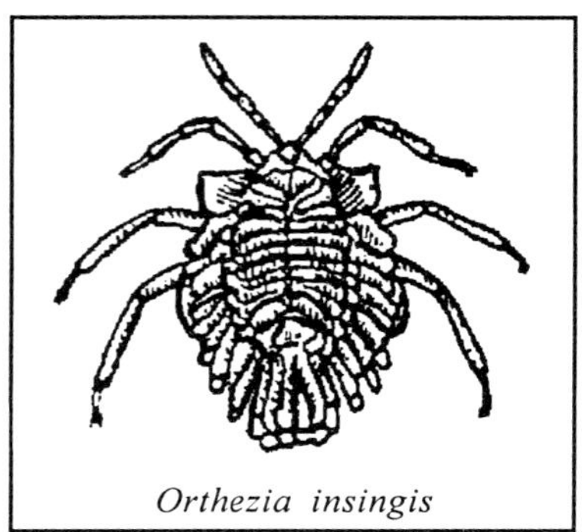

Orthezia insingis

USA, Central and South America (CIE map No. A-73). The main host of this coccid is lantana (*Lantana camara* Linnaeus), an extremely obnoxious weed found in abundance on the hills in India. The pest keeps the weed under check. Besides lantana, it has been recorded feeding on a number of economic crops, including eggplant (Hill, 1983). Both nymphs and adults suck the cell sap from leaves and stems.

Adult females are olive green in colour and have well developed antennae and legs. Segmentation of body is distinct with white waxy plate or laminae at the extremity and the egg-sac is situated between these waxon plates. According to Hill (1983) usefulness of this insect heavily outweighs its nuisance value of damaging the flowering trees; so no control measures are adopted against this scale.

Effective chemical control of scale insects is still a far cry. Fumigation with HCN gas was earlier suggested but this, though affective, is very expensive and requires expert handling and adequate technical knowledge, hence it is neither practicable nor is practiced on fruit trees.

In Australia, UK and USA integrated pest management (IPM) with major emphasis on biological control has proved very promising against these scale insects and is being practiced extensively especially in citrus orchards. In India, not much work has been done on parasites and predators. However, in nature, *Coccidohystrix insolita* is parasitised by *Cheiloneurus latiscapus* Compere and *Leptomastrix nigrocoxalis* Compere and is preyed upon by *Hyperaspis maindronia* Sic. and *Nephus regularis* Soc. *Aspidiotus destructor* is parasitised by *Aneristus ceroplastae* Girault, *Comperiella bifasciata* Howard, *Chartocerus* species, *Chrysonotomyia* species and *Thomsonisca desantisiellus* Shaffee.

Thrips : In addition to eggplant thrips *Scricothrips solnifolii* Shumsher, onion thrips *Thrips tabaci* Lindemann, groundnut thrips *Caliothrips* *indicus* (Bagnall) and chilli thrips *Scirtothrips dorsalis* Hood (Thripidae) have been found infesting and damaging eggplant leaves.

These minute fragile insects are often seen in very large number feeding on ventral leaf surface as the epidermis of ventral side is more tender and thinner than that of dorsal side. Both nymphs and adults lacerate the leaf tissues and lap the oozing sap; as a result white silvery sheens appear on the infested leaves. In case of severe infestation, the leaves start drying from tips downwards.

Caliothrips indicus has been reported feeding on cabbage, cauliflower, knol-khol, chillies, onion, potato, cotton, rice, wheat, linseed etc. it has uniformly blackish-brown body with broad fore wings having 4 dark and 3 pale cross bands and dark pointed apex with a pre-apical pale band. A single life cycle is 11 to 14 days with several generations in a year.

Frankliniella schultzei (Trybom) and *Thrips apicatus* Priesner, the common blossom thrips, have also been reported feeding on flowers of eggplant. A severe infestation causes drying and premature shedding of flowers and as a result fruit setting is adversely affected.

If and when necessary spray 0.03% endosulfan or lindane, or 0.1% malathion.

Bud Worm : *Scrobipalpa blasigona* (Meyrick) (Gelechiidae) is a minor pest of eggplant. Eggs are laid on leaves. Newly hatched caterpillars bore inside the flower buds and feed within. As a result, the affected buds often fall down which ultimately affects the fruiting capacity of the plant. Caterpillars are small, cylindrical and pale whitish with pinkish tinge. They pupate inside the damaged buds. Adults are small moths with heavily fringed wings, measuring 14 to 18 mm across the wings.

If and when this pest appears in large number, dust 5% HCH or spray 0.05% dichlorvos, malathion or endosulfan.

Leaf Miners : *Scrobipalpa ergasima* (Meyrick) (Gelechiidae) and *Phthorimoca operculella* (Zeller) (Gelechiidae) have been recorded as minor pests from Northern India, while *S. heliopa* (Lower) has been reported from South India. *P. operculella* is cosmopolitan in

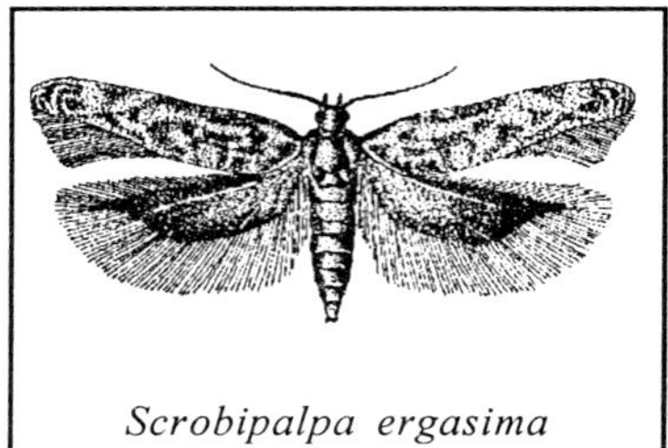
Scrobipalpa ergasima

distribution as the notorious pest of potato.

Scrobipalpa ergasima is active all the year round, maximum during January to March. The caterpillars mine the leaves causing small blotches, especially near the leaf tips. Caterpillars come out of one mine and again start another mine on the same leaf. Pupation takes place either inside the mine or on ventral leaf surface, where the pupa is fastened by a silken thread. Caterpillars are small, about 7 mm long and green in colour. Adults are also small moths having wing expanse of 12 to 15 mm.

Scrobipalpa heliopa - tobacco leaf miner/stem borer - originally described from Australia is widely distributed in South-east Asia. Eggs are laid at night on ventral surface of upper leaves. A female lays 50 to 80 eggs. On hatching, the caterpillars bore into the veins of leaves, midribs, leaf stalks and finally the top shoots and feed within. As a result, gall-like swelling is formed on the plant and the burrows are filled with black excreta pellets. Soon, the affected leaves and shoots start drying up. Pupation takes place within brownish silken cocoons, usually in the stems but occasionally in the midribs. Eggs are elongate-oval, 0.5 mm long, greenish in colour changing to orange-yellow. Full-grown caterpillars are cylindrical in shape, about 10 mm long, transluscent and grayish-white in colour. Puapae are brown in colour, 6 to 7 mm long, cylindrical in shape with tapering posteriorly and small gray cremastral hooks. Adults are small brown moths. Incubation, caterpillar and pupal periods last for 4, 15 to 22 and 6 to 8 days respectively (Nair, 1975). Fletcher (1920) reported that in Bihar, egg, caterpillar and pupal periods and total life cycle takes about 13 to 19, 43 to 67, 18 to 30 and 87 to 115 days respectively.

Chemical control measures suggested are same as those for bud worm. *Chelonus heliopae* Gupta, has been recorded as larval parasite of *Scrobipalpa heliopa.*

Eggplant Leaf Webber : *Psara bipunctalis* (Fabricius) (Pyraustidae) - a minor pest of eggplant, is found throughout the Indian sub-continent. Eggs are laid in batches on under surface of leaves. On hatching, the caterpillars scrap and feed on epidermal tissues, and later, they web the leaves with silken strands and feed on the ventral surface of leaves, skeletonising the same completely. Pupation takes place in the soil. Full-grown caterpillars are stout, greenish in colour and 24 to 28 mm long. Moths are straw coloured having black dots and wavy lines on all the wings. Mechanical control by removing and destroying the webbed leaves can check the damage of this pest.

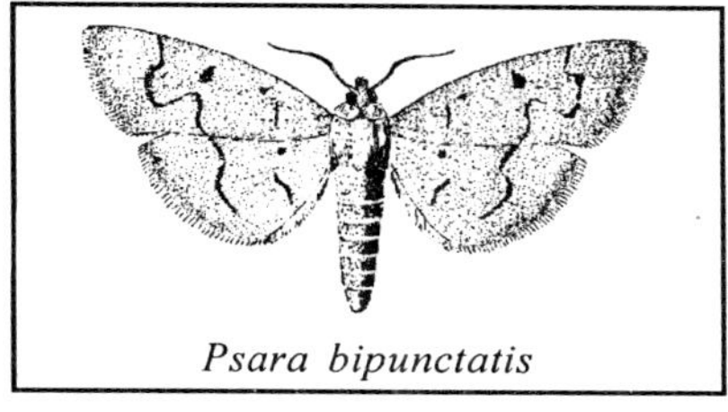
Psara bipunctatis

Hairy Caterpillars : *Selepa docilis* Butler, *S. celtis* Moore and *S. rabdota* Hampson (Noctuidae) have been recorded as minor pests of eggplant. Of these, the first one is comparatively more common, especially in Tamil Nadu. Besides eggplant, it has been reported damaging fig, litchi and mango trees. Eggs are laid in clusters of 4 to 12 on shoots and tender leaves. The caterpillars are yellowish in colour, having a lateral reddish line with a series of black spots. They feed gregariously on dorsal surface of leaves. Moths are medium-sized with head and thorax of pale brown colour and abdomen fuscious. Forewings are pale brown with purplish-gray tinge while hind wings are whitish suffused with fuscous towards the outer margin. Wing expanse is 22 to 26 mm. Incubation, caterpillar and pupal periods last for 3,7 to 10 and 7 to 11 days respectively. As these are minor pests, no control measures are generally required. In nature, however, the caterpillars are parasitised by *Euplectrus euplexiae* Rowher (Cherian and Pillai, 1942).

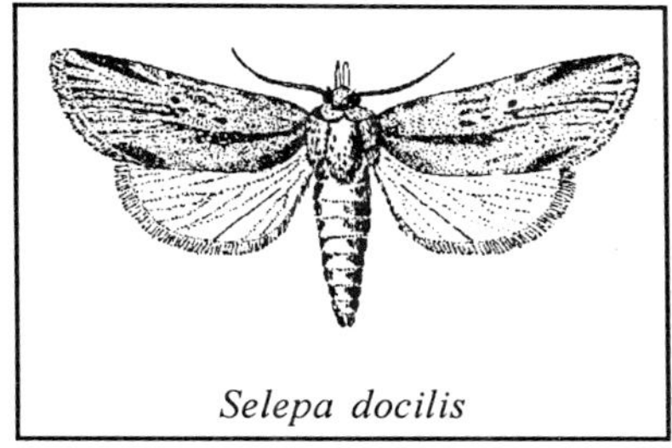
Selepa docilis

Ants : Black ant, *Camponotus infuscus* Forel (Formicidae) is sometimes found nibbling and tunneling the outer tissues of the main stem and shoots of the plants near the ground level and constructing its nest near or around the plants. Their activity is more prominent on plants infested by aphids or scale insects and the ants may be seen going up and down the plants to feed on honeydew secreted by the aphids, scale insects, leafhoppers etc. These ants do not cause much damage to the crop but have plenty of nuisance value. They even carry aphids and scale insects to their own nests or to safety in case of danger and in the process cause dispersal which results in spread of infestation of the aphids and scale insects.

Brown ant, *Solenopsis geminate* (Fabricius) (Formiciade) is another subterranean insect that nests in burrows in soil around the base of plants. It feeds by nibbling the roots of young seedlings and tunneling the plants around the roots, occasionally causing in death of the seedlings. The ants also climb up the plants to lick the honeydew, secreted by other insects.

To control these ants, stir the soil around the plants and mix thoroughly with the soil 5% HCH, chlorpyrifos dust, @ 20-22 kg per hectare.

For the effective management of insect pests and diseases of eggplant Sardana *et al.,* 2004 recommendly under IPM modules of neem cake @ 250 kg/ha in soil and 3-4 sprays NSKE (Neem seed kernel emulsion) @ 5% intermittently with need based application of chemical pesticides.

For insect resistance in brinjal modified Bt Cry genes has been widely used to obtain transgenic plants resistant to Colorado beetle (CBB) and shoot and fruit borer (SFB). Wild type Cry genes were unable to provide sufficient insect control because of their low expression, transgenic eggplants resistant to SFB and CPB by expression of codon-optimized Cry1 and Cry3 genes have been obtained, representing a very effective means of pest control (Rotino *et al.*, 2002). A modified Bt gene of *Bacillus thuringiensis* var. *tolworthi* encoding a coleopteran insect-specific Cry 3B toxin was mobilized to the female parent of the commercial F1 hybrid 'Rimina'. A synthetic Cry1 Ab gene was utilized to perform genetic transformation and a transgenic line had fruits totally protected from larval injury was selected.

6

TOMATO

TOMATO, *Lycopersicum esculentum* Milliere (Solanaceae) - a native of Peru and Ecuador - is now grown all over the Tropical and Temperate regions of the World. This is a tender herbaceous perennial plant but is cultivated as annual. It is procumbently branched and partly erect, having bisexual flowers and bearing fruits in clusters. Fruits are of various shapes and sizes; flattened, round to pear-shaped; segmented and green when immature becoming bright golden to blood red when ripe. It is a warm season crop and is not only highly sensitive to frost but it does not thrive at low, non-freezing temperatures. On the other hand, high temperatures, coupled with low humidity and dry winds frequently damage the floral parts and there is no fruit setting. The plant also cannot tolerate high rainfall and waterlogging. Tomato fruits are rich in vitamin content - most important being vitamin C, followed by vitamins A, B_1 and B_2 (Nath, 1976). The fruit is eaten raw, baked or stewed. It is a common ingredient of salad and is also made into syrup, pickles, ketchup, sauce and juice. The latest improved varieties for Northern India include 'Pusa hybrid-1' and 'Pusa hybrid-2' (Tewari *et al*, 1996). Pakistan imported 5000 tonnes of tomatoes from India in 1995 (Hindu, 7-12-1995).

Tomato fruits

INSECT PESTS

Tomato plants are attacked by a number of insects, nematodes and diseases that reduce the yield and spoil the quality of fruits. The major insect pests reported from Southeast Asia include gram pod borer, tobacco caterpillar, hadda beetles, mealybug, aphids and whitefly.

Fruit Borers: Gram pod borer, *Helicoverpa (Heliothis) amrigera* Hubner (Noctuidae) is a major pest of tomato. It is widely distributed in the Tropics, subtropics and warmer Temperate regions of the World, extending as far North as Japan and Germany (C1E map No. A-15). Some workers have confused this species with the New World species, commonly known as American bollworm, *Heliothis zea* (Boddie) (=*H. obsoleta* Fabricius).

Another less known species *H. assulata* Guenee has also been reported on tomato. These species are polyphagous and have been recorded on a wide range of host plants. The main host of *H. armigera* is chick-pea gram, though it also attacks castor, cotton, Citrus, hemp, indigo, cowpea, groundnut, linseed, okra, millets, safflower, tomato, tobacco etc.

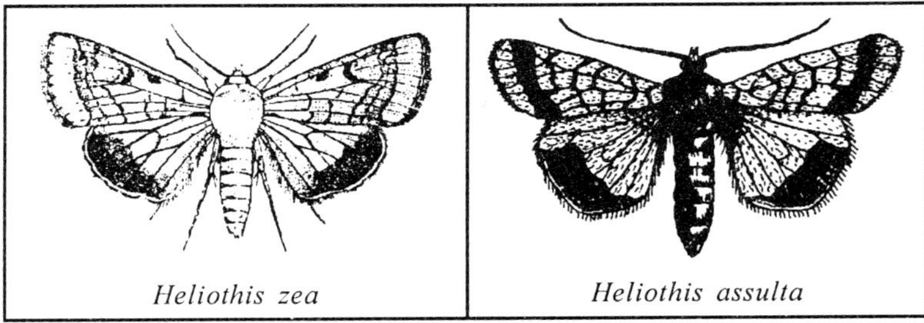

Heliothis zea | *Heliothis assulta*

Eggs are laid singly, generally on leaves and flowers but sometimes on fruits as well. A female lays about 500 to 3000 eggs (usually more than 1000). On hatching, the young caterpillars feed on tender foliage; advanced stage caterpillars (fourth instar onwards) attack the fruits. They bore circular holes and thrust only a part of their body inside the fruit and cat the inner contents. If the fruit is bigger in size, it is only partly damaged by the caterpillar

Tamato fruits damaged by *Helicoverpa armigera*

Tamato fruits damaged by *Helicoverpa armigera*

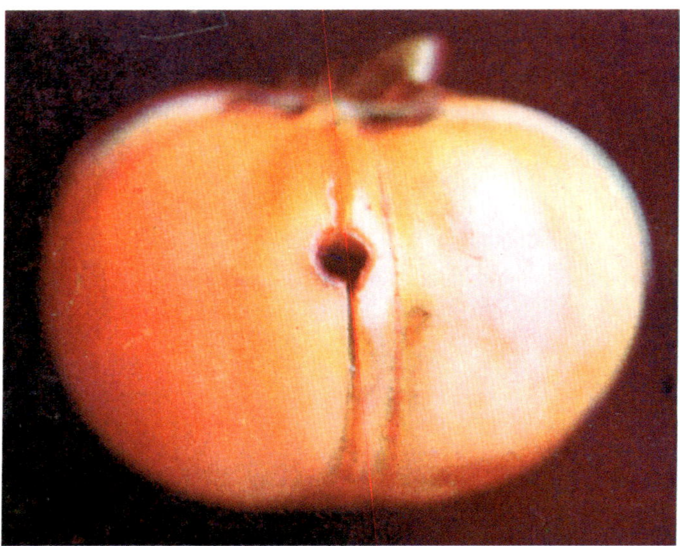

Tamato fruits damaged by *Helicoverpa armigera*

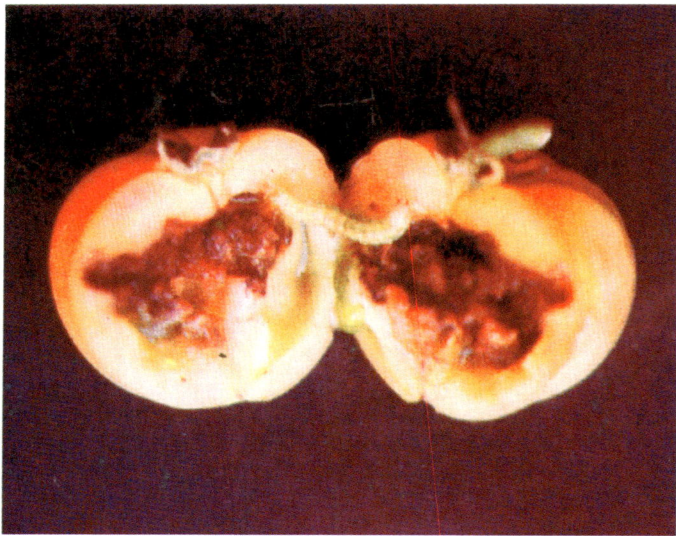

Tamato fruits damaged by *Helicoverpa armigera*

but subsequently, it is invariably invaded by fungi, bacteria etc. and spoiled completely. The caterpillars move from one fruit to another and one caterpillar may eat and destroy 2 to 8 fruits. The caterpillars are cannibalistic - freshly hatched ones may even feed on the eggs, if the suitable host is not readily available, while the older caterpillars prey on younger ones; sometimes these also attack caterpillars of other noctuid moths. Full grown caterpillars drop down from the plants to burrow in the soil and pupate therein. Moths remain active during day and are seen at sunset visiting flowers for feeding on nectar.

The eggs are yellowish-white, ribbed and dome-shaped, 0.4 to 0.5 mm in diameter. Freshly hatched caterpillars are yellowish-white in colour and gradually acquire greenish tinge. Full-grown caterpillars are 40 to 48 mm long, apple-green in colour with whitish and dark-gray broken longitudinal stripes. Pupae are dark brown in colour, 11 to 14 mm long and have a sharp spine at the anal end.

Moths are medium-sized, stout, ochreous with pale brown or reddish-brown tinge; forewings are olive green to pale brown in colour with a dark brown circular spot in the centre and indistinct

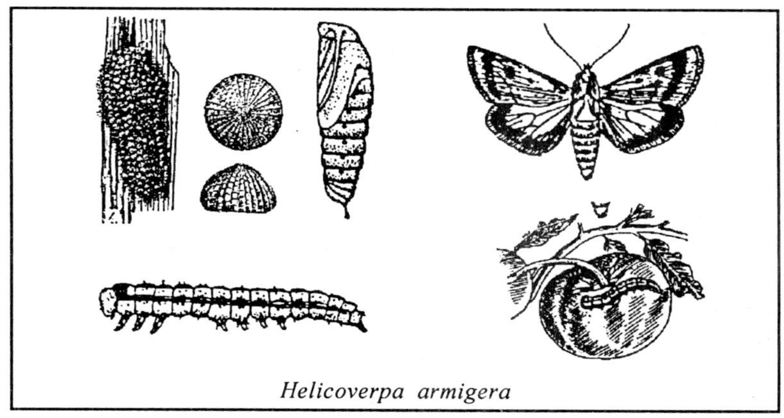

Helicoverpa armigera

double waved antemedial lines; hind wings are pale smoky-white with a broad blackish outer border. Wing expanse measures from 30 to 44 mm. Fecundity of females is rather high ranging between

1200 to 1600 eggs. Incubation, caterpillar and pupal stages last for 2 to 4, 15 to 24 and 10 to 14 days respectively (Butani, 1979 a) and the entire life cycle may be completed in 4 to 6 weeks with 5 to 8 generations in a year depending upon environmental conditions and availability of suitable hosts.

Hand-picking of caterpillars and their mechanical destruction in the early stage of infestation can keep the population under check. In case of severe attack, 5% dust or 0.2% spray of carbaryl or 0.04% lindane or 0.05% endosulfan have been reported to be effective.

A new insecticide emametin 5 sg is reported effective for the control of tomato fruit borer *H. armigera*. Further Suganya Kanna *et al.* (2005) observed that emametis formulation 10.0 gai/ ha and 8.75 gai/ha were very effective as compared to profenofos 50 Ec. (750 g.a.i./ha) and lambda cyhalothrin 5Ec (30 g.a.i./ha) which was comparable to spinsosad 2.5 EC (12.5g.a.i./ha).

Essential oils components Geijerene and pregeijerene isolated from leaves and stems of plant *Chloroxylon surientenic* DC. displayed maximum feeding deterrence DC50 89.8 and 99.6 mg/sq. cm against *H. armigera* larvae (Kiran *et al.,* 2007)

Significantly higher mortality of *H. armigera* larvae observed on gram and pea treated with *Beauveria bassiana* on 8th day (Masarrat Haseeb, 2007).

A strain of Nucleopolyhedrsis virus (NPV) developed at ICRISAT proved to be a powerful means to kill the larvae of *H. armigera* (Anonymous 2007). HEAR-NPV (PAV-1 strain) also reported effective against *H. armigera* (Singh and Battu, 2007) in Pubjab.

In nature, caterpillars are parasitised by *Apanteles souros* Nixon, *Microbracon brevicornis* Wesmael, *Campolites perdistinctus* (Viereck), *Chelonus narayani* Subba Rao, *Horogenes fenestralis* Holmgren, *Tetrastichus israeli* Maniand Kurian, *Exorista fallax* Meigen and *Strobliomyia orba* La Wiedemann.

Fruit Sucking Moths : *Othreis fullonia* (Clerck) and *O. materna* Linnaeus occasionally attack the tomato fruits; the former species is widely distributed in the Oriental region extending from Africa to Australia whereas the latter is confined to South-east Asia.

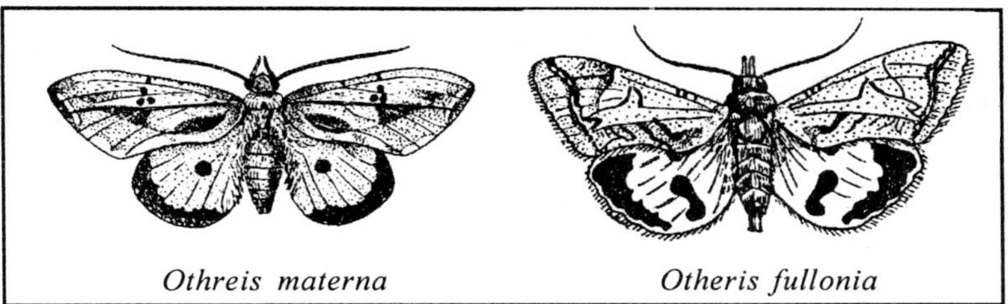

Othreis materna *Otheris fullonia*

Both are polyphagous but fruits of *Citrus* species are favoured by these moths (Butani and Jotwani, 1975).

Eggs are laid on various wild creepers. On hatching, caterpillars feed on foliage of these wild plants. It is the moth stage that causes the economic loss. Moth are nocturnal in habit and appear soon after dusk and cause damage by puncturing fruits and sucking their juice. The infested fruits shrink, shrivel, rot and ultimately fall down.

Eggs are slightly oval in shape and shining pale green in colour. Full-grown caterpillars are stout, 50 to 70 mm long, velvety blue in colour and ornamented with yellow coloured patterns. Pupae arc reddish-brown and 16 to 22 mm long; those of *O. materna* are slightly bigger than *O.fullonia.* Moths are quite large in size and stout, having pale orange body with greenish tinge. Forewings of *O. fullonia* are reddish-brown and those of *O. materna* are greenish-gray. Hind wings are orange in colour; those of *O. fullonia* have marginal black patch towards the apex and a kidney-shaped black spot in centre, while *O. materna* have a round black spot in the centre and a marginal black band. Wing expanse varies between 80 and 110 mm; females are slightly bigger than males and *inter se* two species, *O. fullonia* are slightly smaller than those of *O. materna.*

The egg, larval and pupal stages last for 8 to 10 days, 4 to 5 weeks and 14 to 18 days respectively in South India (Ayyar. 1943). In Northern India, the duration of each stage is little less and there are three generations in a year in the South and four in the North.

As the pests are of minor importance on tomato, generally no control measures are adopted against them.

Fruit flies: Melon fruit fly, *Bactrocera (Dacus) cucurbitae* (Coquillett) and tomato fly, *Acritochaeta excise* Thomson are minor pests of tomato. *B. cucurbitae* is a polyphagous pest having a very wide range of host plants - melons and other cucurbit fruits being the main hosts (Butani, 1975b). Adult female fly punctures the ripening fruits with its ovipositor and thrusts a few eggs inside the fruit. On hatching, the maggots feed on pulp of the fruits. Affected fruits rot and fall down.

Acritochaeta excisa attack only the decaying fruits and vegetable matter but occasionally the maggots bore into the stems as well. Eggs hatch in 28 to 36 hours whereas maggot and pupal development takes 6 to 7 days each (Ballard and Rao, 1924).

Leaf-eating Caterpillars: Tobacco caterpillar, *Spodoptera littoralis* (Fabricius) is widely distributed in the Old World tropics (CIE map No. A-61). Though tobacco and tomato are its major hosts, the pest has been recorded on banana, Citrus, cabbage, cauliflower, colocasia, cowpea, gram, castor, cotton, hemp, maize, millets, mulberry, okra, peas, rice, sorghum, yam etc.

Eggs are laid in clusters usually on ventral side of tender leaves and covered with brown hair. A single female lays on an average 400 eggs (maximum 2000) in 3 to 4 clusters, each of 80 to 150 eggs. Freshly hatched caterpillars feed gregariously, scrapping the leaves from ventral surface but later these caterpillars disperse and show biding propensities as typical cutworms and feeding voraciously at night on the foliage. During severe infestation, entire crop may be defoliated overnight. Pupation takes place in soil in rough earthen cocoons.

Eggs are dirty-white in colour and round in shape. Caterpillars are stout, cylindrical, 40 to 50 mm long when full grown and pale brown in colour with a greenish to violet tinge. They possess a submarginal series of narrow yellow spots having black lunules above them and a lateral series of purplish-black spots and scattered short setae. Pupae are reddish-brown. Moths are stout, pale ochreous suffused with dark-brown. Forewings hind wings are

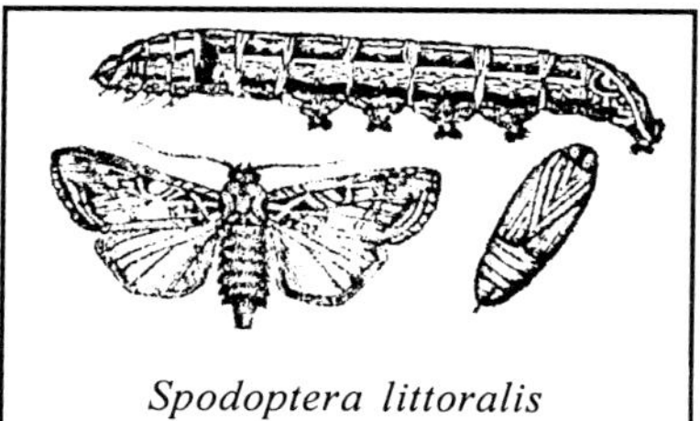
Spodoptera littoralis

opalescent and semi-hyaline white, with dark brown marginal line. Wing expanse is 35 to 45 mm. Incubation, caterpillar and pupal periods last for 3 to 5, 20 to 28 and 7 to 11 days, respectively, and may be extended upto 20, 90 and 30 days, respectively, during winter. Adult longevity is 10 to 24 days. The entire life cycle is completed in 30 to 40 days during summer and 126 to 140 days in winter.

Another species, *Spodoptera exigua* (Hubner) - the *ragi* cutworm - has been reported from Australia, South-east Asia, South China, Middle East, Central and Southern Europe and Africa (CIE map No. A-302). This is also a polyphagous pest recorded in India on *ragi* (*Eleusine coracana*), cotton, crucifers, groundnut, indigo, lucerne, hemp, sesamum, sugarbeet, tobacco, tomato etc. A female lays upto 130 eggs in batches of 50 to 220 eggs (maximum 300) on ventral surface of leaves. On hatching, caterpillars feed on mesophyll tissues and destroy completely the tender leaves. The pest is more common in nurseries and on young plants seldom attacks old plants.

In immature stages it is more or less similar in appearance to those of *Spodoptera littoralis*. Adults are smaller in size with wing expanse of 25 to 35 mm and lighter in colour with indistinct wing pattern. Egg, larval and pupal stages last for 2, 15 and 16 to 17 days respectively in South India and pupation takes place on soil surface or 50 to 100 mm below the soil surface.

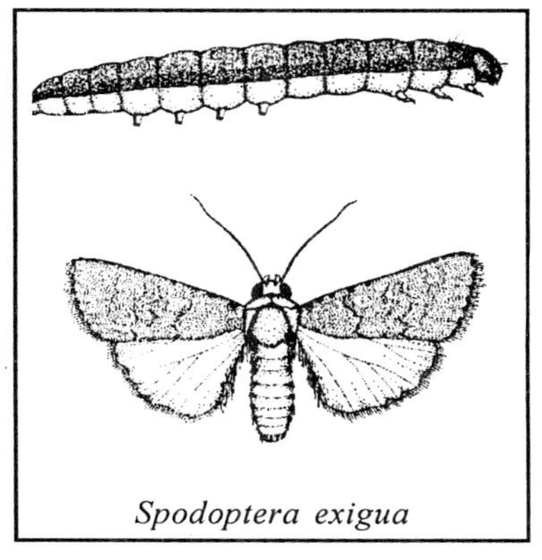

Spodoptera exigua

To check the infestation of these leaf defoliators, fields should be properly ploughed to expose and kill the pupae in the soil. Flood irrigation may be done to drown the hibernating caterpillars. The prominently visible egg-masses, as well as the leaves with young caterpillars feeding gregariously, should be collected and destroyed mechanically (Butani, 1977a). In case of severe infestation, spraying 0.05% dichlorvos or endosulfan has been recommended (Singh, 1970). Dusting 5% carbaryl or 4% endosulfan is also effective.

In nature, caterpillars of *Spodoptera littoralis* are parasitised by *Apanteles prodeniae* Vierveck and preyed upon by pentatomid bug, *Canthecona furcellata* Wolfenstein. Caterpillars of *S. exigua* are parasitised by *Euplectrus gopimohani* Mani, *Sturmia inconspicuella* Baranoy, *Actia monicola* Malloch and nematode *Mermis indica* Linstow. These are predated upon by *Rhynocroris fuscipes* Fabricius.

Cabbage green semilooper, *Trichoplusia ni* (Hubner), is a sporadic pest that feeds on the leaves of tomato plants; eggplant stem borer, *Euzophera peticella* Ragonot occasionally damages the plants by boring into the stems; and potatotuber moth, *Phthorimaea operculella* (Zeller) larvae may sometimes mine the leaves or bore into petioles and terminal shoots. These are major pests of cabbage, eggplant and potato respectively, causing only minor loss to tomato plants.

Tomato pinworm, *Phthorimaea lycopersicella* Busck is one of the most destructive pests in USA (Thomas, 1932; Swank, 1937) and tomato stem caterpillars, *P. melanoplintha* Meyrick and *P. plaesiosema* Turner also cause similar damage (Hyde, 1931; Morgan, 1931). In addition to these, tomato hornworm, *Protoparce quiquenaculata* (Haworth) and tobacco hornworm, *P. sexta* (Johanssen) have also been, reported as major pests of tobacco, tomato, eggplant etc. (Madden and Chamberlin, 1938). Fortunately, these pests have not yet been reported from India.

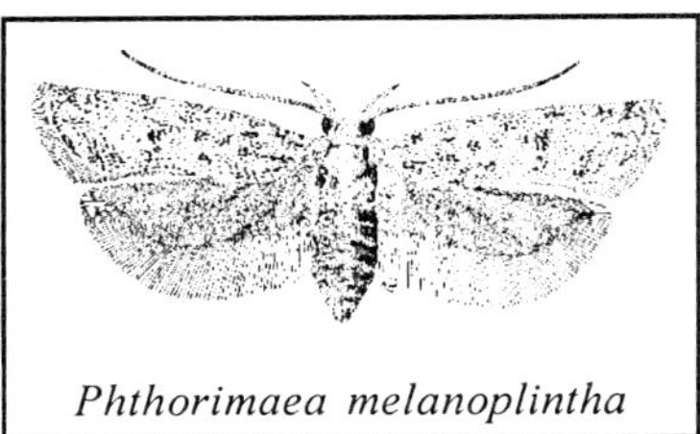

Phthorimaea melanoplintha

Whitefly: *Bemisia tabaci* (Gennadius), the tobacco/cotton whitefly, is found in most of the countries in Tropics and subtropics (CIE map No.A-284). Its main hosts are cotton, tobacco and some Winter vegetables, including tomato. The Infestation on these crops is sporadically severe. White, tiny, scale-like insects may be seen darting about near the plants or crowding in between the veins on ventral surface of leaves, sucking the sap from the infested parts.

Othreis matema *Othreis fullonia*
Othreis ancilla *Othreis cocalus*

Bemisia tabaci on tomato leaf

Tomato leaf curl virus transmitted by whitefly

Bemisia tabaci on tobacco leaf

Mealybug: *Ferrisia virgata* (Cockerell) - the white-tailed or striped mealybug - is a major pest of tomato. It is Pan-tropical in distribution (CIE map No. A-219) and is found all over the Indian sub-continent and South-east Asia. It is polyphagous and has a very wide range of host plants including beans, cashew, cassava, coffee, cocoa, Citrus, cotton, groundnut, guava, jute, sugarcane, sweet potato and tomato (Hill, 1975). The pest is found throughout the year, though, less active during Winter. The population normally increases from February onwards and maximum activity is around middle of April and continues if the weather remains dry. At times of heavy rains, or parasitisation, the population declines and again with the cessation of rains the population increases gradually till November and then again decreases.

Eggs are laid in clusters in cottony ovisac which remains concealed under the female body. A single female lays 100 to 400 eggs. On hatching, the crawlers remain huddled together in cottony nest under the mother. Later, these crawlers become active and wander about, moving swiftly till they find a succulent spot where they puncture the epidermis of host plant, inject their toxic saliva and start sucking the cell sap. The injury thus caused serves as entry for various disease-producing organisms (bacteria and fungi). From second instar onwards, the nymphs secrete honeydew on which sooty mould may develop, which in turn hinders the photosynthetic activity of the plant resulting in stunted growth. Pre-adults and adults secrete a waxy mealy material.

Eggs are pale yellow, cylindrical and about 0.3 mm long. Freshly hatched crawlers are yellowish in colour and become pale white in 2 to 3 days. Adult females are apterous, long, slender slightly oval (3.5 - 4.5 ' 1.5 - 2.0 mm) covered with dusty white waxy secretion and having a pair of conspicuous long glossy wax tassels at the caudal end. Reproduction is sexual as well as parthenogenetic. Incubation period is 15 minutes to 4 hours and the immature stages may last for 20 to 60 days in case of males and 19 to 47 days in case of females (Rawat and Modi, 1969). Longevity of males is 1 to 3 days while the females live for 5 to 7 weeks.

Ferrisia virgata

To check the spread of the pest, remove and destroy mechanically all the affected leaves and twigs as soon as the infestation is first observed. If large areas get infested, spray 0.05% dichlorvos or monocrotophos.

Aphids: Cotton aphid, *Aphis gossypii* Glover and peach green aphid, *Myzus persicae* (Sulzer) have been reported as minor pests of tomato plants, albeit the former is more common than the latter. Nymphs and adults, often found in large number, suck the cell sap and secrete honeydew, which not only attracts the black ants but also favours the growth of sooty mould fungus giving the plants a sickly appearance and hindering their photosynthetic activity.

Spraying 0.03% dimethoate, lindane or methidathion is quite effective in checking the population of aphids. Two to three fortnightly sprayings may be necessary as the surviving population builds up rapidly.

Leafhoppers: Cotton jassid, *Amrasca biguttula biguttula* (Ishida) and potato jassid *Empoasca punjabensis* Pruthi have been occasionally recorded sucking cell sap from the leaves. Both are polyphagous pests, okra and potato being their preferred host. *A. biguttula biguttula* affected leaves turn yellowish, curl and wrinkle while *E. punjabensis* causes hopper burns on the leaves; ultimately in both cases the affected leaves crumple and die away.

Spray 0.03% methidathion, phosphamidon or oxydemeton methyl to check leafhoppcr infestation.

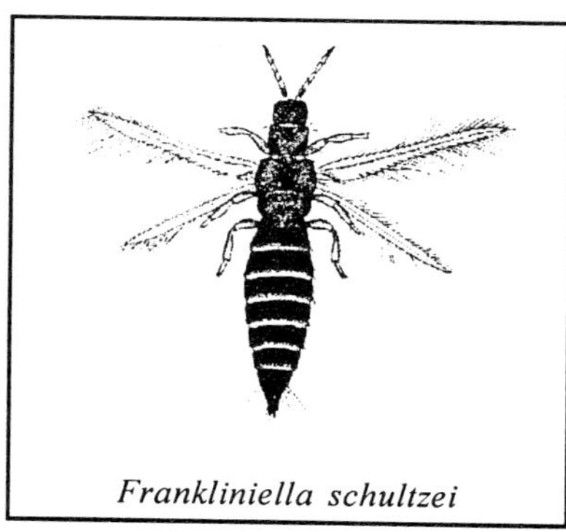

Frankliniella schultzei

Thrips: Onion thrips, *Thrips tabaci* Lindemann and groundnut thrips, *Caliothrips indicus* (Bagnall) occasionally appear in large numbers lacerating the leaf tissues and imbibing the oozing sap. As a result, the tender foliage becomes spotted and pale and silvery sheens appear on the affected leaves. Blossom thrips *Haplothrips ganglbaueri* Schmutz, *Frankliniella schultzei*

(Trybom) (=*dampfi* Priesner), *Megalurrothrips usitatus* (Bagnall) and *Scirtothrips dorsalis* Hood infest the flowers: severely infested flowers wilt, fade and drop prematurely without bearing any fruits.

These thrips are polyphagous in habit and cosmopolitan in distribution. Adults are fragile, slender, minute (about 1 mm long) and have heavily fringed wings. Ratio of males to females is 1: 10 and parthenogenesis is the main mode of reproduction though sexual reproduction also takes place. A female lays on an average 50 eggs. Life-cycle occupies 13 to 33 days. Females live longer than males and there are several overlapping generations in a year.

Frankliniella schultzei is also the vector that transmits the tomato spotted wilt virus (TSWV) causing hud necrosis. The loss caused on this account is much more than by desapping caused by feeding of the thrips.

To control thrips, spray 0.05% methiodathion or endosulfan or any other insecticide that has low residual effect on tomato fruits. Unfortunately by the time these tiny insects are noticed, and control measures adapted, the job of transmitting the virus may have already been accomplished and the loss thus caused is irrepairable.

***Ak* Grasshopper:** *Poekilocerus pictus* (Fabricius) (Acrididae) though a specific pest of *ak* (*Calotropis* spp.), feeds on a number of economic crops; especially when its preferred host, *ak* is not available. Both the hoppers and adults feed voraciously on leaves, eating away the leaf lamina completely.

Dusting 5 to 10% HCH, depending upon the age of the grasshopper, is quite effective in checking the pest population.

Sap Sucking Bugs: *Cyrtopeltis tenuis* (Reuter) (-*Gallabelicus crassicornis* Distant) (Miridae), *Nezara viridula* (Linnaeus) (Pentatomidae) and *Tricentrus bicolor* Distant (Membracidae) appear sometimes in swarms and suck the cell sap from foliage. All these are polyphagous and are considered as minor pests of tomato.

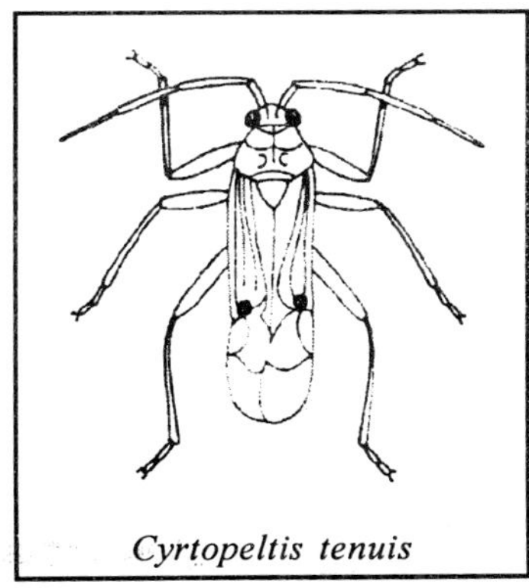
Cyrtopeltis tenuis

Cyrtopeltis tennis – tobacco mirid – is occasionally found feeding on tomato leaves. Adults are active, small, elongated, mm long, greenish-yellow bugs; stramineous; head is robust and vertically deflected in front; mesonotum is exposed and scutellum is triangular. Hernelytra are with leteral margin, straight, cuneus. longer than broad, membrane passing considerably beyond abdominal apex and there is a black spot at the apex of corium and another at the apex of cuneus. An allied species Engytatus geniculatus Reuter is a pest of tomato in USA (Jones, 1931, but not in India.

Tricentrus bicolor has also been reported damaging eggplant, cole crops, okra, potato etc. *Nezara viridula* is a cosmopolitan pest that breeds on a large number of host plants including potato, sweet potato and tomato.

Serpentine leaf miner, *Liriomyza trifolii* Burgers is an introduced pest infesting tomato crop causing damage to leaves by feeding on mesophyll, Severe infestation reduces photosynthetic activity of the leaves which influencing the plant growth, flowering and fruiting, Hemalatha and Maheswari (2004) screened seven cultivars and reported minimum number of leves in mines in 'Hybrid cross - 18' and 'Hybrid cross - 17' ranging from 1.7 to 2.0 followed by *Ruchi* with 2.4 mined leaves / branch.

MITES

Two red spider mites, *Tetranychus neocaledonicus* Andre and *T. cinnabarinus* (Boisduval) (Tetranichadae) and one Eriophyid mite *Aceria lycopersici* (Wolffenstein) have been found feeding on tomato leaves. The last one is occasionally found feeding on twigs also. Affected parts become reddish-brown and bronzy, wither and dry away. A severe infestation affects the flower and fruit formation.

The survey in Ranchi (Jharkhand) revealed presence of Eriophijid mite *Aceria lycopersicae* Wolff on brinjal and tomato (Rabindra Prasad, 2007).

Normally, no control measures are required against these mites on tomato. Clip off affected leaves and burn or bury these in soil. However, in case of severe infestation, dust line sulphur powder or spray the crop with wettable sulphur.

7

POTATO

POTATO, *Solanum tuberosum* Linnaeus (Solanaceae) is a native of Andes mountains of South America. It is a bushy herb with numerous branches and about half to one metre tall. The characteristic feature of the plant is the enlarged tips of underground stems (stolons) bearing edible tubers (potatoes). Roots are adventitious, stem generally winged, leaves odd-pinnate with large terminal leaflet. Flowers are borne on cymose panicles, white suffused slightly with pink or violet; fruits are globose berries, 20 to 40 mm in diameter with thin flattened oval seeds. This plant was introduced in India as early as 17th century, but its cultivation spread rather slowly and at present there are about five hundred thousand hectares under its cultivation. Of all the vegetables potato has the maximum per capita consumption. It is an excellent source of carbohydrates (16.1%) and the food value per lOOg edible portion is: energy 71 calories, proteins 17 g, calcium 9.2 g, vitamin A 12g, thiamine 11 ing, riboflavin 0.06 and niacin 1.18 mg (Singh,1970).

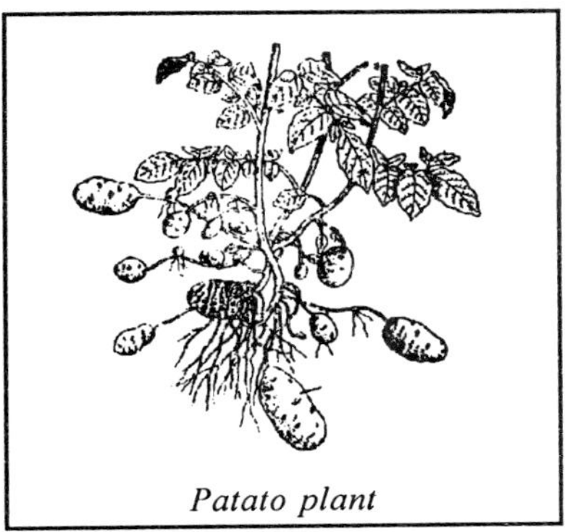

Patato plant

INSECT PESTS

Potato, like any other crop, is vulnerable to attack of pests, both in fields as well as in godowns. Insect pests of major importance include potato tuber moth, cutworms and aphids; those sporadically severe are crickets and occasionally termites whereas red ants, ground beetle bugs, leafhoppers, whiteflies, thrips, beetle grubs, leaf eating caterpillars. Leaf eating beetles and weevils are some of the minor pests. Among the non-insect pests, mites, nematodes and rats cause economic loss.

Potato Tuber Moth : *Phthorimaea opeculella* (Zeller) (Gelechiidae) is the most destructive pest of potato. It is a cosmopoliton pest, found in warmer countries throughout the world (CIE map No. A-10]. Its main host is potato while the alternate hosts include, eggplant, sugar-beet, tobacco, tomato and other solanaceous plants. In India, it was introduced from Italy about a hundred years ago. Its infestation usually starts in fields, goes on increasing and continues in godowns. In plains the pest is active throughout the year; during November to March it is found in fields, mining the leaves or boring into petioles and terminal shoots as well as tubers underground; from April to November, it is active in godowns causing damage to tubers. The loss of tubers may be as high as 30 to 70 per cent (Nair, 1975). Adults are nocturnal in habit. With the onset of winter, the moths fly from godowns to fields and lay eggs singly, generally near the eyes of exposed tubers and sometimes on the ventral surface of leaves. A single female can lay from 150 to 250 eggs.

Eggs are minute, oval (0.5 x 0.4 mm) and yellowish in colour. Full-grown caterpillars are 14 to 20 mm long and pinkish-white to pale greenish in colour. Pupation takes place in rough silken cocoons.

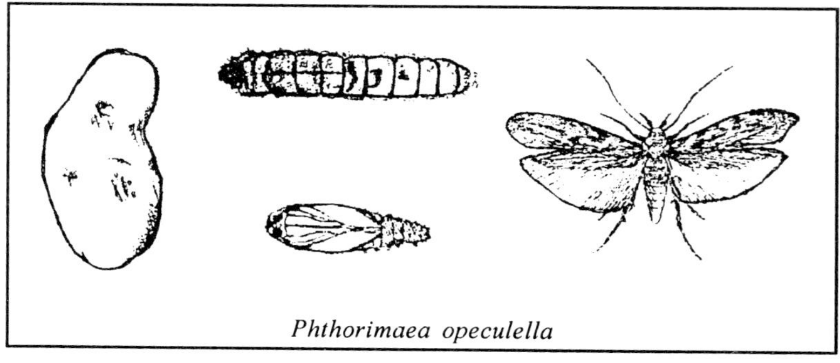

Phthorimaea opeculella

Adults are small with narrow mottled wings that are grayish brown to dark brown in colour. Wing expanse varies from 12 to 18 mm. Egg, caterpillar and pupal periods last for 3 to 4, 7 to 14 and 7 days respectively with at least 5 to 7 generations in a year. Ecological factors influencing the behaviour and development of various immature stages of this insect have been studied by Lall (1949) and Lal and Gupta (1951).

To prevent the infestation by this pest, Lal (1991) found inter-cropping with chillies, onion or pea to be very effective. Besides, earthing-up should be done properly so that tubers are not exposed to ovipositing female moths. All the infested tubers should be removed and destroyed; only good and clean tubers should be stored in well-ventilated, cool, dry place with temperature not exceeding 21°C. Cold storage is the best method or else fumigation may be carried out in airtight godowns with carbon disulphide (CS_2) or a mixture of carbon disulphide and carbon tetrachloride or methyl bromide. Singh (1970) suggested that potatoes kept for seed purpose (not for consumption as food) may be treated with 5% DDT dust @ 150 g per quintal of seed material and in case of infestation, those may be sprayed with 0.05% malathion. In case of field infestation, 2 to 3 fortnightly sprayings with 0.2% carbaryl @ 800 litres per hectare will control the pest. At the time of sowing, the seeds (cut or whole) should be disinfected by dipping the them in pyrogallol (Srivastava, 1993).

Thompson (1946) has listed a number of parasites of this pest but most of them have been reported from USA and other countries. Those recorded in India include, *Habrobracon hebetor* (Say), *H. gelechiae* (Ashmead), *Neochelonvlla* species, *Dioctes* species and *Campoplex* species as also a nematode, *Hexamennis* species on caterpillars of this moth. Unfortunately, none of these parasites have proved to be effective in cheking the pest population.

Cutworms : *Agrotis* species (Noctuidee) are polyphagous pests having a wide range of host plants belonging to different plant families. These are specially destructive at seedling stage. *Agrotis ipsilon* (Hufnagel) is one of the most serious pest of potato. Cosmopolitan in distribution (CIE map No. A-26). it has been reported from China, Northern Europe, Canada, Japan down to

South America and New Zealand. This is generally a cool climate pest. In plains, it is active from October onwards and with the onset of Summer, it migrates to hilly regions. In India, it is more serious in Northern region than in the South. Besides potato, it also feeds on barley, beet-root, cole crops, gram, okra, linseed, lucerne, millets, oats, peas, poppy, pulses, tobacco, wheat etc.

Moths appear soon after dusk, mate and lay eggs on ventral surface of leaves or moist soil; freshly polughed fields are preferred for oviposition. A female lays on an average 300 to 450 eggs (maximum 1800 - Hill, 1975) in 10 to 15 clusters, each consisting of 30 to 50 eggs. On hatching, the tiny caterpillars feed gregariously on foliage for a few days and then segregate and enter into the soil. The caterpillars are nocturnal in habit and hide during day in cracks and crevices in the soil or under the clods or debris around the plants. At night they come out to feed, for which they cut the seedlings near ground level and eat only the tender parts - thus the loss caused is much more than what is actually eaten by these caterpillars. Occasionally, the caterpillars may also nibble tubers. According to Chaudhuri (1953) and Nirula (1961) the loss in badly infested fields may be as high as 30 to 35 per cent. The damage is more pronounced in low lying areas that often remain waterlogged. The caterpillars are also cannibalistic, feeding not only younger caterpillars but at times they bite their own body (Butani, 1979a). The full grown caterpillars enter the soil and pupate in earthen cocoons.

Eggs are globular in shape, 0.5 mm in diameter, ribbed and whitish in colour. Caterpillars are smooth, stout, cylindrical, 40 to 50 mm long, blackish-brown dorsaily and grayish-green laterally with dark stripes. These are greasy to touch, that's why they are often referred to as greasy cutworms. They coil-up at the slightest touch. Pupae are 18 to 22 mm long and reddish-brown in colour. Moths are medium-sized, stout, dark greenish-brown with reddish tinge and have grayish-brown wavy lines and

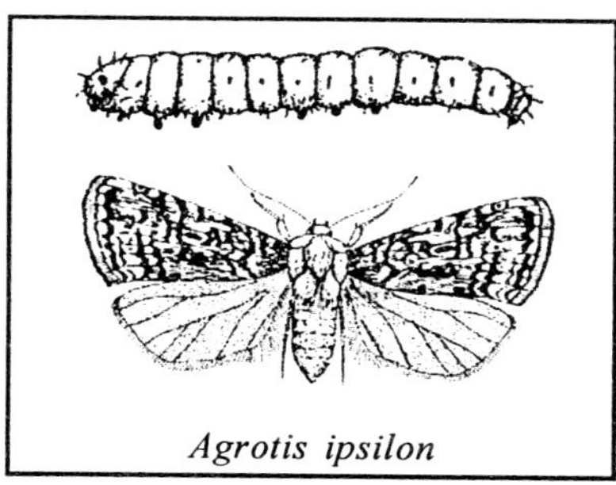

Agrotis ipsilon

spots on forewings; hind wings are hyaline having dark terminal fringe, which is darker in females than in males. Wing expanse is 45 to 55 mm. Incubation, caterpillar and pupal stages last for 2 to 13, 10 to 30 and 10 to 30 days, respectively. Total life cycle occupies 30 to 68 days depending upon the climatic conditions (David and Kumaraswami, 1975).

Other species of cutworms recorded damaging potato in India include, *Agrotis segetum* (Dems and Schiffer-muller), *A. intracta* Walker, *A. flammatra* Schiffer-muller and *A. spinijera* (Hubner) (Srivastava and Khan, 1962). Of these *A. segetum* is comparatively common. Its moths are pale whitish-brown with forewings ochreous-brown and having double waved sub-basal ante and post medial lines and marginal series of specks; hind wings are iridescent white with dark marginal line. Wing expanse is 40 to 48 mm. Moths of *A. intracta* are more or less similar in appearance and size as *A. segetum*. The caterpillars of both these species are generally found feeding on roots and tubers only. Moths of *A. flammatra* are much bigger in size (average wing expanse 56 mm); forewings having a broad pale costal fascia and hind wings fuscous-brown. The caterpillars of this species as also those of *A. sptntfera* cause damage similar to that caused by *A. ipsilon.*

To prevent the attack by these cutworms, soil application of aldrin or heptachlor dust is reported to be quite effective (Nirula, 1961). Nirula and Kumar (1963) suggested mixing of heptachlor (1.5 kg a.i. per hectare) or aldrin (2.5 kg a.i. per hectare) with fertiliser and its application in furrows at the time of planting. This application, followed by clean cultivation and regular stirring of soil, can help in keeping the pest at bay. In case of infestation, hand-picking and mechanical destruction of gregarious caterpillars at early stage coupled with application of 5% HCH or chlorpyrifos or 10% phorate granules, on soil around the plants and raking the soil thereafter is effective (Butani and Verma, 1976 c).

In nature, *Agrotis ipsilon* caterpillars are parasitised by *Microgoster* species, *Microplitis dimilis* Lyle, *Bracon kitcheneri* Will, and *Dliearita ruficuda* Cameron whereaas *Broscus punclalus* Klug and *Liogryllils bimaculatus* have been recorded as predators of this cutworm.

Aphids : Potato aphid *Aulacorthum solani* (Kaltenbach) are the most common and destructive species. Besides these other aphids recorded on potato include, *Acyrthosiphon pisum* (Harris), green peach aphid *Myzus persicae* (Sulzer), *Aphidula nasturtii* (Keltenbach), *Aphis fabae* Scopoli, *A.medicaginis* Koch, *A. rhamni* Koch, *Brevicoryne brassicae* (Linnaeus), *Lipaphis erysimi* (Kaltenbach), *Macrosiphum euphorbiae* (Thomas), *M. rosae* (Linnaeus), *Myzus ornatus* Laing. *Rhopalosiphoninns latysiphon* (Davidson) and *Rhopalosiphum rufiabominalis* (Sasaki).

Aulacorthum solani is a polyphagous pest, widely distributed in Europe, Africa, Middle East, Indian subcontinent, Australia, New Zealand, Fiji, Hawaii, Peru and North America (CIE map No. A-86). Colonies of nymphs and adults may be seen on ventral surface of leaves and shoots sucking the sap therefrom. The infested leaves become yellowish, wrinkled and cupped, while the tender shoots also turn yellowish and die away: The aphids also excrete honeydew on which sooty mould develops covering the affected parts with a thin superficial black coating that hinders the photosynthetic activity of leaves resulting in stunted growth of the plants. In addition, they also act as vectors, for transmitting several viral diseases.

The adults are pale green in colour with long conspicuous cornicles. Both alate and apterous forms can be observed in a colony. Apterous forms have a dark green patch at the base of each cornicle. Winged forms have transverse black spots or broken lines on abdomen. Both the forms are viviparous, giving birth to pale green young ones parthenogenetically. Males are rare. One generation takes about 2 weeks and there are several overlapping generations in a year.

To control aphids, spray 0.03% dimethoate or oxydemeton methyl or 0.05% dichlorvos or thiometon. Repeat the spraying, if necessary, after 10 to 12 days.

Mole Cricket : *Gryllotalpa africana* Palisot de Beauyoes (Gyllotalpidse) is a sporadically severe pest repórted from Bengal (Banerjee, 1955), especially of young seedlings and particularly in moist soils. The species is widely distributed in warm region of Old World (CIE map No. A-293). Eggs are laid during Rainy season,

100 to 150 mm deep in the soil, in earthen chambers prepared by the females. A female constructs 3 to 4 chambers in her lifetime and in each chamber 20 to 30 eggs are laid. Nymphs live underground in ramifying burrows and feed on roots of culvitated and wild plants and also tunnel the newly planted seed material. Both nymphs and adults come out of soil during night to forage for food.

Eggs are oval in shape, about 1.5 mm long and brown in colour. Adults are 22 to 28 mm long, brown with short wings folded over the abdomen and not Covering the abdomen completely. The conspicuous broad and curved fore legs are adapted for digging.

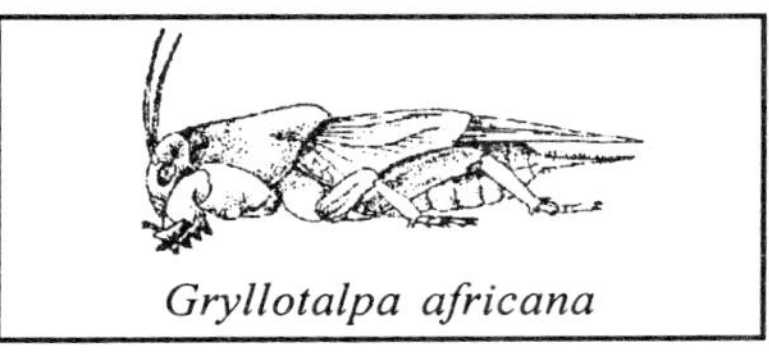

Gryllotalpa africana

Generally, no control measures arc adopted against this pest. If necessary poison baiting may be done; moist crumbly mixture of wheat bran and 10% HCH dust may be broadcast in between the crop rows during evening hours (Butani and Verma, 1976c).

Subterranean Pests : Among the insect pests inhabiting soil, termites are considered to be the most important economically. There are several species of termites reported as pests of vegetables. *Eremotermes* species, *Microtermes obesi* Holmgren and *Odontotermes obsus* (Rambur) have been reported damaging potato crop especially in sandy soil or in fields having bad aeration and poor drainage (Butani and Verma, 1976c). These cause sporadically severe infestation. The first two alone are reported to cause in Bihar a reduction of 4.7 to 6.5 per cent in yield of tubers (Kumar. 1965). The termites feed on roots and tubers, the latter become hollow and are often filled with earth. As a result the leaves of such plants start yellowing and wilting and ultimately dry up.

Gonocephalum hoffmanseggi

Ground Beetles: *Gonocephalum hoffmanseggi* (Steven) (Tenehionidae) and *Hapatroides seriatoporus* Fairmare (Tenebionidae) are thin beetles reported from Karnataka feeding on roots of potato seedlings. As a result the seedlings are outrightly killed. The beetles cause little damage on grown-up crop.

Wireworm : *Draslerius* species (Elateridae) is found feeding on the roots of potato plants (Nair, 1975) but it has not been reported to cause any serious damage.

Red Ants : *Dorylus Orientalis* Westwood (Formicidae) have been reported from Assam and *D. labiatus* Shuckard from Andhra Pradesh nibbling the tubers but the damage caused is negligible. These ants have plenty of nuisance value. They live in the burrows in soil around the host plants and are very aggressive by nature. Barlow (1900) has quoted Forel who has stated that all the species of *Doryius* are exclusively carnivorous and this species is certainly not a pest of potato but must be looking around for worms and other insects. Green (1903) however, most emphatically contradicted this statement stating that workers of this species live entirely underground and are confirmed vegetarians. Lefroy and Howlett (1909) pointed that only this species of *Dorylus* has termite like habit of attacking plants underground. The workers of this ant have also been observed to attack the workers of *Pheideol indica* and carry them off to their nests for feeding of the young ones.

Adults of *Dorylus orientalis* are pale testaceous with a long shiny silky pubescence; head is reddish, thorax silky and shining and abdomen with a rich satiny reflection. Sexual organ is found protruding at the apex of terminal segment and is fringed (Stebbing, 1903).

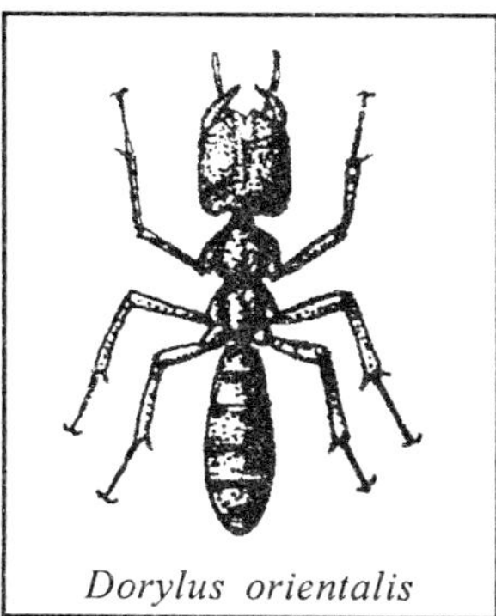

Dorylus orientalis

To avoid damage by these subterranean pests, avoid raising potatoes in sandy soils or areas infested with soil-borne pests. Soil application of 5% HCH, chlorpyrifos or heptachlor @ 20 to 22 kg per hectare can check the damage by these pests. In case the infestation is observed after sowing the seed, add lindane or chlorpyrifos emulsion in the irrigation water @ one kg a.i. per hectare.

Capsid Bug: *Creontiades pallidijer* Walker (Miridac) is a polyphagous pest with a wide range of host plants including eggplant, crucifers. melons, okra and potato. Eggs are laid within tender tissues of growing points, petioles and axils of branches. Nymphs and adults suck the cell sap from leaves and cause small irregular brown spots on young leaves and growing tips; gradually, the affected leaves droop and dry up. This bug has been found to breed throughout the year in Delhi area; during January to April, it feeds on eggplant, peas and potato: April to June on melons and July to September on other cucurbits including bottle gourd and sponge gourd; during October it is found on maize and pulse and during November-December, the bug migrates and attacks cabbage, cauliflower, knolkhol, radish, turnip etc. (Ghulam Ullah, 1940).

Eggs have a conspicuous white ridge and tag-like prolongation. Adults are delicate bugs, 7 mm long, ochreous-green in colour with transparent light green wings (hemi-elytra). A female lays 100 to 200 eggs and these hatch in 4 to 5 days. Nymphal period is on an average 18 days and total life cycle takes about 22 days. Adult longevity during September is 5 and 15 days in case of males and females respectively.

Green Sting Bug : *Nezara viridula* (Linnaeus) is cosmopolitan in distribution (CIE map No. A-27) and has been recorded from South Europe and Japan down to Australia and South Africa. It is a polyphagous pest and though its main hosts are eastor and coriander (Iyer, 1922), it also breeds on coffee, Citrus, cotton, millets, pulses, potato, rice, indigo, tomato, wheat etc. A female lays up to about 300 eggs in clusters of 50 to 60 eggs each, stuck together in rafts, on dorsal surface of leaves. Freshly hatched nymphs remain clustered around the egg-raft and it is only after the first moult that the nymphs disperse and start active feeding. Nymphs and adults suck the cell sap from tender leaves, and shoots, thereby devitalising the plants.

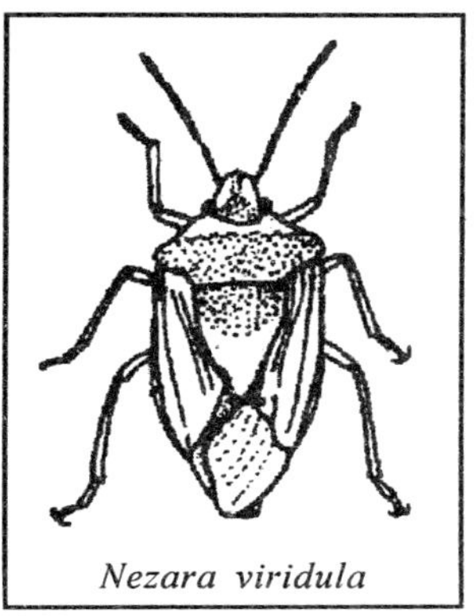

Nezara viridula

Eggs are barrel-snapped, little over one mm in length, whitish in colour, turning pink with age. Adults are medium-sized, about 15 mm long bugs, green to reddish-brown in colour.

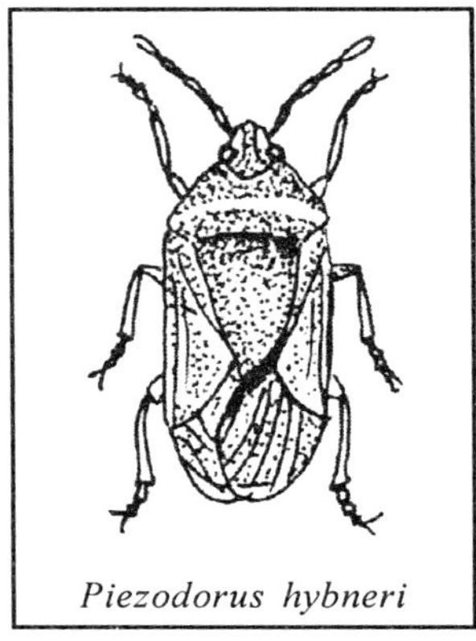

Piezodorus hybneri

Soybean Bug : *Piezodorus hybneri* (Gmelen) (Coreidse) has been observed feeding on potato leaves. Eggs are laid in clusters of 25 to 30 eggs each, on dorsal surface of leaves. These hatch in 3 to 4 days and nymphal development takes 22 to 26 days.

No separate control measures are generally required for these bugs. However, when severely attacked, dusting with 5 to 10% HCH or 4% endosulfan dust can effectively control the pest population.

Leafhoppers : *Amrasca biguttula biguttula* (Ishida) (Cicadellidas) a major pest of cotton and okra is a polyphagous pest, also attacks eggplant and potato. Some *Empoasca* species have also been recorded sucking the sap from potato leaves. The infested leaves first turn yellow, then brown, get wrinkled and gradually erinkle and die away.

Empoasca kerri motti Pruthi (Cicadcllidae) breeds throughout the year but is most active during October to March. Nymphs and adults suck the sap from veins and veinlets on ventral surface of leaves. As a result tips of the affected leaves become brown, turn upwards and get dried. During severe infestation brown spots appear on leaf margins which also get rolled up and subsequently die away. Beside potato, this jassid also attacks eggplant, chilli, cowpea and tomato. In absence of these hosts, it migrates to castor, lucerne and barseem (Rattan Lal, 1946).

Adults are 3 mm long and yellowish-green in colour; vertex is flat and greenish-yellow, smaller than pronotum which is also flat and smooth; hemi-elytra are long, narrow, semi-transparent and pale green in colour; green at costal and gray at distal regions (Pruthi, 1940). A female lays 25 to 60 eggs in 25 to 30 days, generally one egg per sometimes 2 and rarely 3 to 4 eggs per day. Incubation period is 4 to 11 days; nymphal on an average 25 days and adult longevity is about 15 days (males) 90 days (females). Preoviposition period varies between 6 and 16 days.

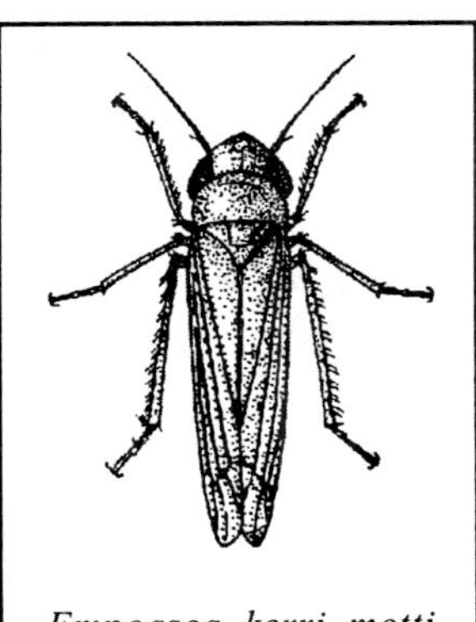

Empoasca kerri motti

Empoasca punjabensis Pruthi (Cicadellidae) causes hopper-burn of leaves. The symptoms are etiolated spots and patches on leaves, browning and rotting of leaf margins and tips and drying of leaves. Eggs are laid in veins. ·Adults are slightly bigger than those of *E. kerri motti*. on an average 3.5 mm long and yellowish-green in colour. Vertex is flat, smooth and pointed; pronotum is transparent and longer than vertex with yellow abdomen and ovipositor stout and green in colour. Hemi-elytra are longer than body and transparent greenish-yellow in colour (costal margin deep green). Eggs hatch in 4 to 9 days; nyrnphal development takes 19to21 days and adult longevity is 7 to 15 days (Vevai, 1942).

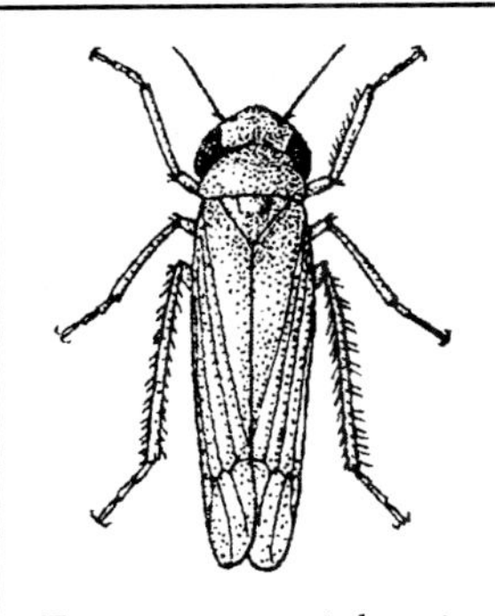

Empoasca punjabensis

Empoasca solanifolia Pruthi (Cieadellidoe) causes similar damage as *E. punjabensis*. Its adults are bigger in size being about 4 mm long, robust and pale brown in colour; vertex flat and slightly raised; pronotum is 1½ times longer than vertex: abdomen is tinged with yellow; ovipositor is stout and pygopher covered with a few minute hairs. Hemi-elytra are transparent and twice as long as abdomen having thin distinct veins (Pruthi, 1940).

Ernpoasca fabae Harris (Cicadellidae) - potato leafhopper is more common outside

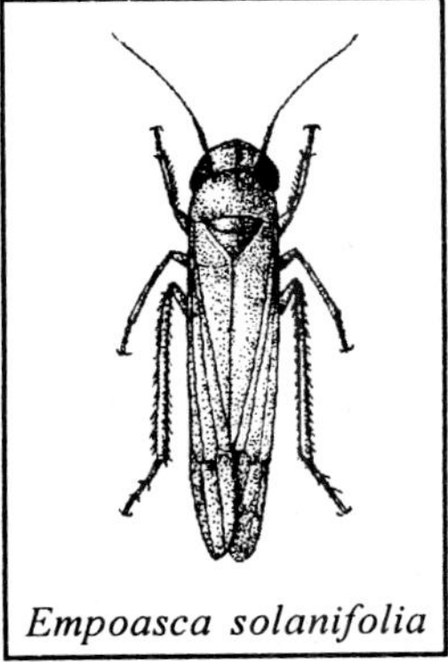

Empoasca solanifolia

India. This also produces hopper-burn, stunting and crinkling of leaves. Adults are wedge-shaped, 3 to 4 rnm long having a row of six round light spots along the anterior margin of pronotum.

All these leafhoppers besides devitalising the plants and thereby affecting the production adversely also spread the virus diseases. The loss caused by desapping may not be so severe, but the damage done by transmitting the viral diseases is irrepairable. While only large population of jassids can cause significant damage by sucking plant sap, a single viruliferrous jassid can affect, and destroy numerous plants by migrating from plant to plant. It is therefore suggested to give prophylactic spraying 0.05% dimethoate or formathion or 0.03% phosphamidon or oxydemton-methyl to prevent the jassid activity. If need be, the spraying may be repeated once or twice at an interval of 10 to 14 days to obtain complete check of the pest.

In nature, black ants, *Camponotus* spp. and a spider *Diclyha albida* have been observed feeding on these leafhoppers but none of these exercises any appreciable check on the pest, perhaps due to rapid rate of multiplication and migration from neighbouring areas.

Whiteflies : *Bemisia tabaci* (Gennadius), the cotton or tobacco whitefly and *Trialeurodes vaporariorum* (Westwood), the greenhouse whitefly, have been occasionally reported feeding on potato foliage; both are polyphagous pests. *B. tabaci* is a serious pest of cotton, tobacco and some winter vegetables. It has been recorded as a pest of bitter gourd, eggplant, okra, potato, sweet potato and tomato. These tiny insects suck the sap usually from ventral surface of leaves and devitalise the plants, but in case of potato, infestation is seldom severe. In addition, these insects also act as vectors for transmitting the virus diseases and as such are of great potential importance. The chemical control measures suggested for leafhoppers are effective against the whiteflies also.

Thrips : *Selenothrips indicus* Bagnall (Thripidse) has been found damaging potato in India especially in the South (Ayyar, 1929). These arc tiny, slender, fragile insects; adults have fringed wings. Both nymphs and adults scrape the epidermal tissues of leaves usually near the tips and rasp the oozing sap. The affected tips get curled and dry up. The loss caused is usually not heavy except when the insects appear in large number. As such, normally no separate control measures are adopted against this pest. However, in case of severe infestation spray 0.2% monocrotophos or thiometon to check the pest population.

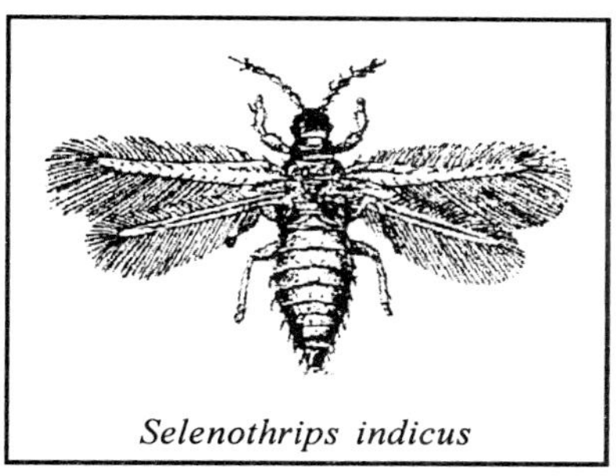

Selenothrips indicus

Lace-wing Bug : *Recaredus* species (Tingidae) attacks stored tubers in India. Nymphs and adults suck sap from critical tissues of tubers. Eggs are laid on tubers. A life cycle is completed in about a month under favourable conditions and as many as 7 generations have been recorded in a year. Adult longevity is 7 to 8 months and the insect hibernates in adult stage.

Being a minor pest, generally no control measures are adopted against this pest. Red ant, *Monomorium indicum* Forel has been reported attacking this bug (Dutt, 1914).

Beetle Grubs : *Anomala* species, *Holotrichia conferta* Sharp, *H. (Lachnosterna)* coriacea* Hope, *H.(L.) loigipennis* Blanchard, *H. (L.) serrata* Fabricius (Melonthidae) and *Xylotmpes gideon* Linnaeus (Dynastidae) infest the potato crop but are considered as minor pests.

These are all polyphagous pests. Eggs are laid in the soil near the host plants. On hatching, the grubs feed on developing roots of potato as well as other grasses growing around. When full-fed, the grubs overwinter deep down in the soil. Adults emerge

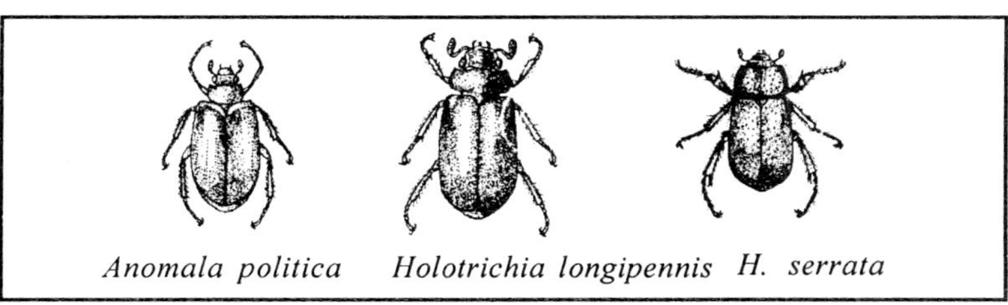

Anomala politica *Holotrichia longipennis* *H. serrata*

* According to Frey (1971) Genus *Lachnosterna* is confined to Europe.

as soon as the temperature starts rising, but continue to remain in the soil till the onset of monsoon, when they come out and feed on the foliage during night.

Hadda Beetles : *Epilachna dodecastigma* Mulsant, *E. oceuata* Redtenbacher, *Henosepilachna vigintioctopunctata* Fabricius and *Epitrix cucwneris* Harrow, have been reported from various parts of India. Both grubs and adults feed on leaf tissues, often skeletonising the leaves completely. Of these, *Epilachna* spp. occasionally cause severe damage. *Epilachna philippinensis* Dieke and *Epitrix tuberis* Gent, are major pests of potato in Philippines, but these species have not yet been reported from India.

Green Leaf Beetle : *Chalaenosoma metallium* Fabricius (Chrysomelidse) as also the flea beetle, *Monolepta orientalis* Jacoby (Chrysomelidse) have also been oceasionally found feeding on potato leaves in South India.

Leaf Eating Weevils : Adults of *Atmeionychus peregrinus* Olivier (Curculionidse) from Orissa and *Myllocerus subjasciatus* Guerin (Curculionidae) from Tamil Nadu have been reported (Nair, 1975). nibbling the leaves of potato plants along the margins. However, the damage caused is rarely serious.

To check damage, if and when these pests appear, collect and destroy them mechanically in the initial stage of attack. Spraying with 0.05% dichlorvos or endosulfan is also effective in controlling these beetles and weevils. Spraying chlorpyrifos 20 EC thrice with 0.5 kg a.i. /ha (each time) gives maximum protection against *Holotrichia* spp. (Mishra, 1995).

Leaf Eating Caterpillars : *Spilarctia obliqua* (Walker) (Lymantriidee) - the notorious Bihar hairy caterpillar and *Spodoptera exigua* (Hubner). the lucerne caterpillar, are both polyphagous pests attacking a wide range of cultivated crops including potato. Among vegetables the preferred host of *S. obliqua* is sweet potato and that of *S. exigua* is tomato. Eggs arc laid in clusters on leaves. On hatching, the caterpillars feed gregariously for some time then disperse and feed voraciously on older leaves. However, the damage caused to potato plants is generally not severe.

Besides, caterpillar of tussock moth *Dasychira mendosa* (Hubner) (Lyrnaniriidge) have also been occasionally recorded feeding on leaves. This too is a polyphagous pest, attacking a number of fruit trees, potato, cauliflower etc. Full-grown caterpillar is 38 to 44 mm long, bluish-brown to blackish in colour having red stripes on its head. Body of this caterpillar is densely hairy and has compact tufts of long hair on certain abdominal segments. Head, thorax and abdomen of the adult are pale brown; males are with bipectinate antennae and females have tufts of hair at their anal end. Forewings are uniformly brown with black specks and a pale patch outside the sub-basal line. Hind wings are pale grayish-brown having outer area slightly darker. Wing expanse of male and female moths is 34 to 42 and 46 to 54 mm respectively.

If and when these caterpillars appear, collect and destroy the same mechanically during the early gregarious stage.

Eggplant Shoot Borer : *Leucinodes orbonalis* Guenee (Pyraustidae) - a major pest of eggplant, has also been recorded boring into shoots of potato plants. Though reported from all over India, it is a minor pest. The affected shoots wilt and droop down. To check the build-up of pest -population, clip off and destroy all the infested shoots by burning or deep burning.

Potato : Kawehuk (2002) reviewed transgenics in potato. For insect resistance transgenic potato lines containing *Bacillus thuringiensis cry*IIIA δ-endotoxin gene controlled first instar Colorado potato beetle (CPB) larvae. Good correlation between insect control and the levels of δ-endotoxin RNA and protein has been observed. Cultivars resistant to CPB are commercially available in Canada and the United States. Resistance to the tuber moth (*Phthorimaea operculella*) in potato plants expressing the cryV Bt transgene has also been reported.

8

CHILLIES

CHILLIES, *Capsicum spp.* (Solanacese) though native of Mexico (Central America) and Peru (South America) is widely grown in Tropics and subtropics as also under glasshouse conditions in Temperate regions. In India, it was introduced by Portuguese in 17th century and is by now grown all over the country, especially Andhra Pradesh, Karnataka, Tamil Nadu and Maharashtra.

Capsicum annum Linnaeus is the commonly grown species in India. This is a shruby, much branched annual plant that grows upto 500 to 600 mm; the flowers having white corolla are borne singly at nodes. Fruits are 10 to 200 mm long, thin with conical tips. The plants are sensitive to excessive rainfall, waterlogging and frost. The fruits are pungent and the pungency is due to an active principle capsaicin contained in the skin and septa of the fruits. The bright red colour at the ripe stage is due to the pigment, capsanthin. *Capsicum annum grossum* Sendtner is usually cultivated in the hilly tracts in India. This is also called bell pepper or sweet pepper. It yields large-sized fruits, 25 to 40 mm in diameter, having thick, smooth and glossy pericarp and is mildly pungent. It is mainly used in vegetable preparations, or is pickled. *Capsicum frutescens* Linnaeus is another species grown in limited areas. This is a shrubby perennial plant, 100 to 125 cm tall with 2 or more flowers in a leaf axil; corolla are greenish and fruits are small, 12 to 20 mm long, conical in shape and are exlremely pungent. This species occurs wild and semi-wild in the Tropics.

Among the vegetables, chilli is a rich source of vitamins A and C. It is an important condiment crop. Fruits are used green as

Aphid and mites on pigeonpea leaves

Myzus persicae on chillies leaves

Myzus persicae on chillies leaves

Myzus persicae on chillies leaf

well as ripe and dried to impart pungency to various foods. The fruits are also pickled and used in chutney, sauce etc.: in some parts of the country green chillies are even taken raw with food. *Capsicum* preparations are used as counter-irritants in lumbago, neuraligia and rheumatic disorders. It has a tonic and carminative action and is especially useful in atonic dyspepsia (CSIR, 1950). Chilli can prevent blood cancer. *L*-asparaginase - an anti-cancer enzyme used in the treatment of acute lymphatic leukaemias in children - has been isolated in a purified form from chilli (Anonymous, 1980).

INSECT PESTS

Over 25 insect species have been recorded from South-east Asia damaging both leaves and fruits of chillies. Except thrips and sometimes aphids, both infesting the leaves, no other insect species is reported to cause significant yield loss so as to classify it as pest.

Thrips : Chilli thrips, *Scirtothrips dorsalis* Hood (Thripidae) is a polyphagous pest having a wide range of host plants (Ananthakrisrman, 1969). Eggs are laid on or just under leaf tissues. Both nymphs and adults lacerate the leaf tissues and imbibe the oozing sap; sometimes even the buds and flowers arc attacked. Tender leaves and growing shoots are the preferred. The pest activity increases during dry weather. The infested leaves start curling and crumbling and are ultimately shed whereas buds become brittle and drop down. If there are no rains, the entire plant may dry and wither away. Besides the damage by their feeding, this thrips is also responsible for transmitting leaf curl disease. Under the conditions of severe infestation 30 to 50% crop may be lost (Nagaraja Rao, 1955). Chillies grown after sorghum are more susceptible to attack by these thrips (Ayyar *et al.*, 1935). Similarly, in mixed cropping of onion and chillies, both the crops suffer badly (Butani and Jotwani, 1984).

Eggs are minute and dirty white in colour. Nymphs and adults are also very small, slender, fragile and yellowish-straw in colour. Adults have heavily fringed wings that are uniformly gray in colour. Reproduction is both sexual and partheno-genetic. In

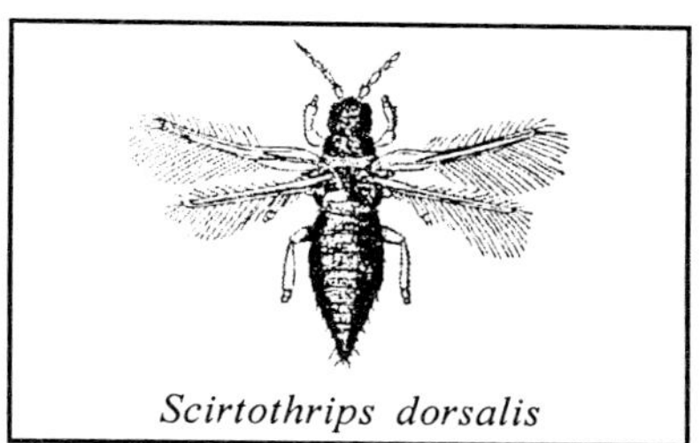
Scirtothrips dorsalis

case of sexual reproduction, oviposition period lasts for about a month during which a female lays on an average 100 eggs @ 2 to 4 eggs per day. Entire life-cycle is completed in 2 to 2½ weeks, with as many as 25 overlapping generations in a year (Raizada, 1965).

In addition to this thrips, groundnut thrips *Caliothrips indicus* (Bagnall) and blossom thrip *Frankliniella schultzei* (Trybom) have also been reported infesting the leaves and flowers, respectively. A severe infestation of the former results in the formation of silvery sheens on the leaves while that of the latter causes drying and premature shedding of flowers. Both the species are highly polyphagous, having a very wide range of host plants.

To control this thrips, spray with 0.03% dirnethoate, phosalone or monocrotophos. Asokan and Venugopal (1992) got cent per cent mortality under laboratory conditions with synthetic pyrethroids like clocythrin 50 ppm and cypermathrin 75 ppm.

Bifenthrm (Talstar) 10 EC was evaluated for control of thrips and mites on chilli. Bifenthrm 10 E.C. @ 80 g., a.,i./ha was found optimum and effective dose in reducing the chilli pest. Complex as insecticide and acaricide under irrigated, ecosystem also reported effective (Thusaliram *et al.*, 2005).

Sequential spray of fenpropathrm 75 g a.i./ha, oxydemeto methyl 25 E.C. @ 200 g.a.i./ha, fenpropathrin 30 EC@ 75 g a.i./ha Dicofol 18.5% @ 425 g.a.i./ha and Dipel 8L@ 1000 ml/ha was found effective 1PM spray seguence in reducing chilli thrips, mites borer damage and recorded higher green chilli fruit under the irrigated eco-system (Hosamani *et al.*, 2005).

Predaceous thrips, *Scolothrips indicus* Priesner and *Frankliniella megalops* Back have been found preying upon *Scirtothrips dorsalis* (Ananthakrishnan, 1971).

Aphids: Cotton aphid *Aphis gossypii* Glover (Aphididye) and peach green aphid *Myzus persicae* (Sulzer) (Aphididae) are commonly found infesting chilli plants, Besides these two species, Isaac (1946) also recorded *Aphis cytisorum* Hartig (=*laburni* Kaltenbach). All these are polyphagous pests.

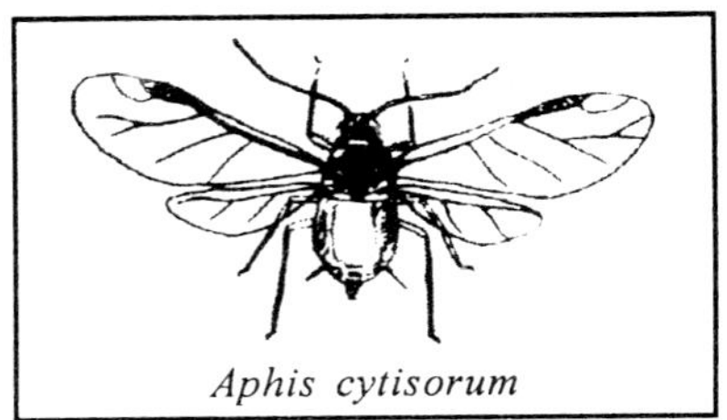
Aphis cytisorum

Small, ovate, soft-bodied nymphs and adults are found in large number on underside of tender leaves and shoots sucking the cell sap from the tissues. The infested leaves curl and dry up. The insects also screte honeydew which attracts ants and favours the development of sooty mould. This black superficial coating, covering the dorsal surface of leaves and twigs, hinders the photo synthetic activity, thus causing further retardation in growth and fruiting capacity of the infested plants. Reproduction being usually parthenogenetic, the aphids mutiply very rapidly. A single female produces about 12 to 24 young ones (nymphs) per day and these mature and start reproducing in about a week's time. Cloudy weather is very conducive for the rapid multiplication of these insects, whereas heavy showers cause a sudden drop in their population.

To check spread of this pest, clip off and destroy promptly the affected shoots and twigs in the initial stage of attack. In case of severe infestation, spray 0.04% monocrotophos or 0.05% endosulfan. Spraying 40% nicotine sulphate (1: 800) is also effective. Two sprayings at an interval of 10 to 14 days can keep the pest under check.

Large Brown Cricket : *Brachytrypes portentosus* (Litchtenstein) (Gryllidse) - nymphs and adults feed on seedlings of chilli and other crops at night. During day, the insects remain in the soil. Damage caused is usually of minor nature and warrants no control measures. Occasionally, the cricket may appear in large number, causing severe damage.

Termites : Roots of chilli plants are occasionally damaged by the termite, *Odontoterrmes obesus* Rambur. This is a highly polyphagous pest, having a very wide range of host plants. The incidence of this pest is more in sandy and sandy-loam soil than in clayey soil.

The pest can be checked by mixing thoroughly in the soil 5% HCH, chlorpyrifos dust @ 20 to 22 kg per hectare.

Beetle Grubs : *Anomala bengalensis* Blanchard, *Holotrichia consanguinea* (Blanchard) and *H. reynaudi* Blanchard (=*insularis* Brenske) (Melolonthidae) have been reported damaging chilli plants.

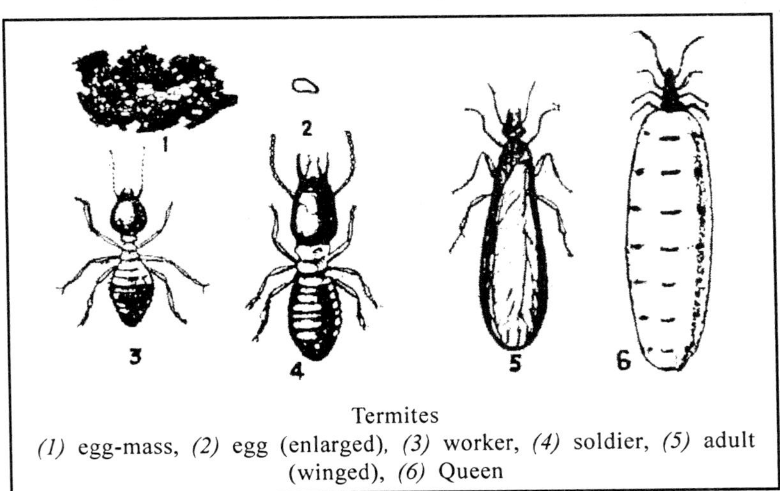

Termites
(1) egg-mass, *(2)* egg (enlarged), *(3)* worker, *(4)* soldier, *(5)* adult (winged), *(6)* Queen

Of these, *H. reynaudi* is comparatively common and more destructive especially in Rajasthan (sandy soil).

Eggs are laid in soil; grubs remain confined to soil and feed on roots, while the adults hide in debris or in the soil during the day but come out at night and feed on leaves. The incubation period of *H. reynaudi* ranges from 8 to 12 days; grubs remain active for 11 to 16 weeks and then overwinter in soil in pupal stage. Sometimes, when the climatic conditions are still favourable, the adults may emerge from the pupae

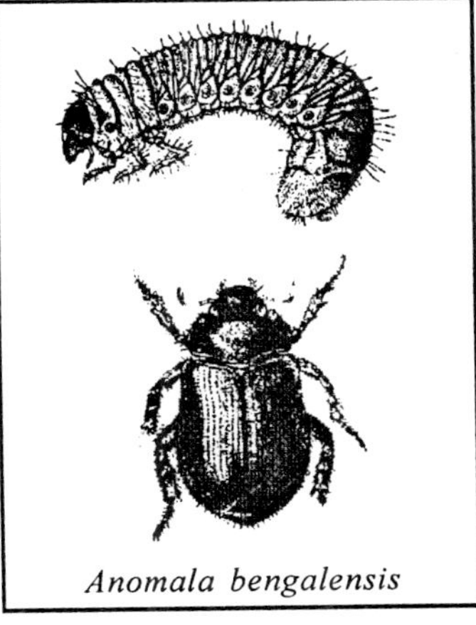

Anomala bengalensis

in 2 to 3 weeks and later overwinter as adults. There is only one generation in a year (Srivastava and Khan, 1963). The soil treatment suggested against termites, controls these pests as well.

Cutworm : *Agrotis ipsilon* (Hufnagel) - cosmopolitan pest, found all over the World is often found damaging chilli seedlings. Caterpillars are nocturnal in habit and hide during day in cracks and crevices in soil. At night they come out, cut the seedlings at ground level and eat only the tender leaves. Older plants are not attacked. Soil treatment with 5% HCH, chlorpyrifos or heptachlor recommended against termites also prevent the occurrence of cutworm.

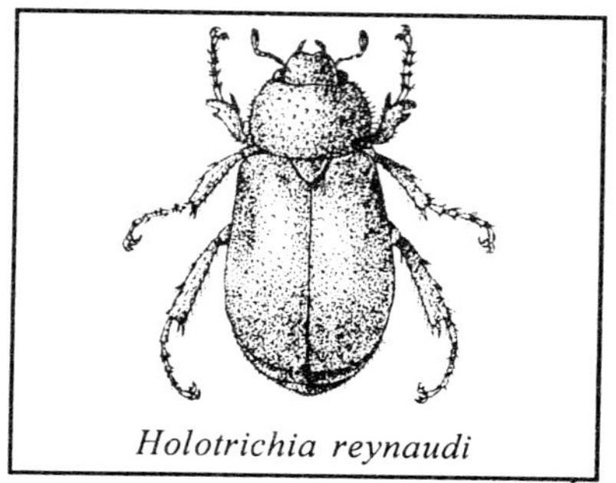

Holotrichia reynaudi

Sap Sucking Insects : Several mealybugs and scale insects, namely, *Ferrisia virgata*(Cockerell), *Saisetia coffeae* (Walker), *Aspidiotus destructor* Signoret and *Lepdosaphes piperis* (Green) have been recorded on chilli plants and may cause damage by sucking the cell sap from the leaves. If and when an infestation is observed, promptly remove and destroy the affected leaves. A lygaeid bug, *Spitostethus pandurus*.(Scopoli), cow bug *Tricentrus bicolor* Distant and whitefly *Bemisia tabaci* (Gennadius) have also been reported sucking vital sap from leaves. The damage done by feeding by these insects is negligible.

Leaf Defoliators : Army worm *Mythimna loreyi* (Duponchel), leaf roller *Archips micaceanus* (Walker), lucerne caterpillar *Spodoptera exigua* (Hubner) and leaf eating beetle *Monolepta signata* Olivier (Chrysomelidae) are the defoliators that cause usually minor damage. Whenever these pests occur in sufficient number, dusting may be done with 10% HCH or 5% carbaryl. Spraying 0.05% dichlorvos or endosulfan is also effective.

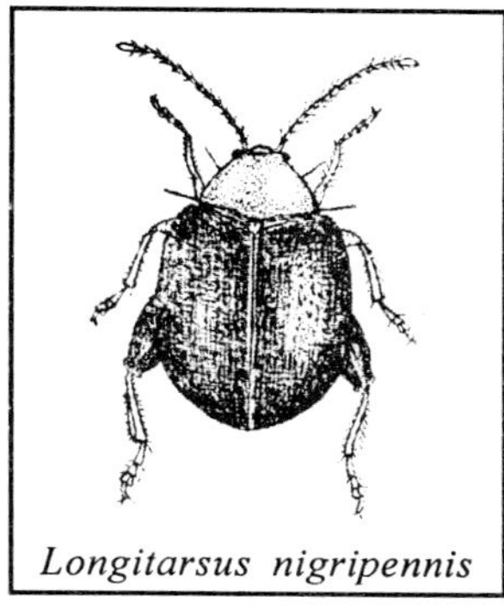
Longitarsus nigripennis

Fruit Borers : *Longitarsus nigripennis* Motschulsky (Chrysomelidae) - a serious pest of pepper, has also been found damaging chillies. Eggs are generally laid on fruits; the grubs bore inside and feed within the fruits while the adults scrap leaves and fruits. Egg, grub and pupal stages last for 5 to 8, 20 to 30 and 6 to 7 days respectively. Total life cycle occupies 40 to 50 days. There are at least 4 overlapping generations in a year, the pest overwinters in adult stage. Spraying 0.1% lindane or 0.2% carbaryl will control this beetle as well as other fruit boring pests.

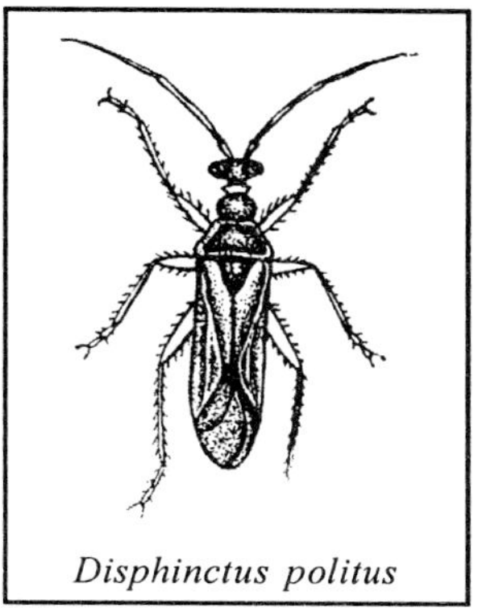
Disphinctus politus

Chilli fruits are also attacked by capsid bug *Disphinctus politus* Walker, tobacco caterpillar *Spodoptera littoralis* (Fabricius), gram caterpillar *Helicoverpa armigera* (Hubner), eggplant stem borer *Euzophera perticella* Ragonot and pepper top shoot borer *Laspeyresia hemidoxa* Meyrick (Eucosmidae). The common names of each pest clearly indicate the main host of these insects. Remove and destroy immediately the affected fruits. If and when infestation is observed and when incidence is high, spray 0.1% lindane or 0.2% carbaryl.

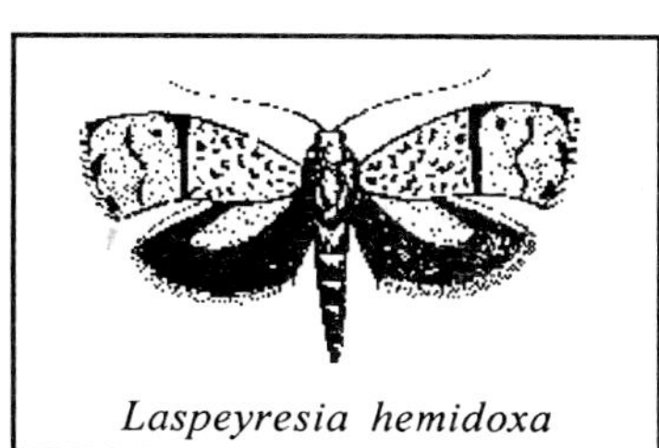
Laspeyresia hemidoxa

In addition to the above *Bactrocera dorsalis* Hendel and *B.cucurbitae* (Coquillett) as also fruit sucking moths *Othreis* spp. and *Ophiusa. coronata* (Fabricius) have been recorded damaging the chilli fruits (Nair, 1975). However, the loss caused is not of any economic importance.

Storage Pests : Dried chillies, when stored, are often attacked by cigarrette beetle, *Lasioderma serricorne* (Fabricius) and drug store beetle *Stegobium paniceum* Linnaeus (Anobiidae). As the names suggest, the primary host of the former is tobacco, cigarettes and cigars while the grubs of the latter feed on stored spices, dry vegetables and animal matter (Nayar *et al.*, 1976).

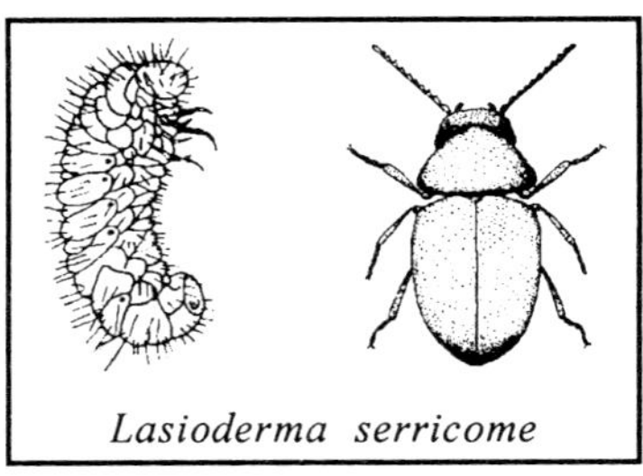
Lasioderma serricome

Grubs of *L. serricorne* are white, fleshy, crescent-shaped with dense hair all over the body; those of *S. paniceum* are also white and fleshy but not hairy. *L. serricorne* beetles are 2 to 3 mm long, robust, oval and reddish-yellow in colour having minute hair on elytra; while those of *S. paniceum* are slightly bigger in size, cylindrical in shape and light brown in colour having striated elytra. Incubation, grub and pupal durations of *L. serricorne* last for 6 to 7, 17 to 29 and 4 to 8 days respectively (Mehta and Verma, 1968).

Arthrodeis species (Tenebrionidas) are small, elongated, flattened, dark brownish-black beetles, nocturnal in habit and feed on decaying vegetables as well as on stored or dried chillies. Loss caused is of minor importance.

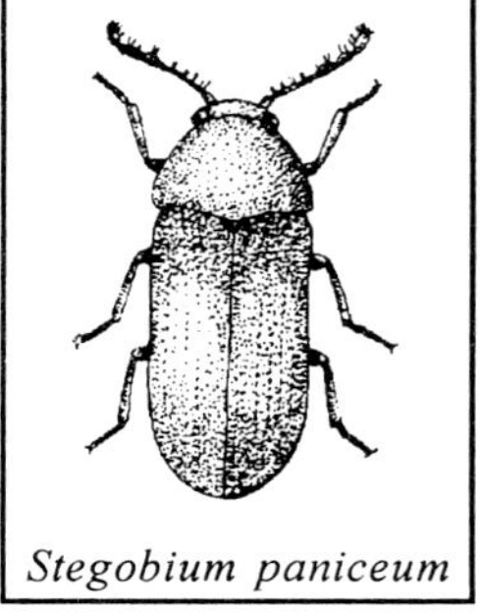
Stegobium paniceum

If the quantity of chillies infested by these storage pests is small, spread them in thin layers under sunshine. If large quantities are infested with these pests, fumigation is the only remedy. Any good fumigant like etylene dibromide or methyl bromide @ one lit per 30 cubic metre space may be used. Fumigate only in airtight containers or chambers.

MITES

Polyphagotarsonemus latus (Banks) (Tarsonemidae) *Tarsonemus translucens* (Green) and *Tetranychus cinnabarinus* (Boisduval) (Tetranichidee) have been reported feeding on leaves of chilli plants.

The tiny creatures are found in large number of ventral side of leaves under a protective cover of fine webs. Both nymphs and adults suck the cell sap and devitalise the plants. Severely infested leaves show brownish patches and the entire foliage may ultimately dry up. *P. latus* is comparatively more common. This mite is cosmopolitan in distribution (CIE map No. C-191) and polyphagous. It also acts as vector for transmitting chilli leaf curl or Murda disease (Kulkarni, 1921; Hirst, 1923; Kar, 1926). Sreeramulu (1976) has suggested spraying 0.1% phosalone to control these mites. Spraying, alternatively, melhomyl and dicofol (2 ml/litre) is also effective (Hindu,3.1.1996). However, wettable sulphur or any other miticide can also be effectively used.

9

OKRA

OKRA, *Abelmoschus (Hibiscus) esculentus* (Linnaeus) Moench (Malvaceae), commonly known as lady's finger, for its shape, is of Tropical African origin, grown extensively in all the Tropical and subtropical regions of the World. It is an annual herb with erect stem, 0.7 to 2.0 metres high, covered with hair. Leaves are cordate, palmate, 3 to 5 lobed and coarsely toothed. Pods or fruits when ripe are 7 to 30 cm long, light green in colour, pyramidal-oblong with longitudinal ridges, smooth or hairy; seeds are round, striate and hairy. Tender fruits (unripe) are used as vegetable as also for thickening soups, stews and gravies due to their high mucilage content. Mucilaginous extract of green stems is used for clarifying sugarcane juice for making Jaggery (*gur*). Root mucilage is used in China as sizing for paper, and leaf mucilage is used as a substitute of soap to remove oil. Leaves are also eaten by the cattle (Chandrasekharan and Ramakrishnan, 1929). In Egypt seed flour is eaten mixed with maize flour. In West Africa, flowers arc eaten in soup. In Malaysia, an infusion of roots is used for treatment of syphilis. Pods are highly nutritive, being rich in pectin and mucilage and a fair source of iron and calcium; besides, they also contain vitamins A, B and C. Ripe seeds contain 16 to 22 per cent edible oil but its extraction is rather difficult (CSIR, 1959). Fibre is white to creamish-yellow in colour, silky, strong and pliant but

Okra

somewhat coarse and stiff; unfortunately extraction and use of this fibre has not been commercially exploited.

An allied species, *Abelmoschus ficulneus* (Linnaeus) Wight & Arnold, is found from Punjab to West Bengal and in South India. Mucilaginous extract of green stem is used for clarifying sugarcane juice; aromatic seeds are used for flavouring purposes and fruits are rich source of vitamin C. (Singh *et al.*, 1983).

INSECT PESTS

Okra plants are attacked by a number of insect pests, mites and nematodes during different stages of growth. Most of the insect pests attacking okra, also attack cotton and therefore work on these pests has mostly been carried out on cotton crop. Due to this reason it is generally emphasised that to protect okra from ravages of these common pests, it is imperative that this crop should not be grown in the vicinity of cotton (Butani and Verma, 1976 d). The major pests of okra include, leafhoppers (jassids), shoot and fruit borers, leaf rollers and red cotton bug. Among the pests of minor significance, mention may be made of aphids, sap sucking bugs, whitefly, thrips, mealybugs, scale insects, leaf-eating beetles and weevils, cutworms, semiloopers and stem borers. Among non-insect pests and red spider mite root knot nematode, *Meloidogyne goldi*, has been reported damaging, respectively, the leaves and roots of this crop.

Leafhoppers : *Amrasca biguttula biguttula* (Ishida) (Cicadellidae), the cotton leafhopper, is a polyphagous pest causing serious damage to cotton, okra, eggplant, beans, castor, cucurbit, hollyhock, potato, sunflower as also other malvaceous plants.

Eggs are laid singly in the tissues of main veins on the under surface of leaves. Both nymphs and adults suck cell sap usually from ventral surface of leaves and while feeding inject their toxic saliva into the plant tissues; affected leaves turn yellowish and curl. In case of heavy infestation the leaves turn dark brick-red, become brittle and crumple. The pest appears with the onset of cloudy weather and their population is adversely affected after heavy monsoon showers.

Eggs are pear-shaped, elongated and yellowish-white in colour. Nymphs are whitish-pale-green, wingless and move in a peculiar fashion, diagonally. Adults are wedge-shaped, 2 to 3 mm long, pale green in colour with a black dot on posterior portion of each forewing. Adults of Winter generation are slightly reddish in colour. A female lays 15 to 30 eggs. Incubation and nymphal periods last for 4 to 10 and 7 to 21 days. respectively; longevity of the adults varies between 5 and 8 weeks and there are 10 to 12 overlapping generations in a year. There is no true hibernation or diapause but the adults have ability to tide over the adverse climatic conditions. Nagpal (1948) has reported that mating takes place 2 to 16 days after emergence and oviposition begins 2 to 7 days after copulation.

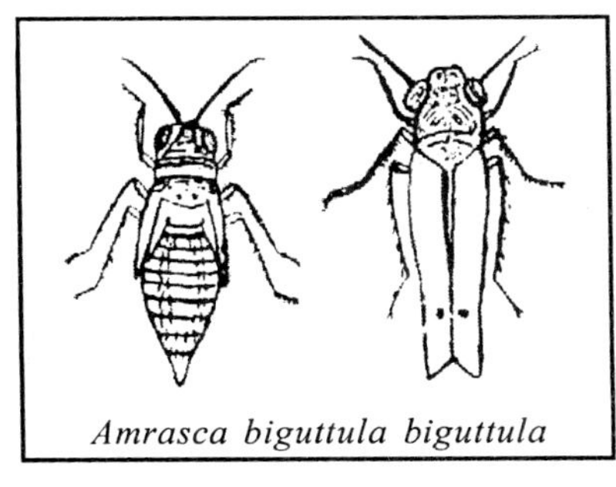
Amrasca biguttula biguttula

Empoasca binotata Pruthi (Cicadellidae) and *Corizus rubicundus* (Distant) (Cicadelhitae) arc the other jassids recorded on okra. Besides, green striped fulgorid *Eurybrachis tomenttos*a Fabricius has been found occasionally on okra leaves in Central and South India (Nair, 1975).

To control these leafhoppers, Katiyar and Bodade (1967) suggested spraying 0.04% phosphamidon while Atwal and Singh (1969) found 0.2% carbaryl to be effective. Dusting 2% methyl parathion has been recommended by Avtar Singh and Butani (1964). Jakhmola and Rawat (1969) as also Rawat and Sahu (1973) found dirnethoate 5% granules applied in seed furrows @ 20 kg per hectare to be effective for 40 days.

Seed treatment of okra with imidacloprid @ 3g/kg and subsequently application of monocrotophos @ 500 g a.i./ha at 55 and 70 days after sowing offered maximum protection against leafhopper (0.53 nymphs/plant) throughout the crop growth stage. The pooled leaf net carbon assimilation rate (NAR) over four crop growth stages varied from 16.66µ mol Co_2 m^{-2} s^{-1} in untreated plots to 34.40µ mol Co_2 m^{-2} s^{-1} in monocrotophos and cypermethrin treated plots. Beta-cyfluthrin provide maximum protection (4.79% fruit damage) against fruit borer, *Earias vitella* Fabricius (Satpathy *et al.*, 2004).

In nature, eggs of *A. biguttula biguttula* are parasitised by *Lymaenon empoascoc* Subba Rao. *Erythmelus empoascae* Subba Rao. *Arescon enocki* (Subba Rao and Kaur), *Anagyrus empoascae* Dozier, *Stethynium empoascae* Subba Rao and *Oligositas* species. A number of predators of this pest have also been recorded. Nasir (1947) reported *Chrysoperla cymbela* Banks; Nagpal (1948) recorded an ant of *Camponotus* species and a spider, *Distina albia*, while Lall (1964) observed *Geocoris tricolour* (Fabricius) arid *G. jacundus* (Fabricius) preying upon this jassid. But generally none of these parasites and predators have been successful in checking the pest population effectively (Butani and Jotwani, 1984).

Shoot and Fruit Borers : Okra shoot and fruit borer, *Earias vittella* (Fabricius)* (auct. *E. fabia* Stollalis) (Noctuidae), commonly known as spotted bollworm of cotton is another major pest of okra. It is widely distributed and is recorded from Pakistan, India, Sri Lanka, Bangladesh, Myanmar, Indonesia, New Guinea, and Fiji (Butani, 1976). It is an oligophagous pest and though okra and cotton are its main hosts, it is also found feeding on a large number of other malvaceous plants, both wild as well as cultivated. The pest is active almost the year round and prefers high humidity and high temperatures; as such it is more abundant in Northern states of India than the Southern states. Summer crop of okra suffers the most from ravages of this pest, after which the pest migrates to cotton.

Eggs are laid singly on buds and flowers and occasionally on fruits: during the early stage of crop's growth, the eggs are laid on shoot tips. When the crop is only a few weeks old, the freshly hatched caterpillars bore into tender downwards, these shoots wither, droop and ultimately the growing points are destroyed: side shoots may arise giving the plants a bushy appearance. With the formation of buds, flowers and fruits, the caterpillars bore

* Stollalis (1782) described *Noctua fabia* as species *nova.* Hampson (1894) redescribed this species as *Earias fabia* (Stollalis). Earlier Cramer (1779) had described a hairy caterpillar as *Eupterote fabia* as species *nova. Earias fabia* is therefore a junior homonym of *Eupterote fabia* Cramer and *Earius vittella* Fabricius (1794) is the next available name.

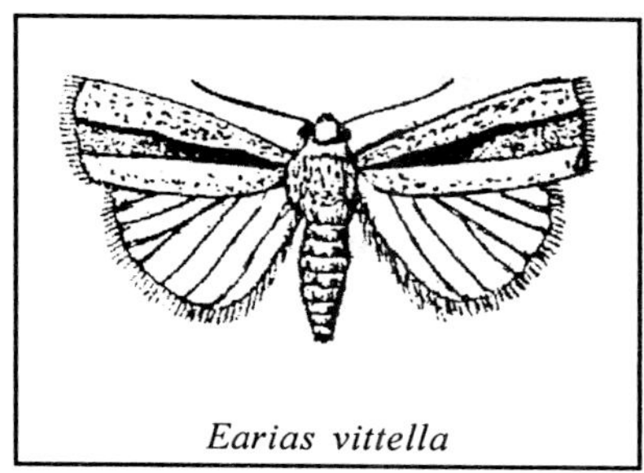
Earias vittella

inside these and feed on inner tissues. They move from bud to bud and fruit to fruit, thus causing damage to a number of fruiting bodies. The damaged buds and flowers wither and fall down without bearing any fruit whereas the affected fruits become deformed in shape and remain stunted in growth. Such fruits have hardly any market value.

Eggs are spherical in shape, about half mm in diameter, light bluish-green in colour and beautifully sculptured with 26 to 32 longitudinal ridges; the alternate ridges project upwards to form a crown, and the eggs look like miniature poppy fruits. Full-grown caterpillars are 18 to 24 mm long, stout, spindle-shaped having long stiff setae. Pupae are 13 to 16 mm long and chocolate-brown in colour, bluntly rounded and enclosed in inverted boat-shaped cocoons. Adults arc medium-sized moths, 13 to 15 mm long, head and thorax ochreous-white; forewings are pale-white with a broad wedge-shaped horizontal green patch in the middle, and hind wings are silvery-creamy-white in colour. Wing spread is 30 to 34 mm.

The moths emerge at dusk; mating takes place 2 to 3 days after emergence, and oviposition commences after another 1 to 5 days. A female lays on an average 400 eggs (65 to 695). Incubation, caterpillar and pupal periods last for 3 to 9, 9 to 20 (50 to 60 during Winter) and 8 to 12 days respectively. A single life cycle takes 22 to 25 days extending upto 74 days during winter and there may be 8 to 12 generations in a year. There is no true hibernation but development and activity is considerably slowed down during Winter.

Another fruit borer, which is also known as spiny bollworm of cotton *Earias insulana* (Boisduval) (Noctnidae), is found damaging okra, especially in drier regions. This is also a major pest of cotton and has been reported from most of the African countries, Malagasy, Mauritius, Mediterranean region. Near and Middle East, Pakistan, India, Sri Lanka, Bangladesh. Myanmar, Japan, Taiwan, and South-east Asia (CIE map No. A-251). Its immature stages are more or less similar in appearance to those of *E. vittella,* except for the size. Full-grown caterpillars are 15 to 18 mm long and pupae are 12 to 14 mm long. The moths are not only smaller in size than those of *E. vitiella* (11 to 13 mm long with wing expanse of 20 to 24 mm) but their heads and thoraxes are pea-green in colour and forewings are uniformly pale yellowish-green, often showing seasonal colour variation (Yathom, 1956). The pattern of damage and other habits of the two species are about the same. Even the two species have been observed mating with each other both under field and laboratory conditions (Khan and Ghani, 1944).

Besides, gram pod borer or American bollworm *Helicoverpa armigera* Hubner (Noctuidae) and tobacco caterpillar *Spodoptera littoralis* (Febricius) (Noctuidae) have also been occasionally reported damaging shoots and tender fruits of okra. Both are polyphagous pests and among the vegetables, tomato is their preferred host.

Satisfactory control of these borers is still a far cry. By adopting cultural and mechanical methods followed by timely application of insecticides, the damage can be considerably reduced and yield of fruits increased. Work on breeding resistant varieties and use of pathogens, parasites and predators as biological control agents is still in experimental stages. To check the infestation, clean cultivation is essential. Remove debris and all the alternate host plants from field, collect and destroy all the infested shoots and fruits. In case of severe infestation, give two to three fortnightly sprayings with 0.05% dichlorvos or endosulfan.

Recently, emamectin benzoate (Proclaim) 5% g., a new insecticide reported effective against fruit borers (*Earias* spp. and *Helicoverpa armigera*). Emamectin benzoate @ 8.50 ga.i./ha recorded lower fruit borers damage and higher fruit yield (Bheenanna *et al.*, 2005)

The natural enemies may play an important role in reducing the pest population in the fields. Eggs are parasitised by *Trichogramma evanescens* Westwood; caterpillar parasites include *Actia aegyptia* Villers, A. *hyalinata* Malloch, *Apanteles* species, *Bassus* species, caterpillars pupate and the moths emerge in early July. Low temperature and high humidity, coupled with cloudy and rainy days, favour *Microbracon brevicornis* (Wesmael), *M. greeni* Ashmead (=*lefroyi* Dudgeon and Gough), *Hobrobracon hebetor* (Say), *Elasmus johnsonii*

Okra fruits damaged by *Earias vittella*

Okra fruits damaged by *Earias vittella*

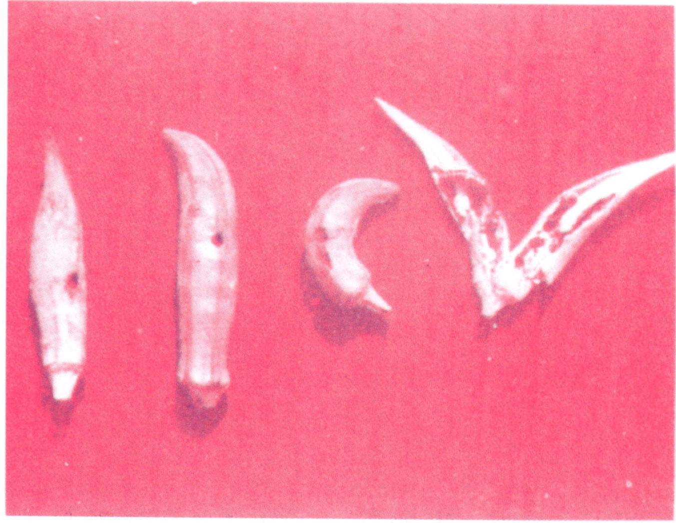

Okra fruits damaged by *Earias vittella*

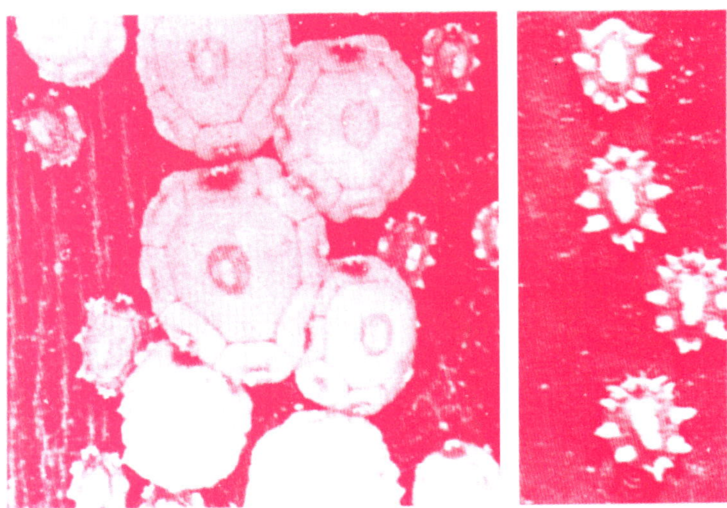

Ceroplasters floridensis on okra leaf

Okra plant infested with jassid and mite

Okra crop damaged by mite

Ferriere, *Rhogus aligharensis* Qadri, *R. testaceus* Spinola, *Phanerotoma hendecasisella* (Cameron), *Polyodaspis* species and *Strobliomyca nana* (Curr): pupae are parasitised by *Brachymeria responsator* (Walker), *B. tachardiae* (Cameron), *Centrochalcis* species, *Chelonus rufus* Lyle, *C.reponsata* Walker and *Goryphus nersei* (Cameron). The shield bug *Catheconidea furcellata* (Wolffenstein) wasp, *Eumenes petiolata* are predaceous on the caterpillars.

Leaf rollers : Cotton leaf roller, *Sylepta derogata.* Fabricius (Pyraustidce) is sporadic pest of American cotton and okra. Besides, it has also been recorded on a large number of other malvaceous plants - cultivated as well as wild. Eggs are laid singly on ventral surface of leaves. On hatching, the caterpillars feed on epidermis of ventral surface of leaves for a couple of days, then they roll the leaves and feed within, eating away large portions of the rolled leaves. More than one caterpillar may be found in a single leaf roll initially but later the caterpillars disperse and attack more and more leaves. Application of nitrogenous fertilisers increases the incidence of this pest. Pupation generally takes place in rolled leaves but sometimes the caterpillars drop down and pupate in the soil.

Eggs are round in shape and yellowish-green in colour. Caterpillars are shiny green in colour and more or less transparent. Pupae are reddish-brown. Moths are yellowish-white with both the fore- and hind-wings having brown lines and distinct markings; wing expanse is 30 to 38 mm. A female lays about 200 to 300 eggs (Sen, 1923). Incubation, caterpillar and pupal periods last for 2 to 3, 15 to 20 and 6 to 12 days, respectively (Rahman, 1940), with 4 to 5 generations in a year. Caterpillars remain overwinter in soil or between the debris from end of November till end of May or early June when these hibernation of caterpillars, resulting in an epidemic during

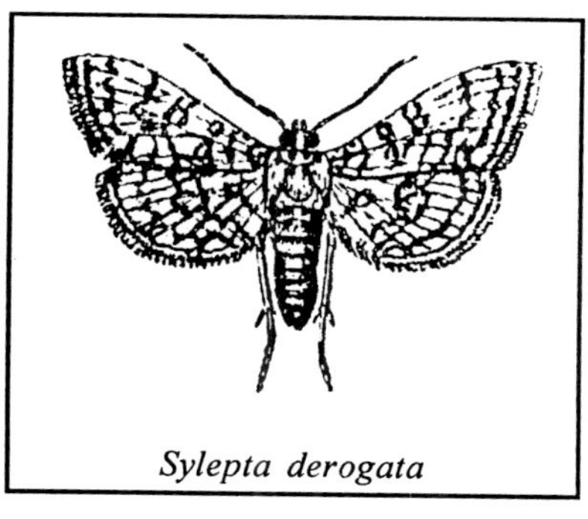

Sylepta derogata

the next season (Lal and Singh, 1951). Warm, damp and cloudy weather is conducive for rapid multiplication of this pest.

Heicysfpgramrna hibisci (Stainton) (Gelechiidae) is another cotton leaf-roller that is a minor pest of cotton and okra. It is widely distributed in South-east Asia. Full grown caterpillars are about 12 mm long, cylindrical in shape slightly tapering posteriorly, head shiny black with scattered minute whitish hair; prothoracic shield also shiny black, body yellowish-green with small dark warts, each emitting a short, white hair. Pupae are brown in colour and cylindrical in shape, broad apically.

To check damage by these leaf rollers, collect and destroy all the rolled leaves with caterpillars inside. The infested fields should be irrigated and ploughed after the harvest of crop to kill the. caterpillars hibernating in the soil. In case of severe infestation, spray 0.2% carbaryl or 0.05% tetrachlorvinphos or fenitrothion. In nature, eggs of *S. derogata* are parasitised by *Trichogromma* spp., caterpillars and pupae by *Apanteles* spp., *Bessa remota* Aldrich, *Brachymeria tachardiae* (Cameron), *Cedria paradoxa* Wilkinson, *Elasmus indicus* Rohwer, *Enicospilus atricornis* Morley, *Goryphus nursei* Cameron, *Microbracon greeni* Ashmead, *Microtoridea lissonota* Viereck, *Neopimploides syleptae* Viereck, *Phenerotoma hendecasiselia* Cameron, *Tricaosplus pupivora* Ferriere and *Xathopimpla punctate* Fabricius. Husain and Bhalla (1937) have listed 86 species of birds preying upon *S. derogata.* None of these parasites or predators have been found exercising appreciable cheek on the pest population.

Red Cotton Bug : *Dysdercus koenigii* (Fabricius) (Pyrrhocoridae) is a very common but economically not so important pest found all over the Indian sub-continent infesting okra and cotton. These bugs have also been found feeding on pearlmillet (*bajra*), maize, sorghum and other millets, wheat, cape gooseberry, mulberry, silk cotton, rose, hemp, musk-melon, Indian mellow, hollyhock and various other malvaceous plants, both cultivated and wild. This species was earlier confused with *Dysdercus cingulalus* Fabricius till Kapur and Vazirani (1956) pointed out its correct identity. Till then, most of the work on this pest has been reported under *D. cingulatus.*

Eggs are laid in loose irregular masses preferably in cracks and crevices in moist soil. A female lays 100 to 150 eggs. Both nymphs and adults suck sap from fruits and to some extent from leaves, thus devitalising the plants. The feeding deprives the plants of carbohydrates, free amino acids and proteins (Saxena, 1955). Pest population usually increases with the advent of cold season. Nymphs sometimes become cannibalistic.

Morphology, anatomy and bionomics of this pest have been studied in detail by Pruthi (1921,1923) as also by Srivastava and Bahadur (1958). Eggs are spherical in shape and bright yellow in colour. Nymphs and adults are very conspicuous, bodies including antennae and legs being red in colour. Adults have a black dot on each forewing. Incubation and nymphal periods occupy 4 to 5 and 30 to 90 days, respectively, depending upon prevailing climatic conditions.

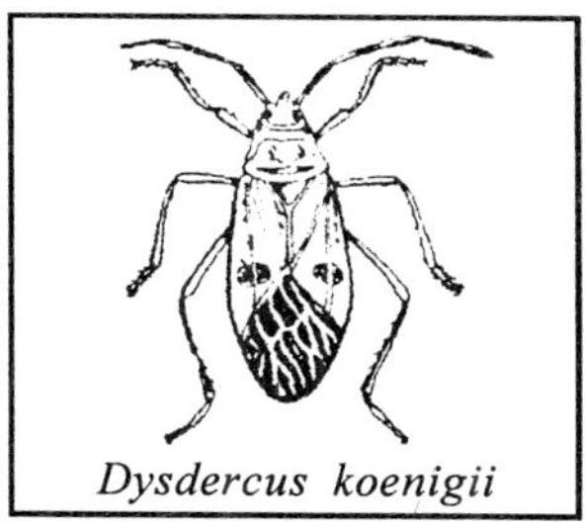
Dysdercus koenigii

High temperature coupled with low humidity keep the pest population under check. If and when there is severe infestation, spray 0.05% dichlorvos, endosulfan or quinalphos.

A tachanid fly parasitises the adults while *Herpactor costalis* Rev. is predaceous on nymphs and adults. Pradhan and Menon (1942) also recorded *Antilochus cocqueberte* Fabricius preying upon this insect. Nagpal (1948) reported the birds, *Crioius rnelanocephalus, Oriolus kurtdoo, Hierococcoyx varius* and *Sitte ferintalis* preying on these bugs.

Dusky Cotton Bug : *Oxycarenus laetus* Kirby (Lygaeidae) is widely distributed all over the Indian sub-continent. It is mainly a pest of cotton plants and only when cotton plants are not available or okra is grown in the vicinity of cotton crop, the pest migrates to okra or any other wild malvaceous plants growing around. Activity of the pest starts soon after monsoon and continues till the advent of cold season. Both nymphs and adults suck the sap usually from the fruits but the loss caused by them is negligible. Egg, and nymphal stages last

Oxycarenus laetus

for 5 to 7 and 30 to 44 days, respectively, and full life cycle is completed in 36 to 50 days.

Lablab Bug : *Coptosoma cribraria* Fabricius (Corimelaenidas) is a major pest of beans, pulses and grams. Nymphs and adults may be seen in large number crowded and feeding on tender shoots and thereby devitalising the plants. In case of severe attack plant growth is adversely affected.

Cow Bugs : Treehopper, *Tricentrus bicolor* Distant (Membracidse) is a polyphagous pest found feeding on cereals, cotton, eggplant, okra, potato, tomato, various fruit trees and wild grasses. It is found throughout the year though maximum activity period is during September-October. Mating usually takes place during warmer days. The mated females cut tender branches, midribs of leaves, petioles, buds and even leaf lamina longitudinally and lay eggs therein. On hatching, the nymphs remain clustered in axils of branches or near midribs of leaves on abaxial sides and suck the cell's sap. Both nymphs and adults exude honeydew which attracts the black ant, *Camponotus compressus.* The damage caused to okra plants is seldom severe.

Eggs are small, about 0.1 mm long and transluscent. Nymphs are sedentary, soft-bodied and shinning, greenish-brown in colour. Adults are black and setose with a prominent sub-triangular pronotum overlapping the entire prothorax, two hollow horn-like recurved suprahurneral arms with apex prolonged posteriorly into a robust posterior spine. Males are darker in colour and more active than

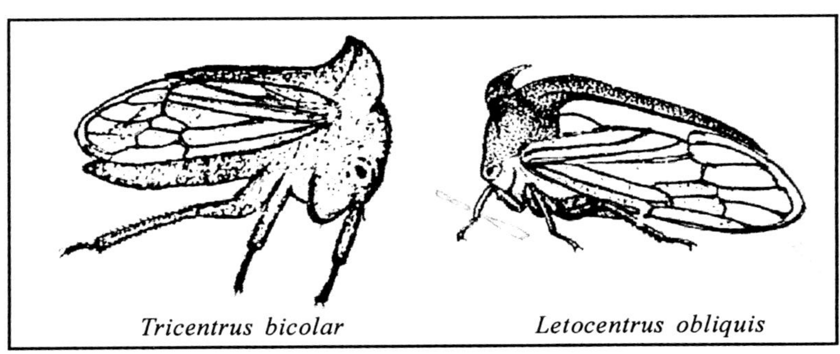

Tricentrus bicolar *Letocentrus obliquis*

females. Incubation period is 8 to 13 days while nymphal development is completed in 33 to 55 days.

Another species of cow-bug, *Leplocentrus obliquis* Walker (Membracidae) is also a minor pest of okra and potato. Its habits and habitat are same as those of *Tricentrus bicolor*.

Capsid Bug : *Creontiades pallidifer* Walker (Miridae), a pest of eggplant and potato, it causes small irregular brown spots on distal portions of young leaves; the affected leaves start drooping and drying but seldom die away.

Lygaeid Bug : *Spilastethus pandrus* (Scopoli) (Lygaeidae), a pest of cole crops has also been reported feeding occasionally on okra leaves. The damage caused by this pest to okra plants is negligible.

These bugs are minor pests and do not need any specific control measures. Nevertheless, when observed in large numbers, spraying 0.05% dichlorvos, endosulfan or quinalphos can prove to be effective in checking the pest population.

Thrips : These are tiny, long, slender and fragile insects that have heavily fringed wings. Because of their small size and hiding habit they are often not detected till the damage symptoms appear. They feed by lacerating plant tissues-and lapping the oozing sap. Ananthakrishnan (1971) reported three species of blossom thrips, namely, *Microcephlothrips abdominalis* (Crawford), *Franklintella schultzei* (Trybom) and *Haplothrips gowdeyi* (Franklin) infesting okra flowers. All these species are cosmopolitan in distribution and highly polyphagous. Their feeding results in drying and premature shedding of flowers, thus adversely affecting the fruit setting.

Microcephalothrips abdominalis is the most common species and attacks a large number of ornamental plants including Chrysanthemum, Dahlia, marigold, safflower and Zinnia. It causes scars on floral parts and in case of severe attack, which is rare on okra, the flowers get discoloured. The adults are 0.6 to 0.8 (males) and 0.8 to 1.0 (females) mm long; yellowish-brown to dark brown in colour; forewings infumate with brownish-gray, darker along the margin and wing setae are very weak. Reproduction is both sexual as well as parthenogenetic. The parthenogenetic females are arrhcnotokons and they produce only males. Fertilised females lay 75 to 100 eggs per day. Life cycle is complete between 9 to 20 days and there are 20 to 25 generations in a year (February to November), only females survive during December to February.

Thrips tabaci Lindemann (Thripidae) - the onion thrips, is sometimes found feeding on tips of okra leaves. This is also a highly polyphagous pest with worldwide distribution. Onion and tobacco are its main hosts and besides okra, it is also found on cole crops, eggplant, etc. The damage caused to okra leaves is generally negligible and only in case of severe infestation, which is rare, the leaf-tips turn pale brown and curl.

Normally, no control measures are required against these thrips on okra. However, when seriously attacked, the crop can easily be protected by spraying with 0.03% phosphamidon or dimethoate or 0.04% monocrotophos.

Semi-loopers : A number of noctuid caterpillars, grouped as semi-loopers due to their peculiar mode of moving, have been recorded feeding on okra leaves: these include, *Anomis flava* Fabricius (= *Cosmophila indica* Guenee). *Xanthodes (Acontia) groellisi* (Feisth), *Acontia intersepta* Guenee. *A. malvae.* Esper, A. *transrversa* Guenee and *Tarache nitidula* (Fabricius). Of these, the first two are comparatively more common and destructive.

Anomis flava, the green semi-looper, is widely distributed in India. It is a sporadic pest of cotton and okra and infests in an epidemic form in years of heavy rainfall. Eggs are laid singly on the upper surface of the leaf. A female lays as many as 600 eggs in 8 to 12 days. Pupation takes place within the leaf folds. Full-grown caterpillars are 25 to 30 mm long, greenish in colour with 5 white longitudinal lines. Adults are medium-sized brown moths; males are more brightly coloured than females. Incubation, caterpillar

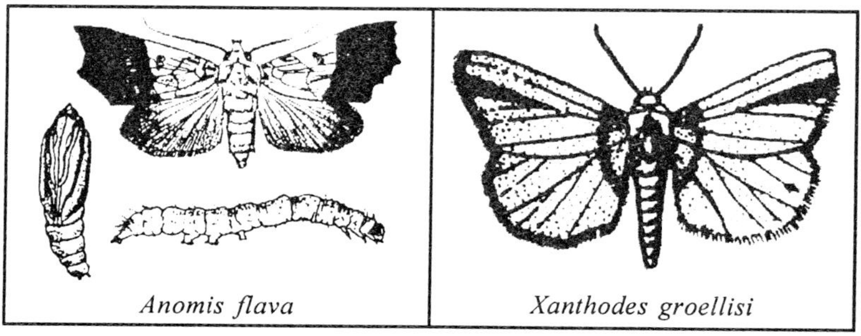

Anomis flava | *Xanthodes groellisi*

and pupal stages last for 4 to 5, 18 to 20 and 8 to 9 days respectively.

Xanthodes groellisi is another sporadic pest that occurs on okra, cotton and other ornamental plants specially hollyhock. The damage caused is similar to that of *Anomis flava*. Continuous wet weather is very favourable for the rapid multiplication of this pest. The caterpillars are green in colour with horse-shoe shaped black markings on each segment. When full-grown they are 20 to 25 mm long. Moths are yellowish in colour with black markings all over.

Tarache nitidula has of late assumed the status of a major pest of cotton in South India. It has been recorded as a minor pest of okra. Eggs are laid singly on tender leaves. On hatching, the caterpillars feed on leaves and in case of severe attack, which is rather rare, they may practically defoliate the entire plant. Incubation, caterpillar and pupal periods last for 3 to 4, 20 to 22 and 10 to 12 days respectively during July-August in South India (Nair, 1975).

As the damage caused by these semi-loopers is of minor nature, generally no control measures are adopted against these caterpillars. Nevertheless, if and when necessary, the infested crop may be dusted with 10% HCH or 5% carbaryl or 4% endosulfan.

Parasites recorded on *Anomis flava* include, *Carcelia hackiana* Townsend, *Sturmia macrophallus* Baranov, *Apariteles* species, *Tetrastichus ayyari* Rohwer, *Trichogramma minutum* Riley, *Actia monticola* Malloch and *Exorista apicalia* Baranov. Puttarudriah and Maheswariah (1958) found that as high as 60% of field collected caterpillars were already parasitised by a tachinid fly (*Isyropa* species).

Cutworms : A group of lepidopterous (Noctuidae) polyphagous pests with worldwide distribution. They are specially devastating at young seedling stage of the crop. The caterpillars are nocturnal in habit and remain hidden in the soil during day time. They come out at night, cut the seedlings at ground level and feed on some of the tender leaves and move from one seedling to another. Thus the actual loss caused by cutting the seedlings in much more than what is eaten by them.

The species reported damaging okra are, *Agrotis ipsilon* (Hufnagel) and *A. flammatra* (Schiffer); the latter is comparatively more common. Besides okra, caterpillars of *A. flammatra* also damage seedlings of cotton, gram, niger, potato, etc. The adult moths have pale-brown head, thorax and abdomen; collar is brown with a deep black triangular mark. Forewings are also pale brown with a broad pale costal fascia from base to postmedial line. Hindwings are fuscous-brown. Wing expanse is 52 to 60 mm. The life cycle is completed in 1 to 2 months depending upon the climatic conditions.

As these cutworms are of minor significance on okra, no control measures are normally adopted against them. However, dusting the seedlings, as also the soil around, with 5% HCH or 4% endosulfan is quite effective in checking their population.

Stem Borers : Three coleopterous species that are minor pest of cotton, have also been reported damaging okra plants. Cotton stem borer, *Sphenoptera gossypii* Cotes (Buprestidae), is a minor pest, widely distributed in Indian sub-continents. Cotton is its main host and okra is a secondary host. The pest prefers warm and humid climate and appears in endemic form during years of drought.

Eggs are laid singly on tender stems and shoots, preferably on lower half of plants. On hatching, the grubs bore into stems and feed within by tunnelling downwards. Pupation takes place in these tunnels. The lower portion of the infested stems often gets swollen and gradually the affected stems wither and die away.

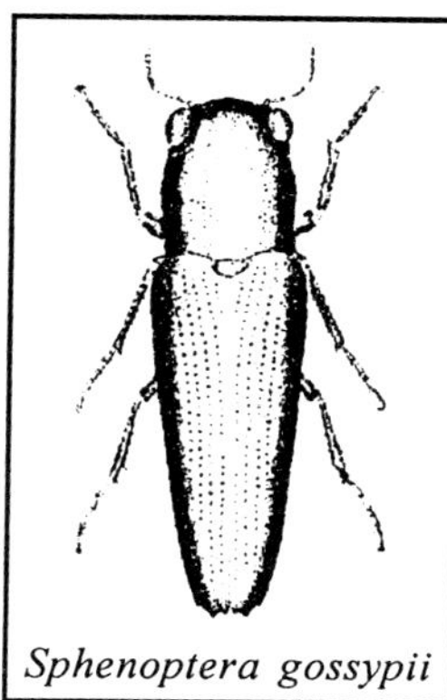
Sphenoptera gossypii

Grubs are stout, fleshy and about 20 to 30 mm long whereas adults are shiny copper brown beetles that measure 7 to 10 mm in length.

Cotton Stem Weevil, *Pempheres affinis* Faust (Cureulionidae) has also been reported from all over the Indian sub-continent. This is also mainly a pest of cotton but in the absence of cotton, it has been recorded on other malvaceous and some wild alliaceous plants, including okra. Females make small cavities in soft and succulent stems, deposit one egg in each

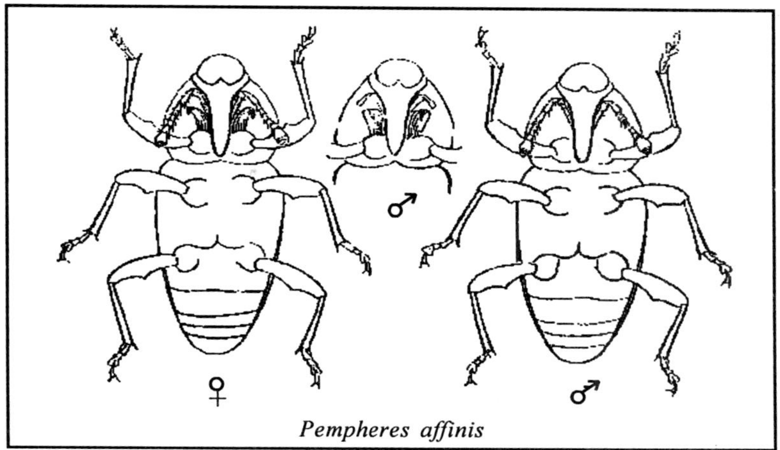

Pempheres affinis

cavity and seal by exudation. The same stem may have a number of such cavities. On hatching, the grubs bore into the stem near the base and cause gall-like swellings. The attacked plants lodge and even break with strong winds. A single female lays 50 to 120 eggs in 60 to 80 days; the eggs hatch in 6 to 10 days; grub stage lasts for 35 to 37 days and adult longevity is 25 to 30 clays. There are 5 generations in a year.

Alcidodes affaber Faust (Curculionilae), a cotton shoot weevil, is widely distributed in Indian sub-continent and attacks a large number of malvaceous plants including okra and also many ornamental plants (*Hibiscus* species). The nature and symptoms of damage are more or less same as those of *Pempheres affinis*. The females of the species excavate cavities on leaf petioles and sterns and lay eggs singly in these pits. On hatching, the grubs bore into the petioles and stems causing gall-like swellings. Pupation takes place within these galls. Adults are comparatively less harmful; they feed on leaf buds and tender terminal shoots. The infested plants become stunted in growth and fruit production is adversely affected. The pest is active from September to December when as many as 12 grubs may be found in the same plant. Eggs are smooth, globular in shape and milky white in colour. Grubs are creamy-yellow in colour and apodous. Adults are dark grayish-brown with pale cross bands on elytra. A female lays about 45 eggs in one

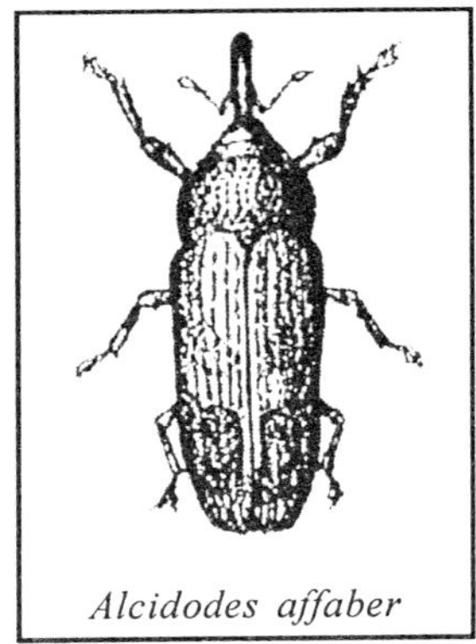
Alcidodes affaber

month. While pre-oviposition period is 8 to 11 days, egg, grub and pupal periods last for 6 to 7, 55 to 62 and 10 to 12 days respectively. Adult longevity is 8 to 32 days and 6 to 43 days in case of males and females respectively (Nayar *et al*, 1976, David and Ananthakrishnan, 2004). Total life cycle occupies 72 to 80 days.

Okra plants should not be allowed to stand in the field after the final picking of fruits. If and when the stem borers appear, destroy the affected shoots.

A number of natural enemies have been recorded on these stern borers. Eggs of *Sphenoptera gossypii* are parasitised by *Ooencyrtus* and *Trichogramma* species; grubs are parasitised by *Euderus gossypii* Ferriere. *Glyptomorpha smeenus* Cameron. *Horminae* species, *Neocatolaccus indicus* Ramakrishna and Margabandhu, *N. sphenopterae* Ferriere and *Vipio* species while pupae are prasitised by *Xanthopimpla punctata* Fabricius (Nair, 1975). The natural enemy complex of *Pempheres affinis* is different on different hosts. *Aplastomorpha catandrae* Howard, *Aximoposis* species, *Bruchrocida orientalis* Crawford, *Dinarmus coimbatorensis* Ferriere, *Euderus pempheriphila* Ayyar and Mani, *Eupelmella pedaloria* Ferriere, *Eupelmus urozonus* Dalman, *Microbracon* species, *Neocatolaecus indicus*, *Rhaconotus cleanthes* Nixon, *R. menippus* Nixon. *Spathius critolaus* Nixon, *S. labadcus* Nixon and the nematode *Geomermis indica* Steiner are some of the reported parasites. *Pediculoides ventricosus* Newport and the ants *Tetraponera* species and *Monomorium* species prey upon the grubs. Grubs of *Alcidoides affaber* are parasitised by *Aphrastohracon alcidophagus* and *Xoridescopus* species.

Grasshoppers : *Ak* grasshopper, *Poekilocerus pictus* (Fabricius] (Acrididae) and small rice grasshopper *Oxya japonica japonica* (Thunberg) have been occasionally recorded feeding on okra plants (Nayar *et at.*, 1976, David and Ananthakrishnan, 2004). Both are minor pests. Nymphs and adults eat. away the leaves and a severe infestation results in complete defoliation of the plants.

Poekilocerus pictus is widely distributed in the Indian sub-continent and Africa. Though it is a primary pest of *ak* (*Calotropis giganta*), it has also been observed defoliating cowpea. eggplant, okra, tomato, various ornamental plants and fruit trees. Eggs are elongated oval in shape, slightly curved and yellowish-orange in colour. Nymphs are yellowish in colour with prominent black and orange dots and stripes all over their body. Adults are yellowish with dark bluish-green stripes on head, thorax and abdomen; antennae are blue-black ringed with yellow. Forewings are bluish-green with yellow veins, and hind wings are hyaline, Pre-oviposition period is of 25 to 30 days. For ovipositing, the gravid female selects a suitable spot on the soil and thrusts its abdomen 150 to 200 mm deep, lays 150 to 200 eggs in a compact mass and covers the same with frothy secretion. Eggs laid in Summer hatch out in about 4 weeks while those laid in Autumn overwinter for 4 months and hatch out in Spring. Nymphal duration varies between 8 and 11 weeks being shorter during Summer, thus there are only 2 generations in a year.

Oxya japonica japonica is widely distributed in Indian sub-continent, South-east Asia, parts of China and Japan. This is also a polyphagous pest, rice being its main host. Both nymphs and adults nibble the leaves. The damage caused to okra plants is generally negligible. Nymphs and adults look alike except that the former has only wing-pads and the Inner fully developed wings. These are pale brownish-green grasshoppers with a brown steak on either side starting from the base of eyes passing through the thorax upto the end of abdomen. Adult males are 18 to 22 mm long and females 26 to 32 mm.

Generally, no control measures are adopted against these grasshoppers. Dusting 5% carbaryl or 1% lindane is quite effective in checking the pest population or spray 0.03% lindane or 0.05% endosulfan.

Pod Fly : *Melanagromyza obtusa* (Malloch) (Agromyzidae) a major pest of red gram, cowpea and soybean, is a minor pest of okra (Venugopal and Venkatramani, 1954). Eggs are laid singly on tender leaves of seedlings. On hatching, the maggots bore inside the tender shoots and leaf petioles. The affected seedlings wither and eventually dry up.

Aphid : *Aphis gossypii* Glover (Aphididaae), the cotton aphid, is a polyphagous pest with Worldwide distribution. Among the vegetable crops it is commonly found on bean, crucifer, chilli, eggplant, gourd, okra, potato and tomato. Besides sucking the cell sap and secreting honeydew, it also acts as a vector, transmitting various viral diseases especially those of beans and potato.

To control aphids. clip off and destroy promptly the infested parts along with colonies of aphids as soon as they are observed. In case of severe infestation, which is not common on okra, spray 0.05% malathion, dimethoate or endosulfan.

Whitefly : *Bemisia tabaci* (Gennadius) (Aleyrodidae), the cotton whitefly, is a polyphagous pest having cotton and tobacco as its main hosts. It also attacks most of the Winter vegetables including cole crops, okra, potato, tomato and ornamental plants. Eggs are laid singly on ventral surface of leaves attached by thin, small whitish stalks. A female lays on an average 50 eggs (maximum 120). Newly hatched nymphs are initially mobile, but later the legs and antennae get degenerated and subsequent instars remain stationary and develop by sucking the cell sap. The damage is caused not only by desapping of the plant and exuding honeydew, but also by their acting as vectors of virus diseases transmitting the leaf curl disease. The activity of the pest is accelerated by scanty rainfall coupled with high temperature. The life cycle is completed in 2 to 3 weeks during summer and takes as long as 15 to 16 weeks in winter.

If and when necessary, spray with any good systemic insecticide, like 0.03% dimethoate, oxydemeton-methyl, pbosphamidon or thiometon.

Mealybugs : A number of mealybugs may sometimes be seen clustered on the terminal shoots and tender leaves of okra sucking the sap from these plant parts. The affected parts turn yellow, wither and get dried. The species damaging okra plants are *Ferrisia virgata* (Cockerell) and *Nipaecoccus viridis* (Newstead) (Pseudococcidae). Both are widely distributed in Tropical and subtropical countries and have a wide range of host plants.

Ferrisia virgata is active during August to November. A prolonged period of drought may result in severe outbreak of this pest (Butani, 1979a). Females are apterous, about 4 to 5 mm long, covered with waxy powder-like substance and having a number of tiny waxy hair all over the body and two long conspicuous waxy filaments at the anal end. Reproduction is oviparous as well as parthenogenetic. the latter being more common. Females mate only once during their life time; eggs are laid in clusters which remain under the female body. It takes from few minutes to 4 hours for these eggs to hatch but the young ones still remain under their mother's body for 1 to 3 days, then they crawl about a little in search of food and settle at a suitable spot. They also secrete a small quantity of honeydew on which sooty moulds develop, covering the affected parts with a thin superficial black coating.

Nipaecoccus viridis is active during November. Its eggs are cylindrical in shape and chestnut-brown in colour. A female lays as many as 400 to 700 eggs. Nymphs are chocolate coloured covered with whitish mealy substance. Adult females are also dark chocolate coloured and covered with white mealy substance. Eggs are laid in ovisacs. Incubation period is 7 to 10 days, while nymphal development of males and females is completed in 15 and 20 days, respectively.

To control these mealy bugs, whenever necessary, spray 0.05%, dichlorvos or dimethoate.

Scale Insects : These are polyphagous pests, cosmopolitan in distribution; one or the other species is found on most of the crops all over the globe. Except a few species, scale insects are generally minor pests becoming sporadically severe. The young ones, as also the adult females (males are rare or not known) devilalise the plants by sucking the vital sap, especially from Ihe tender parts.

The species commonly, found on okra, include *Saissetia coffeae* (Walker) (brown soft scale) and *Parasaissetia nigra* (Nietner) (black scale). The common names suggest the colour of mature females which are hemispherical or domeshaped. Reproduction is usually parthenogenetic. A large number of eggs are laid by the mature female and these remain hidden under her body in a hollow

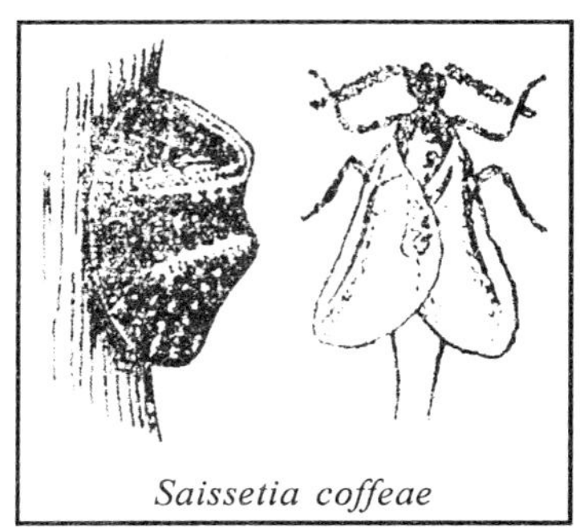
Saissetia coffeae

space specially made for this purpose. After hatching, the crawlers creep out from under the mother scale, move about a bit, search a succulent spot and settle there almost permanently, sucking the cell sap.

Nair (1975) also, reported *Ceroplastes floridensis* Comstock(Coccidae) infesting okra. This soft scale pest occurs widely and sporadically. Its nymphs take about 3 weeks to mature and there are only 3 generations in a year. The scale insects generally do not cause any appreciable damage to okra crop and, therefore, no chemical control measures have been required. However, in case these insects appear in abundance, the infested plant parts should be clipped off and destroyed to prevent their further spread.

In nature, *Saissetia coffeae* is parasitised by *Aneristus ceroplaslae* Howard. *Anysis saissetiae* (Ashmead). *Cardiogaster fusciventris* Motschulsky, *Cheiloneurus paradisicus* Motschulsky, *Cirrospilus cocciuonis* Motschulsky, *Coccophagus flavescens* Howard, *Encyrtus aduslipennis* Motschulsky, *Eucomys rufescens* Motschulsky and *Scutellista cyanea* Motschulsky whereas *S. nigra* is parasitised by *Aneristus ceroplastae* Howard, *Anicetus ceulonensis* Howard, *Anysis saissetiae's* (Ashmead), *Coccophagus lorigiclavatus* Howard, *Encyrtus barbatus* Timberlake, *E. kotinski* (Fullaway), *Eucomys lecanium* (Meyrick), *Marietta leopardina* Motschulsky and *Scutellista cyanea* Motschulsky, besides by a predaceous caterpillar of *Eublemma scitul* Rambur (Nair, 1975). Such a large number of natural enemies may partially be responsible for keeping the population of these scale insects under check in nature.

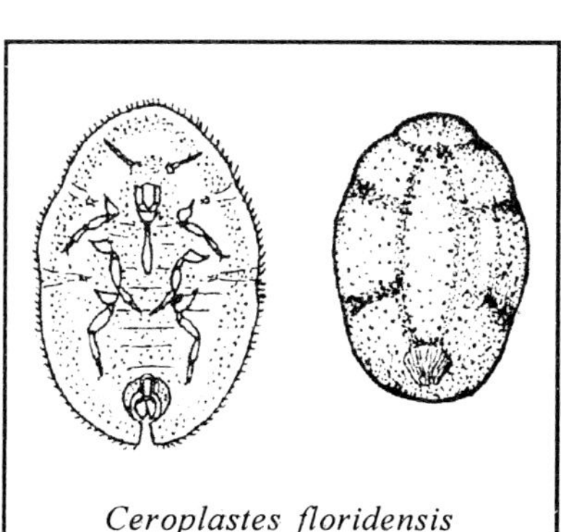
Ceroplastes floridensis

Leaf Miners : *Trachys herilla* Obenberger - a buprestid beetle, is a minor pest of okra, especially in Gujarat. The females make small punctures on leaf margins preferably near the leaf tips and lay eggs singly in the punctured tissues. On hatching, the grubs mine inside the leaf lamina making zigzag galleries. As a number of eggs are laid on a single leaf, there are usually more than one mines on the same leaf; as a result, the apical half of the infested leaf is completely damaged and the leaves thus affected start withering and gradually die way. Incubation, grub and pupal durations are 2 to 3. 6 to 7 and 4 to 5 days, respectively.

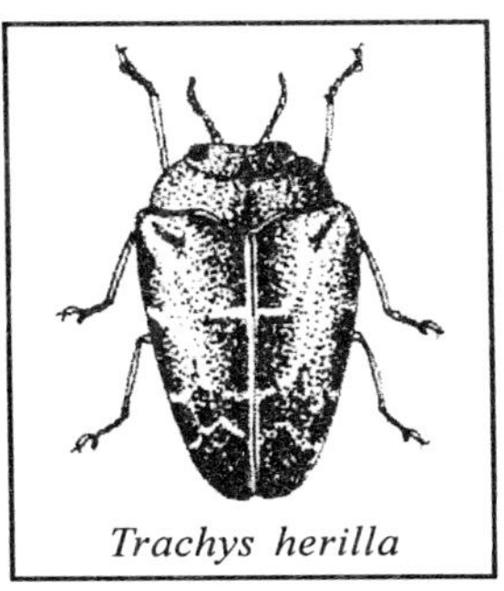
Trachys herilla

Blossom Beetles : *Oxycetonio albopunctata* Fabricius and *O. versicolor* Fabricius (Cetoniidae) are mainly the pests of millets and grasses. *O. albopuctata* has been recorded feeding on rice earheads and flowers of lemon and jute, while *O. versicolor* is a minor pest of groundnut and roses. These beetles also feed on okra flowers and may occasionally gnaw the leaves as well. Adults of *O. versicolor* are 13 to 15 mm long having long head, that is rugosely punctured; elytra are strongly punctate-striate; pygdium is coarsely punctured and setose; there is some colour variation but the base colour of elytra is black with white spots. Adults of *O. albopunctata* are slightly longer than the adults of *O. versicolor* (14 to 16 mm) having long head that is finely and closely punctured; pygidium is finely transverse strigose and elytra are brick-red and opaque.

Blister beetles (Metoidae) *Mylabris pustulata* Thunberg and *M phalerata* Pallas are also occasionally found feeding on okra flowers; their main host being flowers of various gourds.

Flea beetles *Podagria bowringi* Baly is a shining red and blue beetle that feeds on flowers, leaves and tender shoots of okra plants.

These beetles are minor pests of okra and therefore no chemical control measures are usually adopted against these beetles. Only hand-collection and mechanical destruction of adults is suggested to keep the pest population under check.

Leaf Eating Weevils : Okra leaves arc often damaged by a few species of weevils. These include *Myllocerus undecimpustulatus* Desbrochers, *M. variegatus* Boheman, *M. viridans* (Fabricius) and *Ptochus ovulum* Faust. Of these, the first one, known as cotton gray weevil is more common. Eggs are laid in soil about 80 to 100 mm deep. The grubs remain in the soil and feed on roots of grasses and other plants including the host plants. An attack at early stage may therefore kill the seedlings outright. Pupation takes place is soil, 40 to 80 mm deep, in earthen cells. Adults come out of soil and feed on leaf margins; sometimes even the flower burls are attacked and damaged completely.

Myllocerus undecimpustulatus is a polyphagous pest having a very wide range of host plants - cotton being its major host. However, in recent years it has been causing serious damage to a large number of cultivated crops, especially sorghum, pearlmillet and maize.

Eggs are ovoid in shape and light yellow in colour. Grubs are stout, fleshy, apodous, about 7 to 8 mm long, slightly curved and light yellowish in colour. Adults are 8 to 9 mm long, pale green in colour, clothed with scales; elytra with variable dark brownish-gray patches and lines appear punctate in rows owing to absence of scales on these spots. Incubation period is 3 to 11days (average 5 days in summer and 8 days during winter). Grub stage lasts for 28 to 34 days (average 31 days) whereas pupal period takes about 5 to 7 days. Total life cycle is completed in 6 to 8 weeks; there are 5 generations in a year. The pest hibernates in grub stage about 400 to 500 mm deep down in the soil.

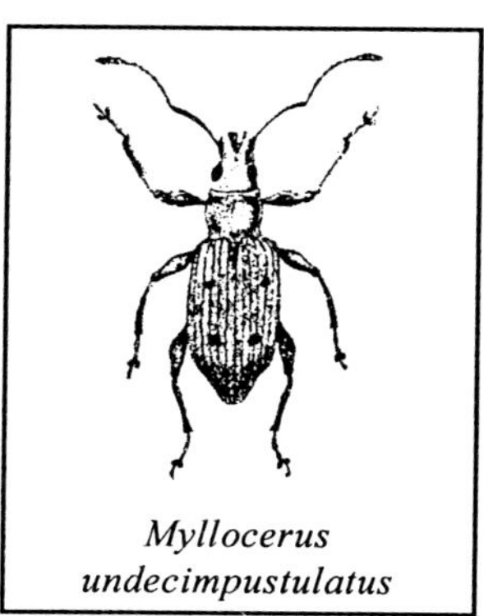

Myllocerus undecimpustulatus

Generally no special control measures are adopted against these weevils on okra. Chemical control measures recommended, against other foliage pests will control these weevils as well. However, it is suggested to interculture the crop (70 to 80 mm deep) regularly to prevent the population build-up and carry-over of these weevils.

In nature, grubs of *Myllocerus undecimpustulatus* are parasitised by *Dinocampus mylloceri* Wilkinson.

Ants : *Myrmicaria brunnea* Saunders, *Phidologiton siverous* (Jerdori) and *Tetramorium smithi* Mayr (Formicidae) have been reported from South India as important pests of okra (Abraham and Remamony, 1978). These ants feed on petals, ovarian tissues and pollen grains in buds and flowers. The infested buds do not open and are often shed while the flowers fade and fall down.

MITES

Mites have often been reported causing serious damage to okra. Colonies of mites comprising larva, nymphs and adults are found feeding on ventral surface of leaves under protective cover of fine silken webs. As a result of their feeding innumerable yellow spots appear on the dorsal surface of leaves and the affected leaves gradually start curling and finally get wrinkled and crumpled. This in turn affects the growth and fruit formation capacity of the plants.

The species recorded on okra from India include, *Tetranychus cinnabarinus* (Boisduval), *T. neoealedonicus* Andre, *T. macmfarlanei* Baker and Pritchard, *Oligonychus coffeae* (Nietner), *Phytoseius minuts* (Narayanan, Kaur and Ghai) and *Amblyseius delhiensis* (Narayanan and Kaur); the first four have been reported from all over India whereas the last two have been recorded at Delhi.

Tetranychus cinnabarinus (Tetranichidae), commonly called red spider mite, is the most common and destructive species. It has a Worldwide distribution and is highly polyphagous having a very wide range of host plants. Its eggs are globular in shape, about 0.1 mm in diameter and whitish in colour. Larvae are about 0.2 mm in length and pinkish in colour. Nymphs are greenish-red in colour and about 3 mm in length. Larvae and nymphs look alike in shape but can be easily distinguished as larvae have 3 pairs of legs while nymphs and adults have 4 pairs or legs. There are only two nymphal stages - protonymphal and deutonymphal. Adults

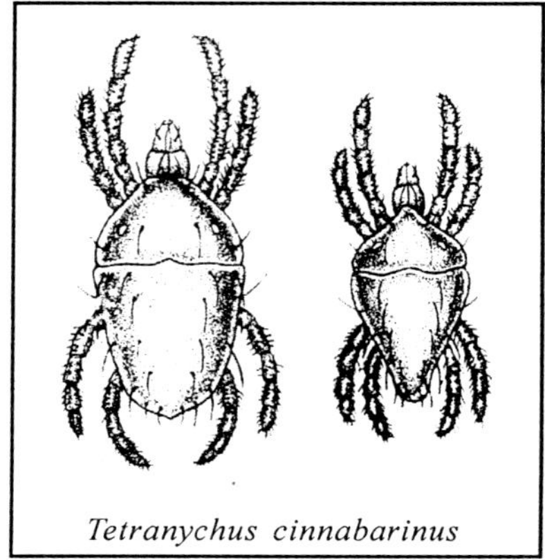
Tetranychus cinnabarinus

are ovate in shape, reddish-brown, in colour and 0.4 mm (males) to 0.5 mm (females) in length with four pairs of legs. Eggs hatch in 4 to 7 days; larval development takes 3 to 5 days; protonymphal and deutonymphal stages last for 3 to 4 days each. Longevity of adult males and females is 4 to 9 and 9 to 18 days respectively. The females aestivate during Summer in Northern India, become active with the onset of monsoon and lay eggs parthenogenetically. These unfertilised eggs give rise to males only, but the subsequent generations are sexual.

Oligonychus cqffeae (Tetranichidae) is another red spider mite commonly called red coffee mite or red tea mite as coffee and tea are its main hosts. It is widely scattered all over the globe (CIE map No. A-165). In India a large number of hosts are reported to be attacked by this species. These include, Java plum (*jamun*), mango, mulberry, peach and several ornamental plants. This mite closely resembles *Tetranychus cinnabarinus* and differs from it only in minute taxonomic characters. The life history and symptoms of damage of both the species are also same except that *O. coffeae* is found on dorsal surface of leaves and is usually active during April to June, after which the activity gradually decreases and the females aestivate till February,

To control these mites, dust fine sulphur or spray wettable sulphur powder. A number of organophosphatic insecticides recommended for the control of aphids and jassids are also found effective in checking the mites.

In nature, the mites are preyed upon by *Clanis soror* Weise *Scymnus* species and *Brumus suturalis* Fabricius.

The major constraints in okra cultivation are occurrence if Yellow Vein Mosaic Virus (VMV), powdery mildew diseases and

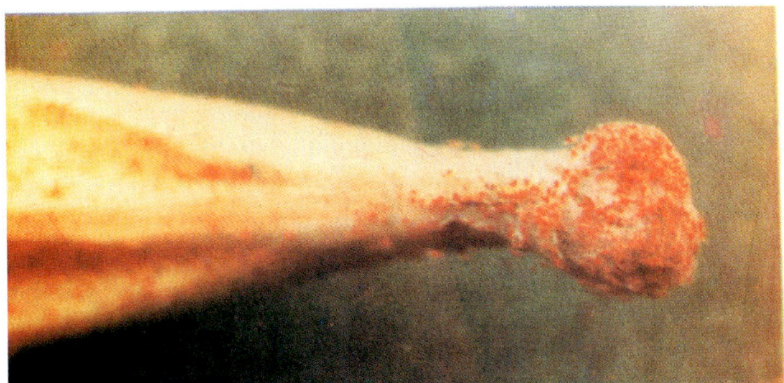

Okra fruit showing abundance of *Tetranychus cinnabarinus*

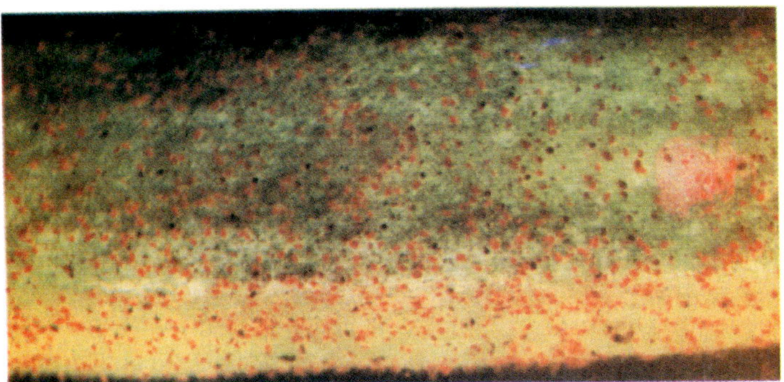

Okra fruit showing abundance of *Tetranychus cinnabarinus*

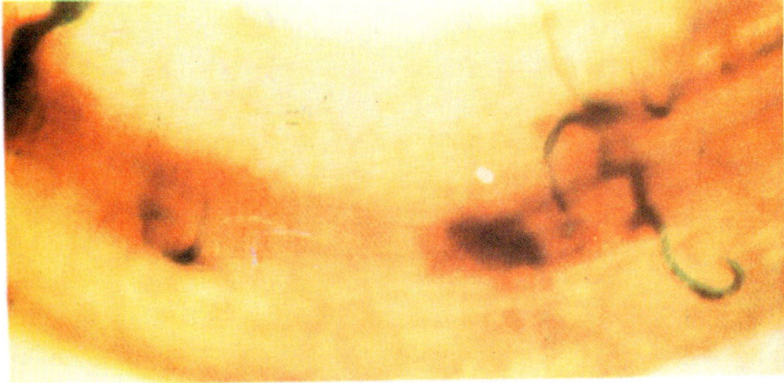

Reniform nematode on okra roots.

Yellow vein mosaic of okra transmitted by whitefly, *B. tabaci*

fruit borer incidence. According to Neerja *et al.*, (2004) revealed that the incidence of VMV was minimum (9.7) in varbha which was at par with HYOH-1 (9.3) KOH-1 (10.4), JNDOH-1 (10.6), AROH-47 (11.2) and MBORH-913 (11.3). The fruit borer incidence ranged from 21.7 per cent in MBROH - 913 to 27.7 per cent in JNDOH-1, HYOH-1 and AROH-47 provided higher marketable fruit yield (70.2) and 69.4 q/ha respectively with less fruit borer incidence.

10

BEANS

BEANS, a group of cultivated crops, occupy a position of considerable importance in our diet, as besides being used as highly nutritious vegetables at green stage, the dried grains are also used as pulses. These are all Tropical legumes and bear on the roots a large number of nitrogen fixing nodules. There are several species of cultivated beans but those commonly grown, for use as vegetables include, French bean, Indian (country) bean, Goa bean and cluster bean.

FRENCH BEAN

Phaseolus vulgaris Linnaeus (Papilionaceae), one of the most popular vegetable, is native of Tropical America. It is sub-erect, twining annual; leaves are trifoliate and flowers are white to violet-purple in colour. Pods are slender. 10 to 25 cm long, straight or slightly curved with prominent beak. Seeds are kidney-shaped, elongate but somewhat compressed and white, red, purple or light black in colour or even mottled. This is a cool weather crop and is sensitive both to frost and very high temperatures. 'Contender' and 'Pusa Parvati' are early varieties whereas 'Kentucky Wonder' is a late variety.

INDIAN BEAN

Lablab purpureus (Linnaeus) (=*Dolichos lablab* Linnaeus) (Papilonaceae), originated in India and its wild forms are found all over the sub-continent. It is a perennial twining or creeping herb but generally cultivated as annual. Most of the cultivated strains are vine type though a few are bushy as well. The leaves are trifoliate; flowers are white, reddish or purple in colour, borne on axillary racemes. Pods are flat or inflated, linear or broad, 25 to 100 mm long and incurved. Seeds are globose, ovate or flattened varying in colour from white to pitch-black. It can sustain cooler climate to some extent. Young pods make an excellent vegetable; dried seeds are nutritious and are eaten after roasting; leaves and flowers are also often eaten in the same way as spinach.

GOABEAN (WINGED BEAN)

Psophocarpus tetragonolobus (Linnaeus) de Candille (Papilionaceae) - A native of New Guinea and South-east Asia is a fast growing perennial climber with tuberous roots. It is cultivated in South India, Bangladesh, Myanmar and South-east Asia. It requires warm and humid climate and can withstand fairly heavy rains. Stems are weak having twining habit; leaves are trifoliate with broad ovate leaflets, 70 to 150 mm long; flowers are large sized and light blue in colour. Pods are 4 angled, 150 to 200 mm long with each angle continued into a much crisped and toothed papery wing, 3 to 6 mm broad. Each pod contains upto 20 seeds. The seeds are smooth shiny, globular (6 mm in diameter) and may be white, brown, black or mottled. The protein content of the seeds is exceptionally high, being 20% as against 1% in potato and sweet potato and 2 to 7% in yams. In India, only the pods are eaten - cooked as vegetable; in Myanmar the tubers are eaten as delicacy (early season potatoes) while in Java ripe seeds are roasted and eaten like peanuts and unripe seeds are used in soup.

CLUSTER BEAN

Cyamopsis tetragonoloba Linnaeus (Papilionaceae) is a native to India. It is hardy,drought resistant, multipurpose crop that is soil restorative. The plant is erect annual, one to three metres high and bears 40 to 100 mm long fleshy pods each containing 5 to 12 seeds. The crop cannot tolerate waterlogging or excessive moisture.

Young and tender pods are used as fodder or silage and the crop is used as green manure. The seeds are highly nutritive, being rich in proteins and are often fed to cattle. Seed-flour is the commercial source of making guar-gum which is.used in food, paper and textile industries. Due to its colloidal nature, the gum acts as a stabiliser and thickener in food products like ice-cream, bakery mixes and salad dressings (CSIR, 1950). Two types of cluster beans are commonly grown in India - dwarf and giant. The former is grown in Punjab, Haryana and Uttar Pradesh while the latter is preferred in Gujarat.

INSECT PESTS

More than 30 species of insects have been reported damaging French bean and Indian bean. The major pests include aphids, sap-sucking bugs, pod borers, leaf miners and stem. fly; as well as pulse beetles, which attack the pods in field and their damage continues in storage.

Cluster bean and winged bean - the poor man's vegetables - are usually not damaged by insect pests. But as these crops are often grown in the vicinity of other leguminous crops the pests of those crops attack these as well, specially in the case of cluster bean. The pests recorded on cluster bean include, aphids, thrips, whitefly, blossom midge, leaf miner, a few species of sucking bugs and leaf feeding weevils. Of these the first two occasionally cause severe damage whereas all others are only of minor importance.

Aphids : Generally polyphagous; are perhaps the most, common pests found all over the World on a vast majority of crops including beans. Both nymphs and adults suck the sap from ventral surface of tender leaves, growing shoots. flower stalks and pods. The infested leaves turn pale yellow, the shoots wither, flower buds fall off whereas the pods shrivel and become malformed. The growth of vines is retarded and ultimately the yield is adversely affected.

A number of aphid species (Aphididae) have been reported damaging beans. The most destructive one is bean aphid *Aphis craccivora* Koch. This is cosmopolitan in distribution (CIE map No. A-99) and has been reported as a severe pest of Indian bean and cluster bean. The other species include, *Acyrthosiphon pisum* (Harris), *Aphis fobae* Scopoli, *A. gossypii Glover. A. rumicis* Linnaeus *A. adusta* Zehntner, *Cervaphis schoutedeniae* van der Goot and *Therioaphis [Chaitophorus] maculata* Buckton. *Aphis medicaginis* Koch, *A. euonymi* Fabricus, *A. glycines* Matsumura, *Macrosiphum euphorbiae* (Thomas) have also been reported on beans from various countries but not from India. These are small-sized insects, about 2 mm long, pear-shaped, green, greenish-brown or greenish-black in colour. Reproduction is usually parthenogenetical and viviparous. Males are rare. Adult females of *Aphis craccivora* are greenish-black in colour and occurs both in alate and apterous forms. Their reproductive life lasts for 5 to 8 days during which a single female gives birth to 15 to 20 off springs. The young ones take about a week to mature and start reproducing at quite a fast rate. In addition to sucking the sap, this aphid also acts as a vector of bean mosaic virus disease, thereby causing greater loss than by feeding.

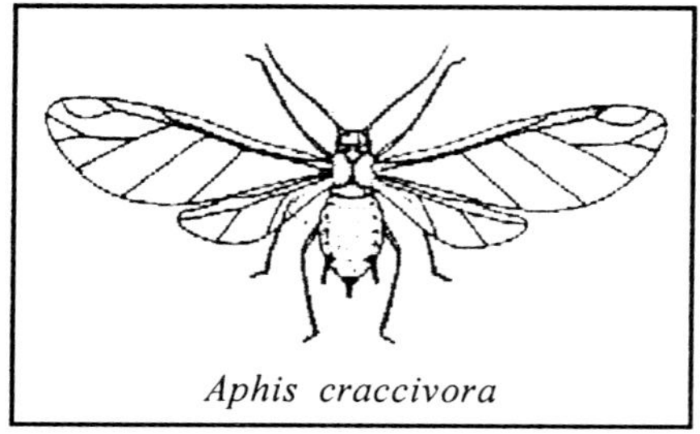
Aphis craccivora

To protect against the damage by aphids, grow resistant varieties and remove and destroy, the infested plant parts during initial stage of attack. If necessary, spray 0.03% phosphamidon or dimethoate when there are no pods on the plants (Butani, 1980). Repeat the spraying after 8 to 10 days. *Menochilus sexmaculatus* (Fabricius), *Coccinella septempunctata* Linnaeus and *Ischiodan scutellaris* Fabricius have been reported as important predators of *Aphis craccivora* and the crop should not be sprayed when these predators are found in sufficient number on the infested crop.

Sap Sucking Bugs : *Anchon pilosum* Walker (Membracidae), *Tricentrus bicolor* Distant (Membracidae). *Anoplocnemis phasiana* Fabricius *Coptosoma cribraria* Fabricius (Corimetaenidae) *C. nizirae* Atkinsoni and *Cyclopelta siccifolia* Westwood (Pentatomidae)are some of the bugs reported feeding on beans. All these are polyphagous pests, leguminous crops being their favoured hosts. The most

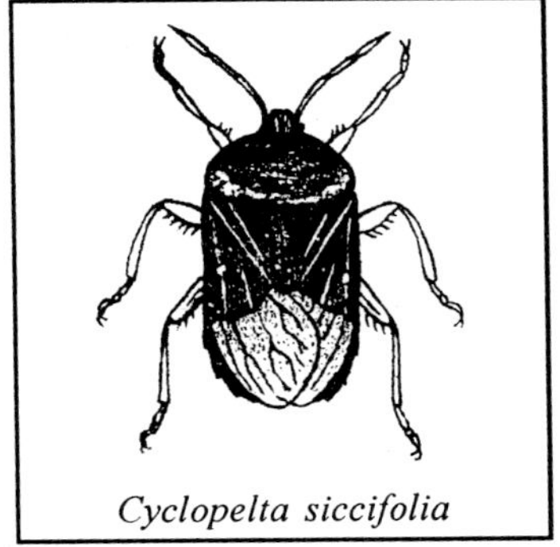

Cyclopelta siccifolia

destructive one to beans is *C. cribraria* - popularly known as lablab bug.

Coptosoma cribraria is found all over the Indian sub-continent. Eggs are usually laid on tender plant parts in two overlapping rows of 30 to 50 eggs. Clusters of nymphs and adults may be seen feeding gregariously on tender shoots as well as ventral surface of leaves. As a result of this desapping, the infested plants become weak and deformed, remain stunted in growth and give poor yield.

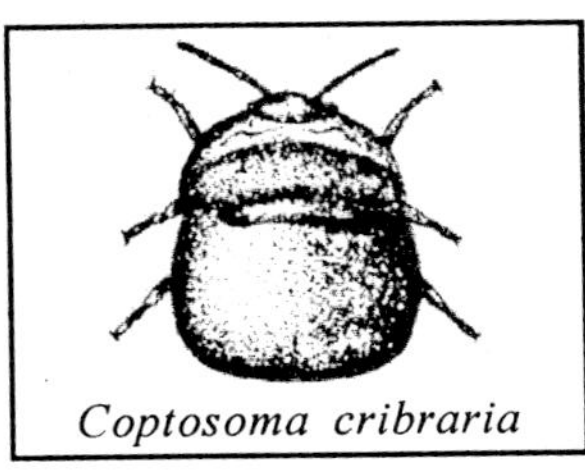

Coptosoma cribraria

Eggs are creamy-white in colour and beautifully sculptured. Nymphs are ovate in shape and pale green in colour. Adults are small oval-shaped greenish bugs, strongly convex with scutellum covering most part of abdomen and wings. These bugs emit characteristic buggy smell. Incubation period is on an average 6 days and and nymphal development takes about 6 weeks.

Anchon piloswn is injurious to Indian bean and cowpea in Kerala. Nymphs and adults arc found in clusters on tender shoots sucking the sap therefrom. Female thrusts its eggs into the shoots singly but close together. The infested shoots become weak. The bugs also secrete honeydew which attracts the common black ant. *Camponotus compressus* Fabricius. Adults are 8 to 12 mm long, characterised by pronotum that extends backwards over the entire abdomen and have two horn-like processes - hence commonly called as cow-bugs. They move by a peculiar jumping motion.

Anchon pilosum

Aphis craccivora on cowpea plant

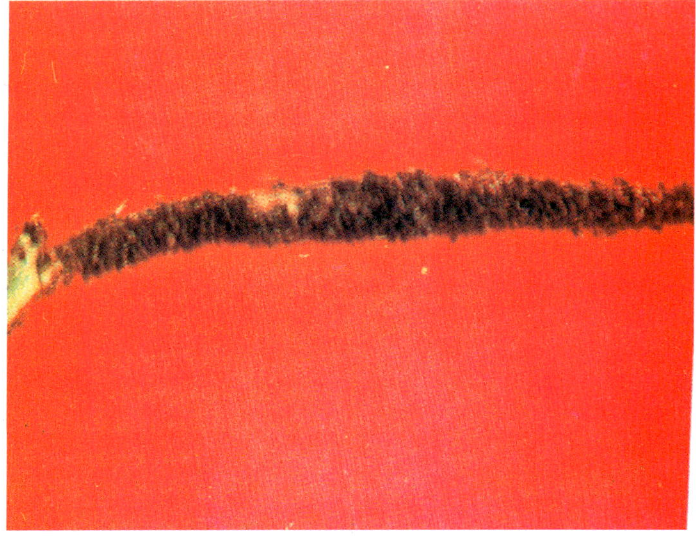

Aphis craccivora on cowpea plant

Aphis craccivora on Indian bean flower

Aphis craccivora on Indian bean pod

Aphis craccivora on cowpea twig

Aphis craccivora on cowpea stem

Mylabris pustulatus on flower of Indian bean

Whitefly on cowpea leaves

Etiella zinckenella (Treitschke), *Adisura atkinosoni* Moore, *Exelastis atomosa* (Walsingham), *Maruca testulalis* (Geyer), *Sphenarches caffer* (Zeller), *Euchrysops cnejus* (Fabricius) and *Lampides (Polyommatus) boeticus* (Linnaeus) have been reported as pests of beans boring tender or mature pods. *H. armigera* is the major pest of tomato. *E. zinckenella* prefers peas and pulses while *A. atkinsoni* and *E. atomosa* are major pests of Indian bean.

Bean Pod Borer : *Adisura atkinsoni* (Noctuidae) is a cold weather pest, widely distributed all over the Indian sub-continent. The eggs are laid singly on leaves, buds, flower spikes and young pods. On hatching, the caterpillars bore inside the developing pods and feed within on ripening seeds. Full fed caterpillars come out of the pods and usually pupate in the soil but occasionally on flower spikes also.

Eggs are small and spherical. Caterpillars are cylindrical, brownish–green in colour, 25 to 33 mm long when full fed. Moths are medium sized and pale yellowish-brown in colour. Forewings arc brown-ochreous with gray tinge and outer margin pinkish: hind wings are straw-yellow in colour with outer area broadly and completely suffused with fuscous: wing expanse is 30 to 35 mm. Egg. caterpillar and pupal durations last for 3, 14 to 15 and 11 days respectively.

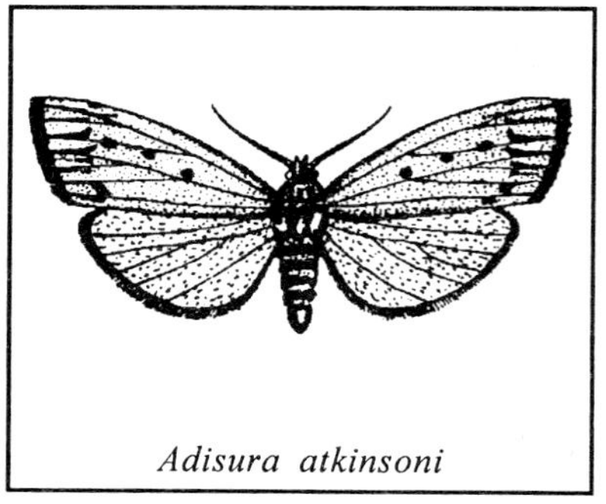

Adisura atkinsoni

Plume Moth : *Exelastis atomosa* (Pterophoridae) was first described from Mumbai (Walsingham, 1885) and by now it has been reported from almost all the Tropical regions of the Old World. It is a major pest of Indian bean, red gram, horse gram, pigeonpea etc. Oviposition takes place usually at night when the eggs are laid singly on the pods, sometimes on flower buds and ocasionally on leaves as well. On hatching the tiny caterpillars bore into the pods and feed on developing seeds; they do not enter the pods completely; their pest population.

Dusting the vines with 4% endosulfan or spraying 0.05% dichlorvos or 0.1% malathion is quite effective in checking he

Pod Borers : Caterpillars of *Helicoverpa armigera* (Hubner). distal abdominal portion remains outside the pods. Sometimes the caterpillars that hatch from eggs laid on flower buds, enter into those buds and feed on inner structures. Pupation takes place outside on the pods and the pupae are attached by means of two circinate spines arising from the 8th and last abdominal segments of pupae.

Eggs are minute in size, about half mm long, oval in outline and pale green in colour. Full grown caterpillars are 10 to 14 mm long, moderately stout, densely covered with short hair and long spines, energing from raised greenish coloured wrats with pinkish or light green stripes on its dorsum, Pupae are cryptically coloured - brownish or pinkish or greenish and are thickly covered with spine-like hair. Egg period lasts for 3 to 4 days during Summer and 5 to 6 days in Winter; caterpillar development is completed in 16 to 21 days and 25 to 30 days during Summer and Winter respectivaly; pupal period lasts for 3 to 5 days in Summer extending upto 7 days during Winter (Fletcher, 1920).

Legume Pod Borer (spotted pod borer) : *Maruca tesiulalis* (Pyraustidae) is widespread in Tropical and subtropical regions of the world (CIE map No. A-351) damaging almost all the varieties of beans and peas. Besides, it has also been reported damaging castor, groundnut, paddy, tobacco etc, Though of regular occurrence, it is usually a minor pest, but of late it has attained the status of major pest in East Africa, In India, the pest has been recorded damaging flowers and pods of Indian bean and cowpea; in Papua New Guinea it has been recorded as a pest of winged bean.

Eggs are laid singly on or near flower buds of host plants. On hatching the young caterpillars feed on reproductive parts of flowers and move from one flower to another. Later they web the inflorescences with the adjacent leaves and developing pods and feed within by boring into the flowers and pods. The

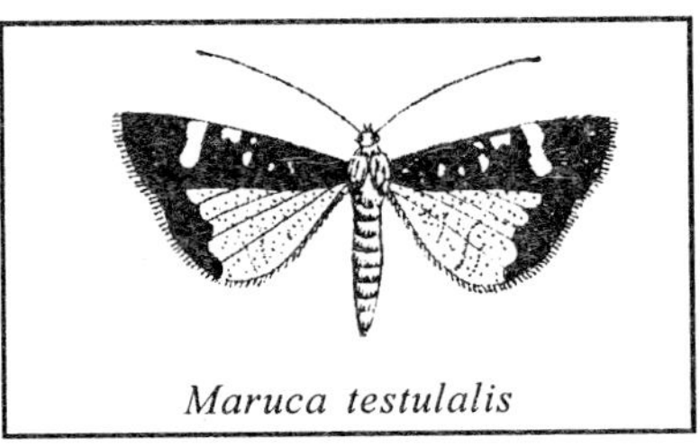

Maruca testulalis

infested flowers do not develop into pods while the affected pods become malformed as these are not able to grow normally due to webbing and feeding of the past. Occasionally the grown up caterpillars also bore into peduncles and stems. Pupation takes place in debris or on soil surface. Moths are nocturnal in habit.

Eggs are elongate-oval in shape, less than one mm in length and light yellow in colour. Full grown caterpillars are on an average 20 mm long, light brown in colour with irregular bronish–black dorsal, lateral and ventral spots. Adults are medium sized slender moths with head, thorax and abdomen fuscous, brown. Forewings are fuscous-brown. with a lunulate white spot at the end of the cell and a maculate semi-hyaline spot beyond the cell. Hind wings are semi-hyaline white with a fulvous-brown distal patch Wing spread is 20 to 30 mm. Incubation period is 2 to 3 days- caterpillar stage lasts for 8 to 14 days (maximum 35) pre-pupal 2 days and pupal 6 to 9 days.

Due to sheltered nature of feeding site, it is rather difficult to control these pod borers. Growing resistant varieties has been recommended. Besides, mechanical destruction of the caterpillar in the initial stage of attack can be easily undertaken. If attack is severe, spray 0.2% carbaryl or 0.05% endosulfan. Spraying *Bacillius thuringiensis* Berliner is also effective.

In nature, the caterpillars of *Adisura atkisoni* are parasitised by *Habrobracon hebetor* (Say), *Microbracon greeni* Ashmead. *M. brevicornis* Wesmael *Corcelia evolans kochiana* Townsend. *Hymenobosmina* species and *Enicosphilus* species (Ramachandra Rao, 1917).

Blue Butterfly : *Euchrysops cnejus* (Fabricius) (Lycaenidse) and Pea blue butterfly (*Lampides boeticus* Linnaeus) are widely distributed in the Indian sub-continent. These are minor pests of Indian bean and have also been recorded damaging various other leguminous crops. Caterpillars bore pods to eat the seeds within. The damage could be detected only after opening the pods. Even the undamaged seeds in the pod are unfit for human consumption due to secondary infection of fungus etc. Eggs are laid on flower buds and pods. On hatching, the caterpillars burrow in the buds, which are usually killed. Later they bore into the tender pods and feed on the developing seeds. The caterpillars are generally found in association with ants which probably imbibe some secretions that are given out by the caterpillars. Pupation usually takes place in the soil.

Caterpillars of *Euchrysops cnejus* are flat, 7 to 10 mm long and pale greenish in colour with dark green dorsal and sub-dorsal lines, entire body surface is covered with minute white tubercles and a few scattered hair. Pupae are pale green in colour and smooth. Adults are typical lycaenid butterflies pale purplish to dark brown in colour; wings having a number of brown spots are edged on either side with white; the tail being dark brown to black in colour. Wing expanse is 26 to 33 mm. *Lampides boeticus* is slightly bigger than *E. cnejus*. Its caterpillars are 10 to 13 mm long and pale green to violet in colour. Pupae are yellowish-green in colour and without any hair. The butterflies have upper side violet-blue to brown in colour and underside pale grayish. Wings are covered with white scale-like hair over the ordinary scaling; the tail is black, tipped with white and wing spread 34 to 38 mm.

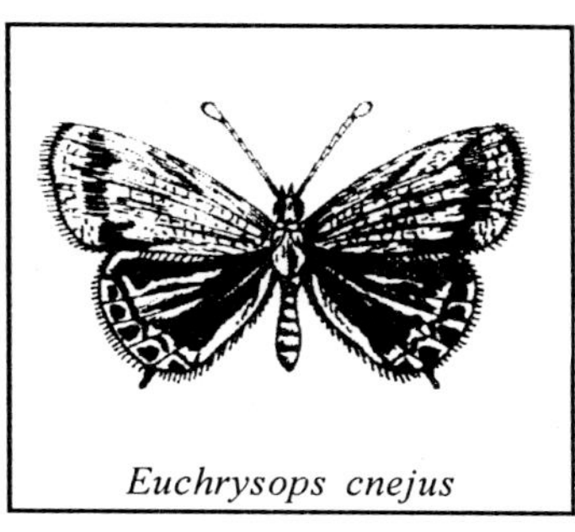

Euchrysops cnejus

Leaf Miners : *Cosmopteryx mimetis* Meyrick, *C. phaeogastra* Meyrick {Cosrnopterygidae) and *Cyphositicha coerulea* Meyrick (Gracilaridae) have been recorded as leaf mining pests of French bean and Indian bean.

Cosmopteryx mimetis is widely distributed in the Indian sub-continent and has also been reported from Australia, New Guinea, Mauritius, the Seychelles and British Guiana (Fletcher, 1920d). The larvae mine either in the middle or apical portion of leaves and the silvery mines run along on either side of midribs. Full grown larvae are about 3 mm long, tapering posteriorly and uniformly light yellow in colour. Wing expanse of moths is 5 to 6 rnm.

Cosmopteryx phaeogastra is found all over the Indian sub-continent. Caterpillars mine bean leaves on the sides of main veins in between the epidermal layers; each constructs cylindrical silken case about 10 mm long and lives in this case. The narrower end of the larval case remains open and black coloured excreta is

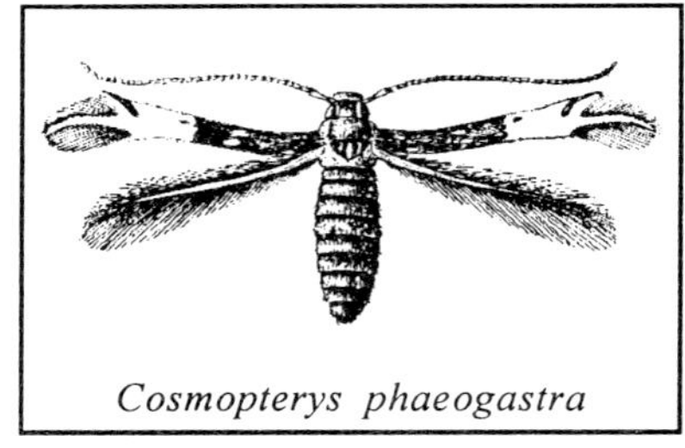
Cosmopterys phaeogastra

extruded through it (Fletcher, 1920). Full grown caterpillars are stout, flattened, about 6 mm long and uniformly pale yellow in colour. Moths are black in colour with golden sheen. Forewings are black with a broad medial creamy white blotch narrowing down towards the distal end which is broadly fringed with long black hair. Hind wings are slender, heavily fringed with dark hair. Wing spread is 6 to 8 mm.

Caterpillars of *Cyphositicha coerulea* mine the dorsal surface of leaves of Indian bean, cowpea and red gram feeding just below the epidermal layer making narrow zigzag line which later widens into a broad-elongated blotch, brown or brownish-white in colour. The full grown caterpillars are 4 to 5 mm long, somewhat flattened, uniformity reddish-brown in colour with a red dot sub-dorsally on each segment. Pupae are 3 to 4 mm long and pale yellow in colour,

Removal and mechanical destruction of infested leaves along with the various immature stages of the pest thereon, checks the pest infestation to a great extent. In nature, caterpillars of *Cyphositicha coerulca* are parasitised by *Euryscotolinx coimbatorensis* Rowher.

Leaf Eating Caterpillars : A number of lepidopterous larvae have been reported feeding on leaves of various beans. These include, *Amsacta* spp., *Acherontia styx* {Westwood), *Spilarctia (Spilosoma) obliqua* (Walker), *Pericallia ricini* (Fabricius). *Estigmene lactinea* Cramer, *Spodoptera exigua* (Hubner) and *S. littoralis*' (Fabricius). These are all polyphagous pests causing severe damage to one or the other crop of economic importance. On beans these have been reported as minor pests except *Amsacta* spp., which cause severe damage, though occasionally.

Red Hairy Caterpillars : *Amsacta* spp. (Tiger moths, Arctiidae), are widely distributed in the Indian sub-continent; though polyphagous, groundnut is their preferred host, but they can consume voraciously most of the *kharif* crops like maize, millets and pulses. The infestation is often severe in the rain-fed areas and in India their ravages have been reported mainly from Gujarat, Rajasthan and South India. Eggs are laid in clusters, mostly on ventral surface of leaves. On hatching the caterpillars feed gregariously on those leaves and even a moderate infestation may result in complete defoliation of vines. The caterpillars are very active and move from field to field in large numbers. Pupation takes place in the soil.

The species commonly met with in India are *Amsacta moorei* (Butler) also known as kutra and *A. albistriga* (Walker). Their habits, habitat, nature of damage and life-history are more or less identical; even the immature stages of the two species look alike and are difficult to differentiate. Eggs are round in shape and cream coloured. Full grown caterpillars are 50 to 70 mm long, densely hairy, pinkish-orange to brown in colour. Moths of *A. moorei* have white coloured head with a crimson line behind it: thorax is white; abdomen is scarlet above with a series of black dorsal spots, white below with 2 series of black lateral spots. Wings are white. Forewings have a scarlet fascia along the costa and a black speck at each angle of cell. Wing expanse is 45 to 55 mm. *A. albistriga* is comparatively smaller in size (wing spread 40 to 50 mm) and differs from *A. moorei* in having yellow line on the head; thorax tinged with ochreous and costal stripe on forewing is of yellow colour.

Amsacta albistriga

Adults emerge out from soil with the first shower of monsoon. A female lays 10 to 15 egg dusters, each of more than 100 eggs. These hatch in 2 to 4 days. The caterpillars become full grown in 3 to 4 weeks. Pupal period lasts for several months extending from October-November to following June-July. Thus there is only one generation in a year.

The mechanical method recommended for the control of these pests is to collect and destroy the conspicuous egg-masses as also the moths. These moths are usually sluggish and therefore easy to collect and kill. This will reduce the pest incidence appreciably if timely and cooperative action is taken. In the case of severe

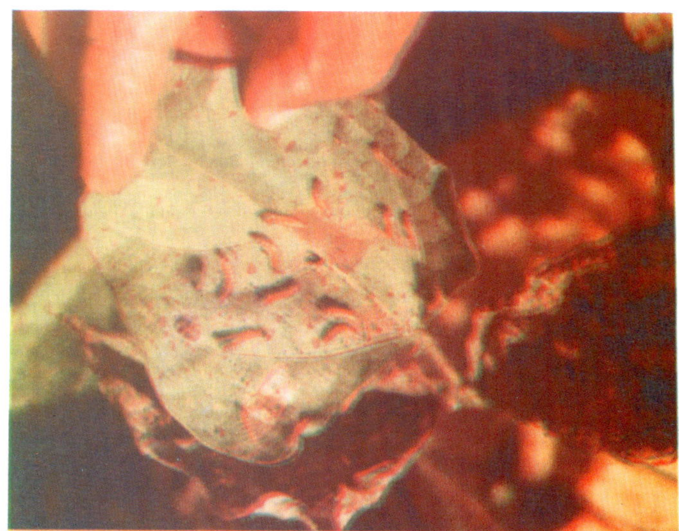

Spilarctia obliqua caterpillars feeding on cowpea leaves

Spilarctia obliqua caterpillars feeding on cowpea leaves

Aphid, *Aphis craccivora* on cowpea flower

Beetle, *Madurasia obscurella* on cowpea leaf

infestation crop should be dusted with 4% endosulfan. In nature, parasites recorded *Apanteles obliqua* Wilkinson, *A. colemani* Vireck, *A. creatonoti* Vireck. *Trichogramma* spp. and *Sturmia inconspicuella* Baranov.

Acherontia styx (Westwoorl) - Sesamum Hawk Moth (Sphingidae), is found throughout the plains of Indian sub continent. Eggs are laid singly on leaves. Caterpillars are voracious feeders and can defoliate the vines in no time. Pupation takes place in soil.

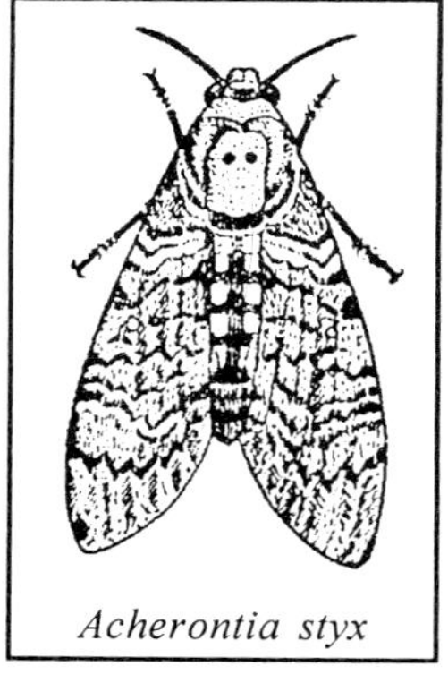
Acherontia styx

Caterpillars sturdy and stout, grow upto 100 mm length, they are greenish in colour with dark green oblique bands laterally bear a conspicuous, curved, yellowish horn-like projection dorsally at the anal end. Pupae are stout, 50 to 60, mm long, chestnut coloured with broadly rounded head and antennae slightly longer than forelegs. Adults arc big. yellowish coloured hawk like moths with skull-like mark on dorsum of thorax. Forewings are blackish covered with yellowish, bluish and gray powdery scales. Hind wings have only the basal portion black. Wings expanse is 90 to 130 mm. Eggs hatch in 2 to 5 days while caterpillars take 14 to 20 days and pupal period lasts for 7 to 11 days.

Hand picking and mechanical destruction of caterpillars in early stage of attack is helpful in checking the pest population. To reduce the population build-up, the infested fields should be deep ploughed immediately after the harvest to expose and kill the pupae in the soil. In nature, the caterpillars are parasitised by *Agiontmatus acherontiaz* Ferriere and *Apanteles acherontiae* Cameron.

Pod Boring Bugs : In addition of the lepidopterous pests, three coreid bugs, *Riptortus fuscus* Fabricius, *R. linearis* Fabricius and *R. pedestris* Fabricius have also been reported as pod borers. Of these *R. pedestris* is comparatively more common. Besides beans, if attacks cowpea, soybean and different types of grams while *R. linearis* has been found damaging maize, sorghum, sweet potato

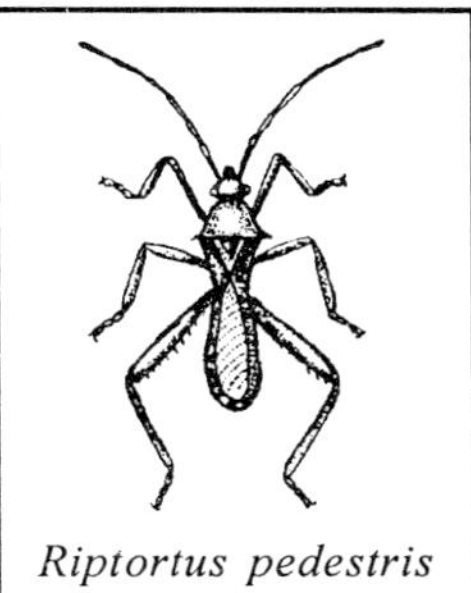
Riptortus pedestris

and fig (Nair, 1975). Both nymphs and adults suck sap from developing seeds within the green pods. the infested pods show light yellow spots, become shrivelled and ultimately die away.

Generally, no separate control measures are required to check the population of these bugs. However, spraying 0. % carbaryl or malathion is quite effective to control these pests,

Thrips : *Ayyaria chaetophora* Karny, *Caliothrips indicus* (Bagnall) and *Sericothrips ramaswamiahi* Karny (Thrlpidae) have been found feeding on leaves of beans - their preferred host being jute, groundnut and indigo respectively. Blossom thrips include *Frankliniella schultzei* (Trybom), *Megalurothrips ustitatas* (Bangall). *Taeniothrips distalis* (Karny). *T. Ieyfroyi* (Bangall) and *T. longistylus* Karny, All these are polyphagous pests causing minor damage to beans. Nymphs and adults lacerate the tender plant parts and imbibe the oozing sap. The blossom thrips infest the flowers and feed on pedicels, sepals, petals and even stigma. The infested flowers get devitalised and fade. Besides, the possible role of *T. distalis* as a vector for transmitting the pod twist disease of beans caused by *Pseudomonas flectans* cannot be over-looked (Ananthakrishnan, 1971).

Ayyaria chaetophora completes its life-cycle on crotons (*Croton tiglium* L.) in 15 to 20 days. Egg, nymphal, pre-pupal and pupal periods being, 4 to 5, 7 to 10, one and 3 to 4 days respectivley (Ananthakrishnan, 1971).

To control the thrips, if and when necessary, spraying with 0.03% dimethoate, phosphamidon, formothion, thiometon or endosulfan has been recommended (Jotwani and Butani, 1977).

Stem Boring Beetle : *Sagra nigrita* Olivier (Chrysomelidas), has been reported as a minor pest of Indian bean in South India. In Karnataka, the pest is active in fields during June to September. Eggs are laid on stems of bean plants. On hatching, The grubs bore into stems and feed on inner tissues for 2 to 3 months. As a result of their feeding large galls are formed on the infested stems. Pupation takes place within these galls and glistening green adult beetles emerge after 4 to 5 months (Nair, 1975). There is only one generation in a year.

Leaf Feeding Beetle : *Platypria hystrix* Fabricius (Hispidae) is a small oval, spiny beetle that is found occasionally feeding on leaves of Indian bean, but the damage caused is negligible and calls for no control measures.

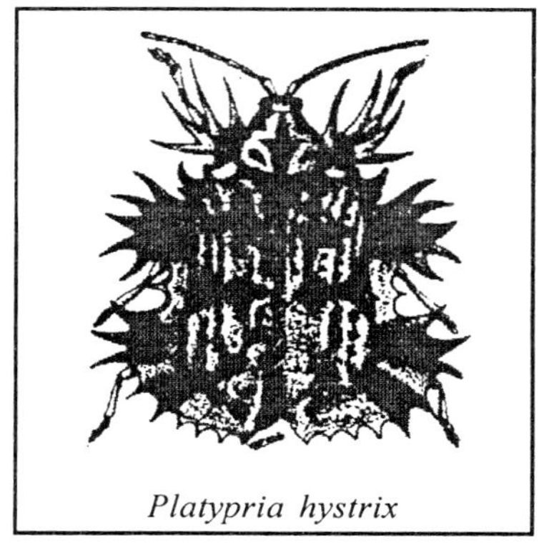
Platypria hystrix

Pulse Beetles : *Callosobruchus* spp. (Bruchidae). are the most important pests off various pulses and devastation caused by these insects is observed more prominently under storage conditions of the harvested pods or grains. These are cosmopolitan in distribution and one or the other species occurs throughtout the Tropics and sub tropics of the world. These beetles have been reported from Africa, Middle and Far East and the Indian sub-continent, both in fields as well as godowns. The infestation usually starts in the fields and continues till the seeds arc consumed - fresh or dried. In fields, eggs are laid on the pods firmly glued on their surface. If the pods have dehisced with the grains partially visible, the eggs are laid directly on the seeds. Grubs bore inside the seeds and complete entire life within the bean or pea seeds. Pupation takes place just under the testa of seed but before pupation the grubs cut a circular

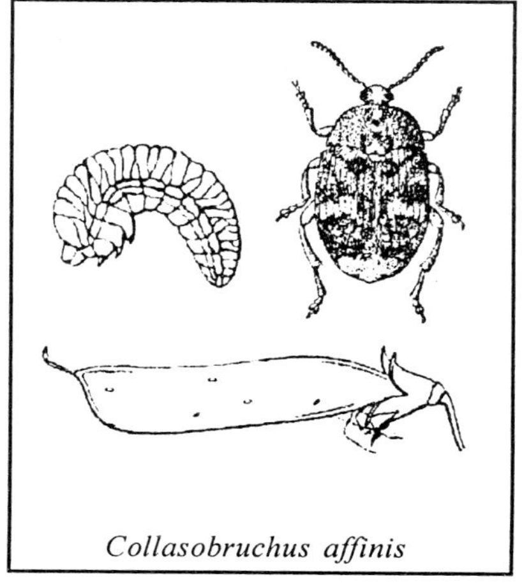
Collasobruchus affinis

hole which remains covered by thin membranous outer coat and through this hole the adult, emerges after completion of the pupal period.

The species found damaging beans and peas in the Indian sub-continent include, *Callosobruchus affinis* Froll, *C. chinensis* (Linnaeus) and *C. quodrimaculatus* (Fabricius). Their habits, symptoms of damage and even the general appearance is more or less species can be differentiated by their specific morphological characters. Grubs are scarabaeiform. Adults are small brownish beetles about 3mm long characteristically emarginated eyes along the inner edges where the antennae are inserted. Distinctive sexual dimorphism exists in the antennal characters. Adults of *C. chinensis* are more or less roundish and their abdomen covered completely with elytra while *C. quadrimaculatus* is elongate with posterior part of abdomen exposed. Egg and grub durations are about 6 and 20 days respectively and a life-cycle is completed in 4 to 5 weeks (Hill, 1975).

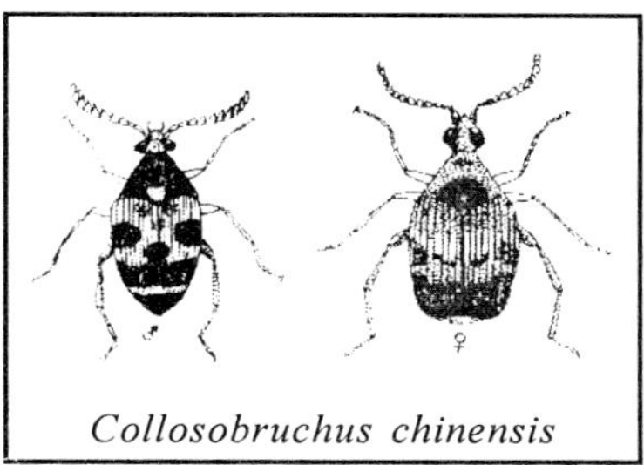

Collosobruchus chinensis

As vegetable pests, these beetles are of minor importance and do not require any control measures in the fields, In stores and godowns however, fumigation with methyl bromide has been recommended.

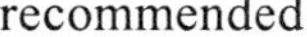

Weevils : several different species of weevils (Curculionidae) have been reported damaging bean vines. Some of these are: *Colobodes dolichotis* Marshall, *Cytozemia cognata* Marshall, *C. dispar* Marshall, *Myllocerus undecimpustulatus* (Desbrochers), *Blosyrus inaequalis* Boheman, *Alcidodes bubo* Fabricius, *A. collaris* Pascoe, A. *fabricii* Fabricius, *A. leopardus* Olivier, *A.*

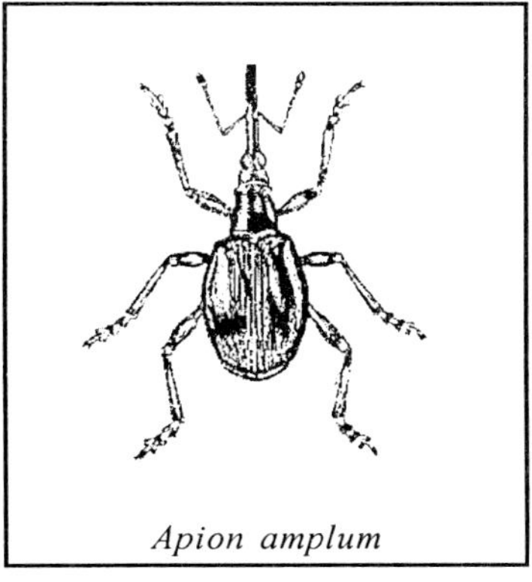

Apion amplum

pictus Boheman, *Episomus lacerta* Fabricius. *Apion amplum* Faust. *Tylopholis ballardi* Marshall and *Desmidophonts* species. Most of these are polyphagous.

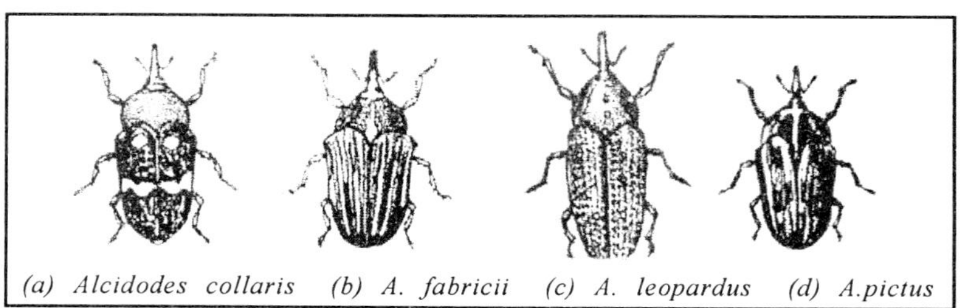
(a) Alcidodes collaris (b) A. fabricii (c) A. leopardus (d) A.pictus

Indian Bean Weevil : *Colobodes dolichotis* originally described by Marshall (1936), was considered to be specific pest of Indian bean, but of late, it has also been reported feeding on cowpea and red gram. Eggs are laid on the main stems near the base of plant. On hatching, the grubs riddle the tissues of main stem and make irregular galleries inside, resulting in formation of gall-like swellings. The effected stems break at these swellings and die away. Adults feed on stem tissues making small excauations for oviposition, but the loss caused by adults is negligible. Adults are 7 to 8 mm long and brownish-black in colour with a V-shaped white mark at the posterior end of elytra. Pre-mating period of female is 3 to 6 days, pre-oviposition 5 to 10 clays and oviposition lasts for 43 to 120 days during which it lays 59 to 64 eggs. Incubation, grub and pupal periods occupy 6 to 7, 59 to 64 and 10 to 12 days respectively (Subramanian, 1959). Longevity of males and females in captivity is 50 to 97 and 54 to 127 days respectively. In nature, the grubs are parasitised by *Tetrostichus* species.

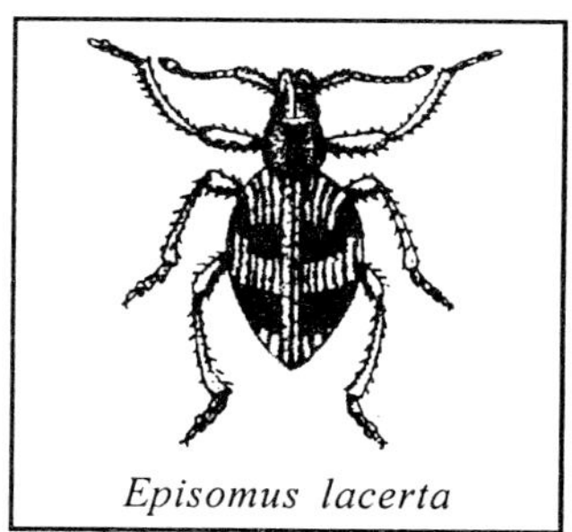
Episomus lacerta

Gray Leaf Weevil : *Episomus lacerta F.* is a serious pest of Indian bean in South India. Eggs are laid in leaf folds specially fabricated by the females for oviposition. A single female can lay over 1000 eggs in batches of 4 to 12 eggs. On hatching, the grubs drop down, enter the soil and feed on roots of host plants causing some minor damage. After completion of the pupal period in the soil the adults emerge out and feed voraciously on leaves; a severe infestation may result in defoliation of the vines.

Stem Gall Weevils : *Alcidodes* spp. have also been reported, as gall forming, from South India. Eggs are laid singly in small excavations made by the females on host stems. On hatching, the grubs burrow inside the stem and a swelling or gall is formed at the point of entry. Incidence of weevils is more in old vines left after harvesting than in the main crop. The most destructive species is *Alcidodes bubo* Fabricius. This is a serious pest of indigo in Bihar and major pest of cluster bean in South India. Adults are small, 8 to 10 mm long, reddish-brown weevils having longitudinal white spots on elytra, incubation period is 5 to 6 days whereas grub and pupal development takes 4 to 5 weeks and one week respectively (Nair, 1975). In nature, the grubs are parasitised by *Eurytoma pigra* Girault and *Metastenomyia juliani* Gtrault. To prevent the attack by these weevils treat the soil with 5% HCH dust @ 20-22 kg per hectare; if the pest population rises, spray 0.2% lindane or carbaryl (Butani, 1980).

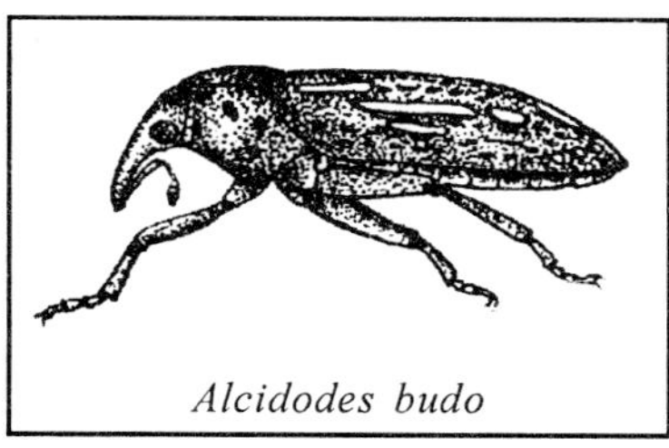

Alcidodes budo

The serpentine leafminer, *Liriomyza trifolii* (Burgess) is posing a serious threat to the French bean (*Phaseolus vulgaris* L.) cultivation. Indiscriminate use of insecticides in crops has created an environment free of natural enemies and resulted in epidemic occurrence of leaf miner besides control failure in the field. Kenadappa *et al.*, (2005) evaluated certain insecticides on larval and pupal stages of leaf miner under greenhouse and laboratory conditions respectively. Acephate 0.15% chlorpyrifos 0.05% and Nagata 0.1125% a.i. were found to be most effective in checking the damage to foliage. In the pupal dip bioassay test, Lamda cyhalothrin @ 0.0075% and quinalphos 0.0625% a.i. observed effective followed by chlorpyrifos 0.05 a.i.

11

PEAS

PEA(S) -- nutritious seeds of self pollinated annual herb (Papilionaceae) Garden Pea is one of the most common and popular vegetable which is grown extensively all over the World. The green tender pods are quite tempting and are even eaten raw while in advanced stage seeds are removed from the pods and cooked. Besides, some varieties having sweet smelling flowers are also grown as ornamentals. The other related pea which is used as vegetable is cowpea; though often cowpea is also used as pulse after drying the mature pods. These plants are capable of utilizing atmospheric nitrogen with the aid of bacteria found in the nodules of their roots.

GARDEN PEA

Pisum sativum arvense Linnaeus - a native of Europe and North-western Asia is being cultivated since prehistoric times. It is a herbaceous annual with vining habit. Its stems are hollow ending in one or more tendrils: leaves are pinnately compound with auricled stripes; flowers are white, bluish or purple in colour, hermaphrodite papilionaceous and borne in cymose inflorescences. Pods may be straight or curved, generally pointed and 30 to 100 mm long containing upto 12 seeds. Seeds are yellowish-green to dark green in

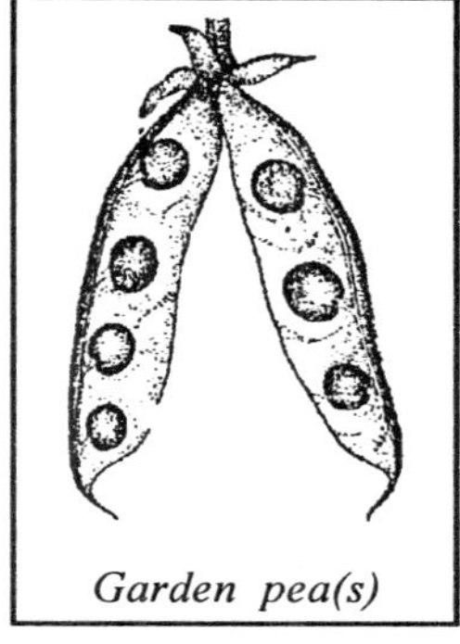

Garden pea(s)

colour, smooth and round being 3 to 5 mm in diameter. Pea is relatively a hardy crop and can be grown under colder climate. The seeds are an excellent source of vitamins A, C and B_1 and are also rich in vitamin G, besides being a good source of phosphorus, iron and carbohydrates (Roe, 1974). In India, these are grown mainly in Punjab, Himachal Pradesh, Haryana, Delhi and Uttar Pradesh. Early varieties like 'Arkle Badger' and 'Meteor' take little over 2 months to mature while Bonneville and NP-20 take 2½ to 3½ months to mature. 'Asauji'- a variety evolved at IARI, New Delhi is a dwarf early variety that takes only 2 months to mature.

COWPEA

Vigna unguiculata (Linnaeus) (=*sinensis* Savi, *catjang* Walp., *cylindrica* Skeels. *sesquipedalis* Fruhw.) is said to have originated from Tropical African region- It is of great antiquity and has been in cultivation for thousands of years in Tropics of the old World. The plant is sub-erect, trailing or climbing-busby annual with glabrous stems; leaves are pinnate, leaflets 75 to 150 mm long, ovate, rhomboidal or broad or narrowly ovate; flowers are in recemes, white, pale violet or purple which subsequently fade to become yellowish in colour. Pods are borne in clusters and are variable in length (maximum 90 mm) and contain 10 to 20 seeds. This is a warm season crop and cannot withstand cold weather nor it can tolerate heavy rainfall. It is mainly grown for fodder or purpose or to be used as green-manure. The improved edible varieties include 'Pusa dofasli' which can be grown in both Spring and rainy seasons.

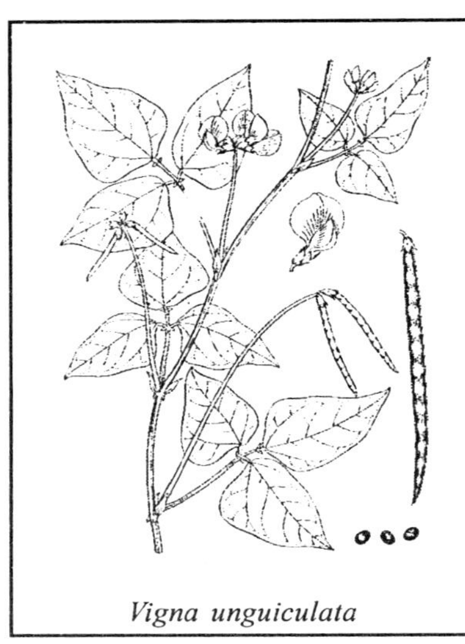

Vigna unguiculata

Garden pea pods damaged by *Lampides boeticus*

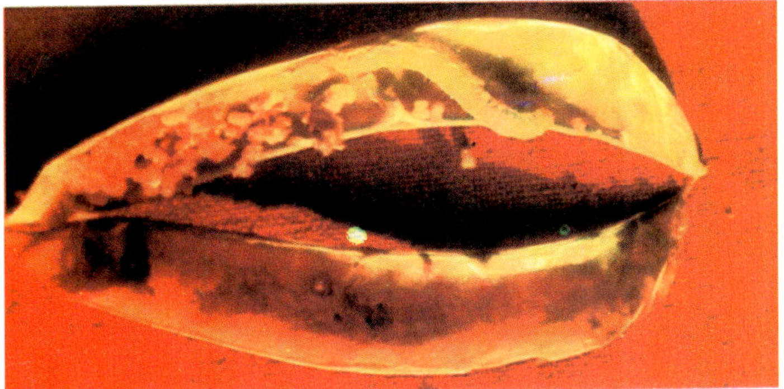

Hellicoverpa armigera on pea pod

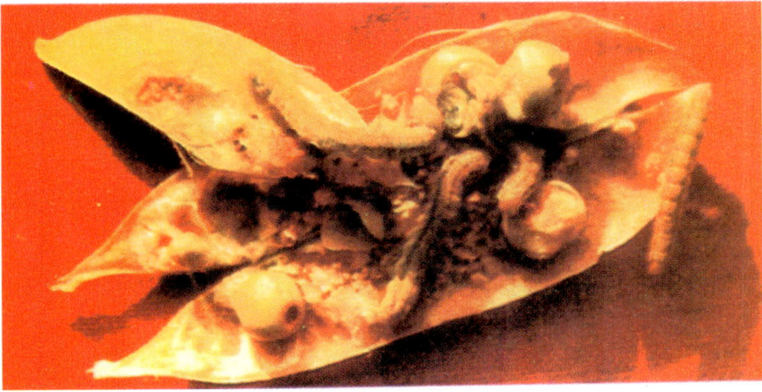

Hellicoverpa armigera on pea pod

Pigeonpea pods damaged by *Helicoverpa armigera*

Root knot nematodes on pea

INSECT PESTS

The major pests of garden pea include, pod borers, pea stem fly, pea leaf miner and pea aphid. Besides, there are more than a dozen minor pests including sap sucking bugs, whitefly, thrips, leaf-eating caterpillars and stem gall weevils. As many as 52 insect species have been reported damaging cowpea but only stem fly and aphids cause substantial loss and are classified as major pests. Minor pests include, grasshopper, leafhoppers, sap sucking bugs, whitefly, leaf-eating caterpillars and beetles.

Pod Borers : A number of lepidopterous larvae have been reported damaging the pods of cowpea and garden pea by boring and feeding inside on the developing seeds. Those commonly found are, *Adisura atkinsoni* Moore (Noctuidas), *Exelastis atomosa* (Walsingham) (Pterophoridae), *Maruca testulalis* (Geyer) (Pyraustidae), *Helicoverpa* (*Heliothis*) *armigera* (Hubner) (Noctuidae), *Etiella zinckenella* (Treitschke) (Phycitidae), *Leguminivora* (*Laspeyresia*) *tricentra* (Meyrick), (Eucosmidae), *Sathrobrota simplex* (Walsingham) (Cosmopterygidae), *Paraspistes palpigera* (Walsingham) (Gelechiidae), *Sphenarches caffer* (Zeller) (Pterophoridae), *Leucinodes orbonalis* Guenee (Pyraustidae), *Anarsia* spp. (Gelechiidae), *Euchrysops* (*Catochrysops*) *cnejus* (Fabricius) (Lycaenidae) and *Lampides boeticus* (Linnaeus) (Lycaenidae). All these are polyphagous pests; the first three prefer beans while *H. armigera* is a major pest of tomato. On peas, *E. zinckenella* is the only pest of major economic importance, the others cause only minor or negligible damage.

Etieila zinckenella originally described from Sicily is by now Pan-tropical in distribution (CIE map No. A-105). It has been reported from USA, Mexico, West Indies, South America, Hungary. France. Austria, Southern Russia, UAE, parts of Africa, Indian sub-continent, Indonesia, Japan and Australia (Bindra and Harcharan, 1969). It is a major pest of various species of beans, grams and is often found damaging peas as well. It is a cold climate pest active throughout Winter.

Eggs are laid singly or in small batches, usually at the junction of calyx and pod or occasionally on young pods. On hatching the caterpillars bore inside the green pods and feed within; generally one caterpillar is found in one pod. When full fed they come out of the pods, drop down and pupate in the soil. Late maturing varieties are comparatively more damaged than early maturing ones.

Eggs are small, 0.5 to 0.7 mm long, elliptical and glistening white in colour. Caterpillars are reddish-pink dorsally and pale greenish ventrally: pronoturn is yellowish with 5 black spots on prothoracic shield. Pupae are 6 to 10 mm long, light green when freshly formed later becoming light brown or amber coloured. Adults are medium-sized moths having purplish-brown or pale rufous body, palpi protruding forward to form a snout (3 mm); forewings grayish-brown suffused with rufous and have a prominent pale white costal fascia and an antemedial yellowish-brown band with a ridge of raised pale reddish scales near its inner edge. Hind wings are semi transparent and fuscous. Wing expanse is 20 to 28 mm. A female lays generally 50 to 180 eggs in 5 to 6 batches. Incubation, larval and pupal periods last for 5.3, 12.7 to 18.4 and 33.1 to 17.8 days respectively; total life-cycle occupies on an average 35.8 days and longevity of moths is 3.9 to 7.4 days (Singh and Dhooria, 1971). Mating takes place 24 to 30 hours after emergence and pre-oviposition period may be one'day or more than a month (44 days). Full fed caterpillars drop down and pupate under the debris or 20 to 40 mm deep in the soil in thin cocoons. Hibernation is in caterpillar stage; these caterpillars remain in filmsy cocoons in soil and pupate with the onset of Spring. The caterpillars which pupate in Autumn are invariably killed during Winter because of low temperature.

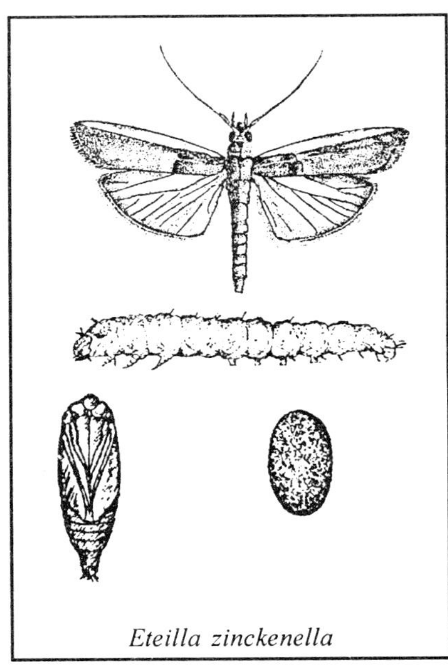
Eteilla zinckenella

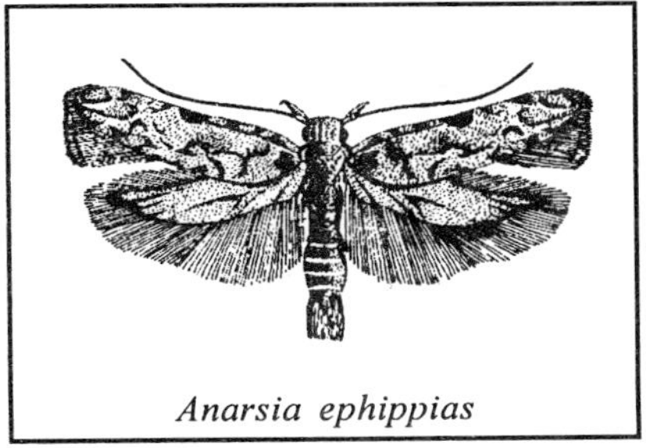
Anarsia ephippias

Anarsia spp. are minor pests of various leguminous crops. The most common species *A. ephippias* Meyrick is widely distributed in the plains of India. It is polyphagous, besides, cowpea and other pulse crops it has been reported damaging groundnut in Punjab and Uttar Pradesh and indigo in Bihar, The caterpillars roll the leaves and web these with the top shoots, then bore inside lower buds and pods and feed inside on seeds. Full grown caterpillars are 9 to 11 mm long, dark red, reddish-brown, deep pinkish-brown, or dark purple in colour; head and body are flattened and segments are distinct. Pupae are about 5 mm long and reddish-brown in colour. Adults are medium-sized active moths with heavily fringed hind wings. Wing expanse is 10 to 12 mm.

Besides *Anarsia ephippias*, *A. acerata* Meyrick has been reported from Karnataka (Coorg) and Tamil Nadu and *A. exallacta* Meyrick from Bihar on various leguminous crops including cowpea and garden pea.

Paraspistes palpigera (Walsingham) was originally described from East Africa by Walsingham (1891) but now it is found in the Indian sub continent, British West Indies, Panama. Bahamas, Seychelles etc. In India, it has been reported from Tamil Nadu and Karnataka as minor pest of Indian bean, cowpea and indigo. The caterpillars bore into pods and feed on developing seeds. The affected pods lose their market value.

Sathrobrota simplex (Walsingham) is a saprophyte and feeds on vegetable refuse and rubbish (Fletcher. 1920 d). It has been recorded feeding on dry pea pods. Eggs are very minute (0-3 x 0.2 mm) and oval in shape. Full grown caterpillars are 7 to 8 mm long; head and thoracic shield are light yellowish-brown; body is yellowish-white with very distinct reddish-pink transverse and narrow lines on each segment. Pupae are smooth, 4 to 6 mm long and yellowish-brown in colour. Adults are small-sized, slender, fragile moths having wing expanse of 7 to 9 mm.

To control the pod borers, spray with endosulfan 2 to 3 times at weekly interval from flowering onwards. Spraying with cypermethrin (25 g a.i. per hectare) or permethrin (200 g a.i. per hectare) is also effective.

Pea Stem Fly: *Ophiomyia phaseoli* (Tryon) (Agromyzidae) is widely distributed and from Africa to Australia (CIE map No. A-130). A polyphagous pest, it feeds on almost all varieties of beans, grams and peas. It has been reported as a pest of cowpea and sweet pea in Uttar Pradesh, often causing severe damage specialty to the seedlings and young plants. The pest breeds on black gram and moth bean from July to October and on different varieties of peas from November to April. Early sown crops are comparatively more prone to the attack of this pest.

The eggs are usually deposited singly under the leaf epidermis but occasionally these are laid on petioles and stems as well. A female lays oil an average 33 eggs (ranging from 14 to 64). On hatching the maggots mine the leaves, bore inside the petioles and tender stems and tunnel down wards. Sometimes adult females also puncture the leaves. The affected leaves turn yellow while the stems drop down and gradually wither away. Pupation takes place in the underground portion of affected stem.

Eggs are slender, oval in shape, less than half mm long and white in colour. Maggots are initially white in colour, later becoming yellowish. These are small in size being less than one mm in length. Pupae are barrel-shaped and brown in colour. Adults are metallic-black flies. 2.0 to 2.5 mm long, having hyaline wings that have a distinct notch in the coastal regions; females are slightly bigger than males, Wing expanse is on an average 5 mm. The flies mate 2 to 6 days after emergence and pre-oviposition period is 2 to 4 days. Egg, maggot and pupal stages last for 2 to 4, 9 to 12 and 18 to 19 days respectively during November-December but during March-April the maggot and pupal

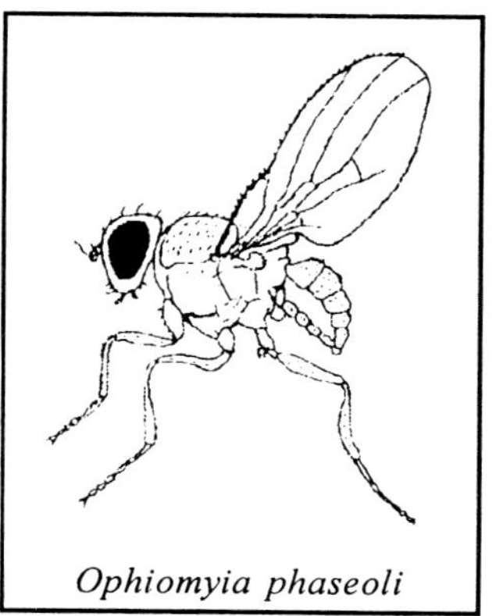

Ophiomyia phaseoli

durations are 6 to 7 and 5 to 9 days respectively (Agarwal and Pandey, 1961). Total life-cycle occupies 2½ to 4 weeks. Longevity of males is on an average 11 days while the females live up to 22 days. There are 8 to 9 generations in a year. The bionomics of this pest has also been studied in Indonesia (Goot, 1930), Philippines (Quesales, 1918), China (Cambell, 1925) and Australia (Froggatt, 1898, Morgan, 1938),

To check the infestation by these flies, remove and destroy all the affected branches during the initial stage of attack. In the case of wide-spread incidence, spray with 0.03% dimethoate, or monocrotophos @ 800 litres per hectare. Alternatively, mix with soil around the root zone of plants, aldicarb 10 g @ 2 kg a.i. per hectare. Seed dressing with phorate or disyston at the time of planting also provides effective protection against these flies for 2 to 3 weeks after germination (Jotwani and Butani, 1977).

Pea Leaf Miner : *Chromatomyia (Phytomyza) horticola* (Goureau) Goureau (=*Atricornis* Meigen) (Agromyzidae) is found throughout the Temperate region of the World. It is polyphagous in habit, having a wide range of host plants (Ahmad and Gupta, 1941). Besides pea, it has also been reported damaging carrot, cole crops, eggplant, potato etc. A female lays about 300 to 400 eggs, thrusting these into leaf tissues with the help of sharp and pointed ovipositor. On hatching the maggots mine the leaves in a zigzag fashion feeding on mesophyll within these mines. The infestation of this fly can easily be recognised by the presence of shiny whitish streaks on the leaves against the green background. Adults puncture the leaves and feed on exuding sap, The flowering and fruiting capacity of infested plants is adversely affected. The pest breeds during Winter and Spring and passes Summer and Autumn as pupa hidden inside the dried shed leaves discarded after harvest.

Adults flies are dark metallic-green in colour about 1.5 to 2.0 mm long; antennae are short and compound eyes big and prominent. The adults mate 36 to 48 hours after emergence. Incubation, maggot and pupal periods last for 2 to 4, 5 to 11 and 6 to 15 days respectively and adult longevity is 15 to 85 days. The adults usually appear in January. It takes less than a month to complete one life-cycle and 4 to 5 generations may be completed by April.

Aestivation begins in May, however the pest remains inactive till January.

To check the pest population, collect and burn infested leaves. Spraying with 0.5% dichlorvos or 0.1% carbaryl, 2 to 3 times at fortnightly interval is also effective. *Solenotus guptai* Subba Rao and *Rhopalotus thakerei* Subba Rao have been reported as larval ectoparasite and endoparsite respectively (Subba Rao. 1957). Larvae are also parasitised by *Neochrysocharis* species and *Opius* species (Nayar *et al.*, 1976, David and Ananthakrishnan, 2004).

Acrocercops (Cyphosticha) coerulea (Meyrick) (Gracillariidae) is another leaf miner widely distributed in the Indian sub-continent. It is a minor pest of cowpea and Indian bean. The caterpillars mine the dorsal surface of leaves, feeding just below the epidermal layer making whitish) elongate blotches, on leaves that later become pale white blisters, The pest is active from August to October. Full grown caterpillars are 4 to 5 mm long somewhat flattened and pale yellow in colour with sub-dorsal reddish dots on each segment giving the caterpillars pink appearance. Pupae are 3 to 5 mm long and pale yellow in colour, Adults are very small delicate moths with narrow long fringed wings.

To control, spray 0.25 to 0.5% dichlorvos or phosphamidon or 0.25% neem cake extract or neem oil emulsion (120 ml in one litre water); two to three fortnightly sprayings may be necessary to check pest population effectively.

Pea Aphid : *Acyrthosiphum pisum* (Harris) is cosmopolitan in distribution in both Palaearctic and Nearctic regions (CIK map No. A-23) and has been recorded from practically all the areas where the peas are grown. Colonies of young and old aphids are found on the young shoots, ventral surface of tender leaves and even on stems. The affected leaves often get cupped or become irregularly distorted, while the shoots become stunted and malformed. Honeydew secreted by the aphids encourages growth of sooty mould and this superficial black coating on leaves and stems hinders the photosynthetic activity of the plant which becomes weak, thus affecting adversely the pod formation.

Adults aphids are large, pear-shaped, green, yellow or pink in colour with long conspicuous cornicles. Both alate as well as apterous forms arc present and these are generally females; male are rare. Winged as well as wingless males have been reported from Europe and USA but not from India. Reproduction is parthenogenetic and viviparous. It takes about a week to complete one generation and there are several overlapping generations in a year.

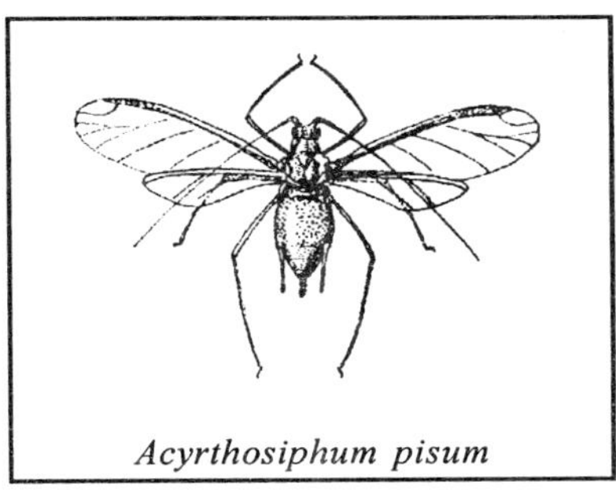

Acyrthosiphum pisum

To control the aphid, spray a non-residual contact or systemic insecticide: 0.3% dimethoate or formothion have proved to be very effective. Two to three sprayings at fortnightly interval may be required to check the pest population effectively.

Neem seed kernel extract formulation 'Achook' reported effective against green peach aphid, *A pisum.*

Deccan Grasshopper : *Colemania sphenarioides* Bolivar (Acrididae) is a serious pest, specially in the Peninsular India, on various minted millets and pulses. It is a minor pest of cowpea. Both hoppers and adults feed on leaves and often on flowers and pods as well.

Adults are wingless, 30 to 40 mm long, greenish-yellow in colour with a reddish. stripe extending from behind the eye and laterally along the thorax; antennae are bluish-black in colour. Eggs are laid in batches. 50 to 80 mm deep in vertical burrows during October-November. These hatch with the onset of monsoon showers (June-July). Nymphal development takes 2 to 3½, months. Thus there is only one generation a year.

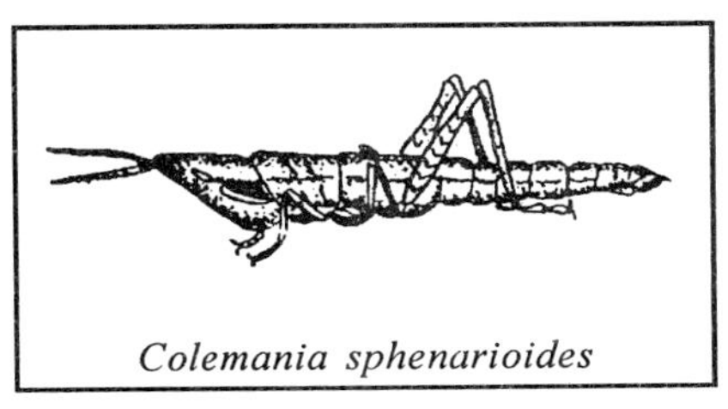

Colemania sphenarioides

The infested fields may be deep ploughed to expose and kill the egg-masses between November and June. Dusting 10% HCH is also effective in controlling the pest population, In nature, the eggs are preyed upon by the maggots of *Sustaechus nivalis* and grubs of *Zonabris* species (Nayar *et al.*, 1976, David and Ananthakrishnan, 2004).

Leafhoppers : *Empoasca kerri motti* Pruthi and *E. binotata* Pruthi (Cicadellidae) have been reported damaging cowpea in India. Both these are polyphagous; the preferred host of *E. kerri* motti is potato and that of *E. binotata* various cucurbits. Nymphs and adults generally remain confined to vental side of leaves near the main veins. They cause damage by sucking the cell sap but the damage caused is usually not serious and therefore needs no control measures.

Whitefly : *Bemisia tabaci* (Gennadius) - cotton or tobacco whitefly (Aleurodidae)- has a wide range of host plants including cowpea and garden pea. Among the vegetables, its preferred host is tomato. The pest not only causes damage by sucking the vital sap from tender plant parts, but also acts as a virus vector, transmitting mosaic disease.

It is quite easy to control whitefly as a pest of peas by giving 1 or 2 sprayings with any good systemic insecticide like 0.03% oxydemeton-methyl or thiometon, but by the time the insect is noticed and control measures adopted, the viruliferous insects may have completed the job of transmitting the mosaic disease and the loss caused thereby is irrepairable.

Sap Sucking Bugs : Sunnhemp bug, *Ragmus importunitas* Distant (Miridae), a major pest of sunnhemp in South India is a minor pest of cowpea and also of young cotton bolls. Swarms of these bugs may be seen infesting cowpea vines. Eggs are thrust into the tissues just under the outer layer oftender plant parts. Both nymphs and adults are very active and feed by sucking the cell sap. The affected plants turn pale and remain stunted reducing adversely the crop yield, Nymphs are 1 to 2 mm long and green in

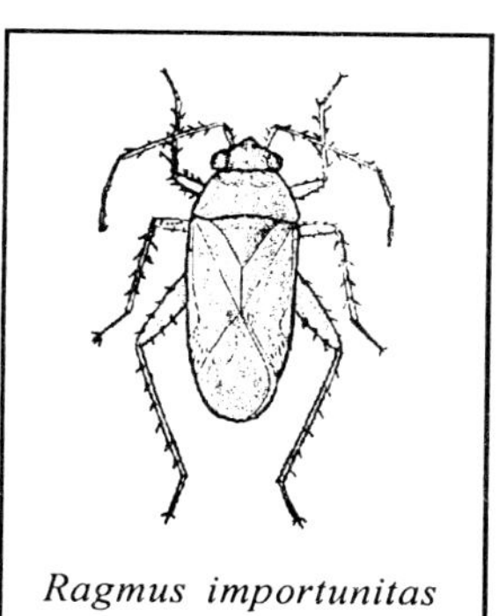

Ragmus importunitas

colour. Adults are 3 to 4 mm long and uniformly grayish-ochraceous in colour; antennas are concolorous and wing membrane is hyaline.

Anoplocnemis phasiana (Fabricius) (Coreidae) is widely distributed in Ethiopian and Oriental regions. This bug has been reported damaging cowpea, eggplant, different varieties of gram, grapevine, indigo etc. Eggs are laid singly or in rows on petioles of leaves.

Nymphs and adults suck sap from tender shoots and in case of cowpea. occasionally from pods also. The infested shoots wither and dry away adversely affecting the crop yield. Colour of nymphs and adults has been found to be variable - castaneous, ochraceous or piceous - in insects collected from different areas and different food plants. However, the colour does not change with the age of nymphs or when nymphs become adults, it only becomes slightly darker. Adults are 22 to 28 mm long and have characteristically curved hind legs, the posterior femora are curved at base thence incrassated, inwardly broadly dentate near apex while its outer margin is linearly serrate, Egg and nymphal stages last for 8 to 9 and 45 to 54 days respectively (Maheswariah and Puttarudriah, 1956).

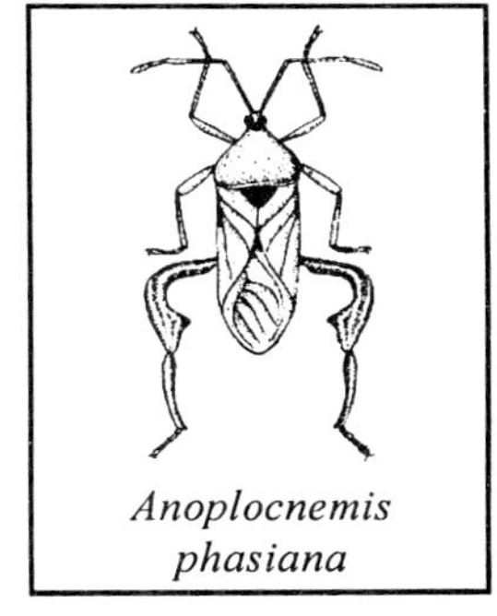
Anoplocnemis phasiana

Generally no control measures are adopted against these bugs on cowpea or garden pea. However, dusting the vines with 4% carbaryl or endosulfan is quite effective in checking the pest population.

Thrips : *Caliothrips indicus* (Bagnall) (Thripirdae) - The groundnut thrip - also attacks cowpea and Indian bean. Singh and Vaishampayan (1967) recorded heavy infestation on garden pea from germination till harvest. A severe infestation results in formation of white silvery sheens all over the leaf surface and the infested leaves may appear completely bleached. Life-cycle is completed in 2 to 4 weeks with several overlapping generations in a year.

The blosom thrips *Ayyaria chaetophora* Karny, *Megalurothrips distalis* (Karny), *Taeniothrips lefroyi* (Bagnall) and *T. longistylus*

Karny have been recorded infesting the flowers of cowpea though the preferred host of these thrips is Indian bean. A severe infestation by these insects affects the pod formation adversely.

Spraying a non-residual contact, or systemic insecticide against any other insect pest will control thrips as well.

Pea Leaf Roller : *Hedylepta (Nacoleia) indicata* (Fabricius) (*vulgaris* Guenee) (Pyraustidae) has been reported as a minor pest of garden pea from South India. Eggs are laid on the leaves in clusters of about 15 eggs each. On hatching the young caterpillars feed on green tissues of leaves: later the advanced stages of caterpillars web together a few leaves and feed within the folds, often skeletonising the leaves completely. The pest is active from August to February in Kerala.

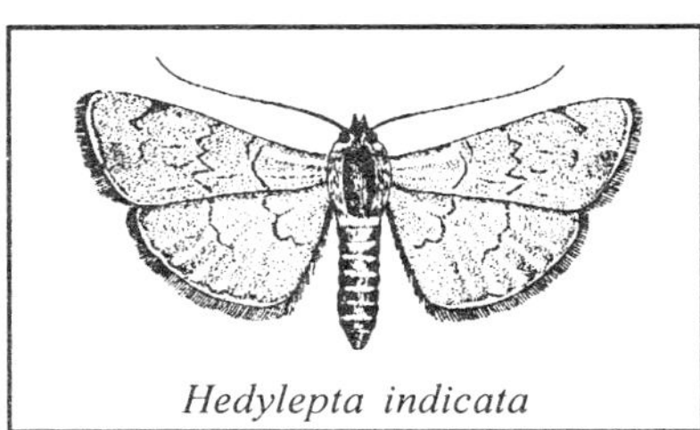

Hedylepta indicata

Eggs are flat and scale-like. Moths are fulvous-yellow with abdomen showing white rings. Wings are suffused with fuscous except costa of forewings and have discocellular spot. Forewings have obliquely curved antemedial black line; hind wings have post medial line bent outwards between vein 5 and 2, then retracted to lower angle of cell. Both wings have marginal black line as well as line at the base of cilia which are fuscous on forewings and white on hind wings (Hampson, 1896). Wings expanse is 20 mm. Incubation, larval and pupal periods last for 4 to 6, 13 to 15 and 5 to 6 days respectively, while the adult longevity is 5 to 6 days (Butani and Jotwani, 1984).

In nature. *Cardiochila fulvus* Cameron and *Xanthopimpla punctata* Linnaeus have been recorded parasitising the caterpillars.

Leaf-eating Caterpillars : Leaves of garden pea are damaged by various lepidopterous larvae. Those commonly found are an arctiid - *Spilarctia obliqua* (Walker) - and a few noctuids – *Diachrysia* spp., *Plusia* spp., *Mythimna separata* (Walker) and *Spodoptera* spp. The leaf defoliators on cowpea include hairy caterpillars -

Euproctis fraterna. (Moore) and *Porthesia scintillans* (Walker). All these insects generally cause minor damage to pea and cowpea. However, periodically these may appear in an epidemic form and then the loss caused is enormous; as the cultivators are caught unaware and get hardly any time to check the damage.

Spilarctia obliqua though prefers sweet potato, has also been recorded damaging various cole crops and root crops as well as cowpea and garden pea.

Plusia criosoma Doubleday and *Diachrysia orinchalcea* (Fabricius) are the two common semi-loopers that feed on leaves of cowpea and garden pea. Both are polyphagous pests and besides peas, these insects also damage different varieties of cole crops.

Mythimna (Leucania, Pseudaletia) separata (Walker) -rice armyworm is found all over the Indian sub-continent, South-east China, Japan. South-east Asia, Korea, Philippines, Indonesia, New Guinea, Eastern Australia and New Zealand (CIE map No. A-230). It is a major pest of rice but being polyphagous, it attacks other cereals, forage crops, grasses and garden pea as well. In recent years this pest has assumed the status of a serious pest of different *kharif* crops in South India. Eggs are laid in batches. Young caterpillars scrap the leaf tissues and skeltonise the leaves while the advance stage caterpillars feed gregariously and voraciously on leaf lamina: consequently even a medium infestation may result in complete defoliation of the vines. Eggs are spherical in shape and greenish-white in colour. Full grown caterpillars are cylindrical in shape and measure 42 to 48 mm in length; these are variable in colour - dark green to greenish-brown - with 4 distinct longitudinal black stripes on either side of mid-dorsal line. Its chaetotaxy has been described in detail by Butani (1960). Pupae are obtect, dark brown in colour and measure on an average 25 mm in length, Moths are pale brown, irrorated with dark specks. Forewings have a minute white mark at lower angle of cell. Hind wings arc pale suffused with fuscous. Wings expanse is 40 to 50 mm. Incubation, larval and pupal periods last for 4 to 12, 14 to 24 and 6 to 26 days respectively. The entire life-cycle is completed in 24 to 62 days and there are 5 generations in a year (Butani and Jotwani, 1984).

Mythimna (Cirphis, Leucania) unipancta (Haworth) has been often confused with *M. separata* as all the stages

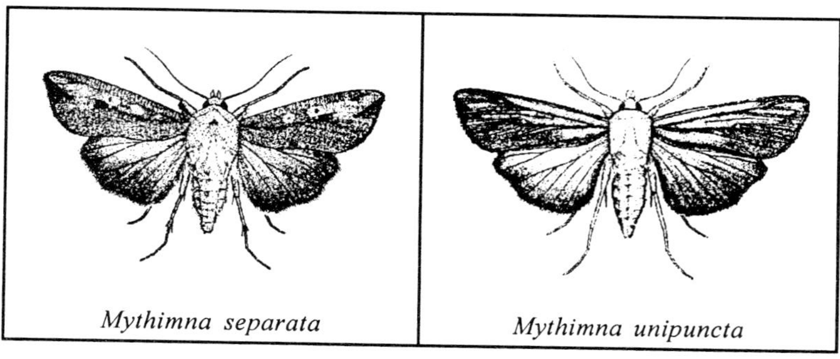

Mythimna separata | *Mythimna unipuncta*

including moths of both the species look alike. Franclement (1951) pointed out that the two are distinct species and later, Ramamani and Subba Rao (1965) confirmed that the Indian species is *Mythimna* (*Pseudaletia*) *separata* (Walker). *M. unipuncta* is found in South Europe, Mediterranean region, West Africa, USA, Central and South America (CIE map No. A-231) and has not been recorded from India.

Spodoptera spp. are highly polyphagous, with a very wide range of host plants. Among the vegetable crops only tomato suffers severe loss while the damage caused to garden pea and cowpea is of minor importance. The species reported damaging cowpea are, *S. exigua* (Hubner) and *S. littoralis* (Fabricius), whereas on garden pea in addition to these two species, *S. exempta* (Walker) has also been recorded.

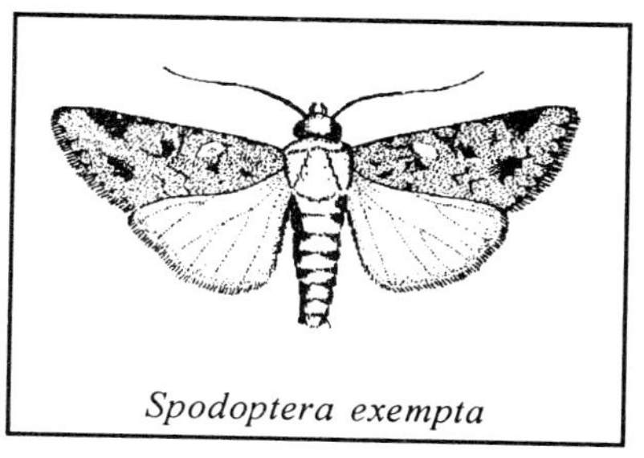

Spodoptera exempta

Spodoptera exempta - African armyworm - is widely distributed in Africa, India, Sri Lanka, Myanmar, Malaysia, Indonesia, Borneo, Celebes. Philippines, Papua New Guinea, Australia and Hawaii (CIE map No. A-53). Eggs are white. Caterpillars are blackish-velvety, measuring 30 to 35 mm when full grown. Pupae are brownish-black in colour and 16 to 20 mm long. Moths are stout having head and thorax ochreous-white, abdomen fuscous; forewings are also ochreous-white with a conspicuous small white mark and hind wings are pale fuscous. Wing expanse is 34 to 38 mm. Biology of this species has been studied in Africa where its egg, larval and pupal periods are 2 to 5, 14 to 32 and 7 to 21 days respectively (Hill, 1975).

Euproctis fraterna (Lyrnantriidae) - The tussock moth - is found all over the Indian sub-continent. Eggs are flat, round in shape and yellow in colour. Full grown caterpillars are 35 to 40 mm long with crimson coloured head and black body; first somite has long lateral tuft of black hair arising from prominent tubercles, the other somites have dorsal and lateral tufts of white hair. Head, thorax and forewings of moths are bright orange-yellow; abdomen tinged with fuscous and the anal tuft orange. Females have an almost complete submarginal scries of black spots. Hind wings are paler and wing expanse is 24 to 28 and 30 to 38 mm in case of males and females respectively.

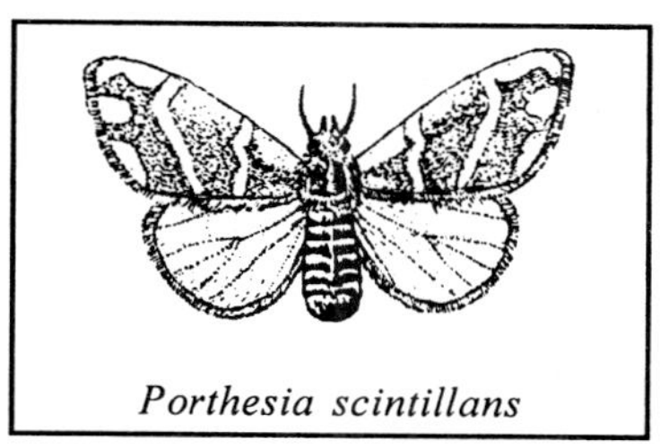
Porthesia scintillans

Porthesia scintillans (Lymatriidae) is another tussock moth widely distributed in the Indian sub-continent. The caterpillars are gregarious in the early stage, when they feed on epidermal layers of leaves. Advanced stage caterpillars are voracious feeders and move from plant to plant.These are dark brown with series of crimson coloured lateral tubercles on a yellow line bearing tufts of gray hair. Moths have yellow coloured head, brown thorax and yellowish-brown abdomen with orange coloured anal tuft in females. Forewings are vinous-brown irrorated with dark scales and hind wings are fuscous-brown with broad yellow margin. Wing expanse is 20 to 30 and 32 to 38 mm in case of males and females respectiveely (Hampson, 1892). Egg, larval and pupal periods occupy 6 to 10, 30 to 40 and 8 to 12 days respectively.

Generally no control measures need be adopted against these caterpillars. Nevertheless, if required, spray with 0.2% carbaryl or 0.05% endosulfan or quinalphos which is quite effective in controlling these pests.

Stem Borers : Pea stem borer *Grapholita (Laspeyresia) torodelta*

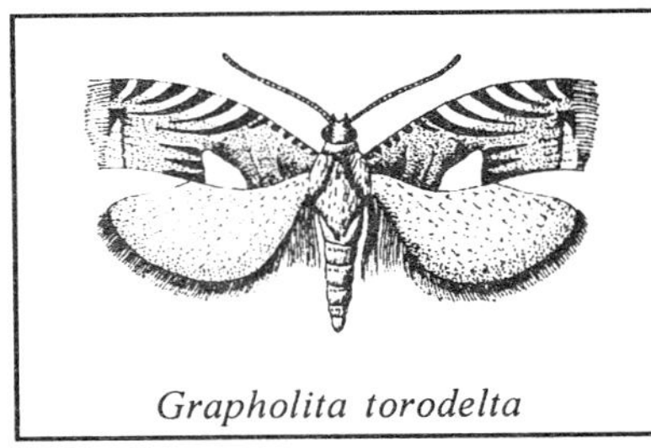
Grapholita torodelta

(Meyrick) (Eucosmidae) has been reported from Tamil Nadu and Kerala. The yound caterpillars bore into the growing tips devouring tender tissues. The terminal portions of the infested stems droop and die away. Pupation takes place in larval burrows. The larvae are slender, 10 to 15 J mm long, pale green in colour with reddish head and short hair scattered all over the body.

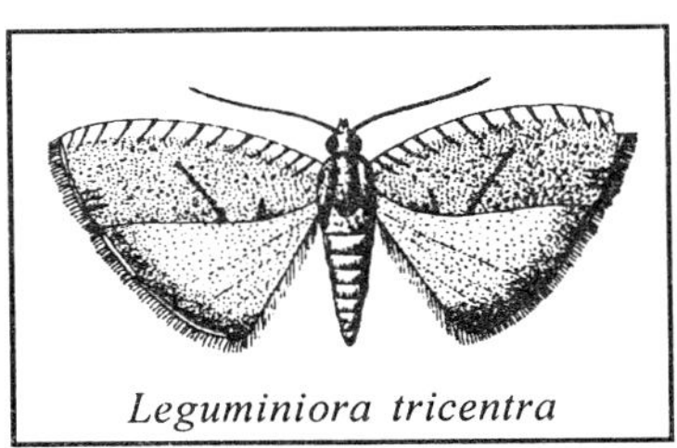
Leguminiora tricentra

Another allied species, *Leguminivora* (*Laspeyresia*) *tricentra* (Meyrick) (Eurosmidae) has been reported damaging cowpea. If is widely distributed in the Indian sub- continent. Caterpillars tunnel into shoots and feed within. Symptoms of damage are similar to those caused by *Grapholita torodelta*. The pest hibernates as caterpillar from November to February and then aestivates from March to June. Full grown caterpillars are 6 to 10 mm long, cylindrical in shape, tapering towards both the sides; these are orange-yellow in colour with head and thoracic shield shiny black. Adults are small-sized delicate moths, grayish-brown in colour and having wing expanse of 8 to 12 mm.

To prevent the spread of these borers, clip and destroy the affected shoots as soon as these pests are observed.

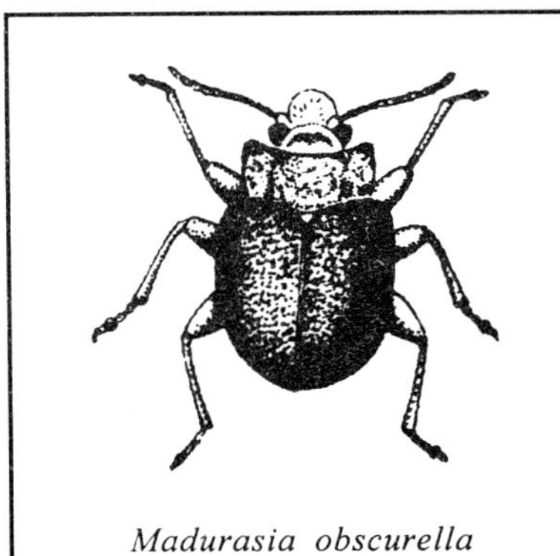
Madurasia obscurella

Leaf-eating Beetles : *Madurasia obscurella* Jacoby is a minor pest of cowpea. It is a polyphagous pest and has also been reported damaging cucurbits in Himachal Pradesh and yam in South India. Eggs are laid in soil around the host plants. The grubs remain in the soil and feed on root hair usually causing minor damage, however, it may facilitate entry and growth of disease causing micro-organisms. The adults come out and feed on follilage.

Soil application with disulfoton @ 1.0 to 1.5 kg a.i. per hectare at the time of sowing (Saxena *et al.*, 1971) or foliar spraying with 0.1% lindane or 0.07% endosulfan (Saxena, 1976) is quite effective in controlling this beetle.

Tanymecus indicus Faust, (Curculionidae) popularly known as *gujhia* weevil, is a polyphagous pest reported from all over India, attacking wheat, barley, gram, pea, papaya, maize, rice etc. Eggs are laid in soil and both the immature stages - grub and pupa - are confined to soil and only adults come out to feed on the aerial parts of the plants. However, adult weevils also hide during bright, warm and sunny hours of the day under clods and loose soil and feed only during dawn and dusk. They cut the seedlings at or below the ground level and feed on tender leaves. The adults are small-sized, snouted beetles, 4 to 6 mm long and grayish-brown in colour. They emerge around end of May and become sexually mature after another 10 to 20 weeks, incubation period is 2 to 3 weeks. Grub development takes 11 to 13 weeks. Incubation period is 2 to 3 weeks, grub development takes 11 to 13 weeks; pupal period during March-April is 4 to 5 weeks. There is only one generation in a year.

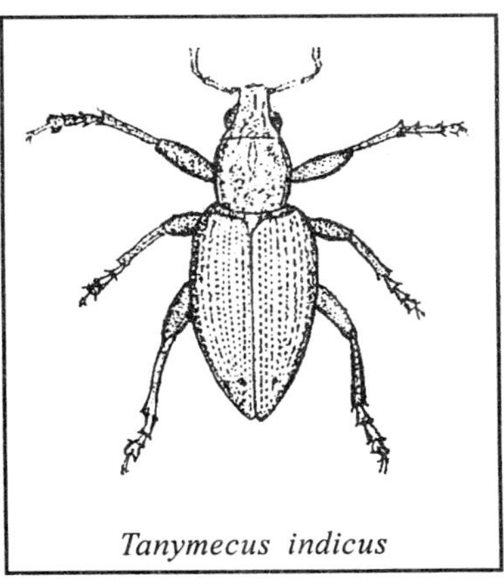

Tanymecus indicus

Soil application with 5% HCH, chlorpyrifos dust @ 20 to 22 kg per hectare will effectively control this weevil. When foliar damage is observed dust the crop with 4% carbaryl or endosulfan @ 18 to 20 kg per hectare.

12

LEAFY VEGETABLES

LEAFY VEGETABLES - the herbs that are grown for their edible greens (leaves). According to Singh and Arora (1978) about 220 species belonging to families, Amaranthaceae, Chenopodiaceae, Araceae, Asteraceae, Convolvulaceae, Malvaceae, Nymphalaceae, Papilionaceae etc. occur in India. While most of these are found growing wild, a few are regularly cultivated and consumed as cooked vegetables; the common one being Amaranthus (*choulai*), spinach (*palak*), lambs-quarters (*bathua*), fenugreek (*methi*), dill (*soya*), agathi (*basna*), Indian saltwart (*Loonuk*) water-coconut (*gol phal*) etc.

AMARANTHUS

Amaranthus is said to be a native of India and is most common Summer leafy vegetable in the plains of the Indian sub-continent. The varieties cultivated include *Amaranthus tricolor* Linnaeus *(gangeticus* Linnaeus*)* and *A. blitum* Linnaeus, (*choulai*) Both are herbaceous annuals, very rich in vitamins A and C, besides containing proteins, carbohydrates and iron. The fresh tender leaves and stems are cooked and eaten as vegetable dish. *A. tricolor* (*ban chulai*, *lal sag*) is ready for cutting after 35 days of sowing. *A. spinosus* Linnaeus (*kanteli choulai*) and *A. viridus* Linnaeus (*jagli choulai*) are found growing wild all over the Indian sub-continent.

The leaves are eaten raw or cooked and taken as vegetable preparation with cereals. In addition, *A. cruentus* Linnaeus is cultivated for its seeds. The seeds (*ramdana*) are yellowish-white with thick round border, are used as food-grain by the poor.

INSECT PESTS

About two dozen insect species have been recorded on *Amaranthus*. Of these, only *Amaranthus* weevil and two species of leaf webbers are of major importance, the rest are minor pests or mere host records.

***Amaranthus* Stem Weevil:** *Hypolixus truncations* (Boheman) (Curculionidae), a specific pest of *Amaranthus* is widely distributed in India and neighbouring countries. It attacks both wild and cultivated crops; leafy varieties with large leaves being comparatively less damaged than those with prominent stems (Butani and Verma, 1977 b). Adult females bite circular holes in the stems, lay eggs singly in each hole then cover the holes with black coloured secretion. A female lays an average of 30 eggs. On hatching, the grubs bite their way into the stems and feed on pith region, making irregular zigzag tunnels, which are filled with excreta. As many as 17-18 grubs have been recorded in a single stem (Ayyar, 1922). The affected stems become weak and often spilt longitudinally due to transpiration and this results in excessive evaporation; the plants get dessicated and ultimately dry up completely, Adults feed on tender leaves and stems but the loss caused by them is negligible. Full-fed grubs cut a small round hole in the stem, leaving thin semi-transparent epidermal tissue intact, then form a grayish-brown hard compact chamber out of the faecal matter and pupate therein. Adults on emergence remain inside the stems for 5 to 6 days, then cut epidermal membrane and emerge out.

Eggs are smooth, oval, one mm long and pale yellow in colour. Grubs are stout, curved, legless, white in colour and about 13 to 17 mm long. Pupae are about 9 mm long and pale yellowish-brown in colour. Adult weevils are ash-gray in colour, 10 to 15 mm long with elbowed antennae and brown elytra. Incubation period

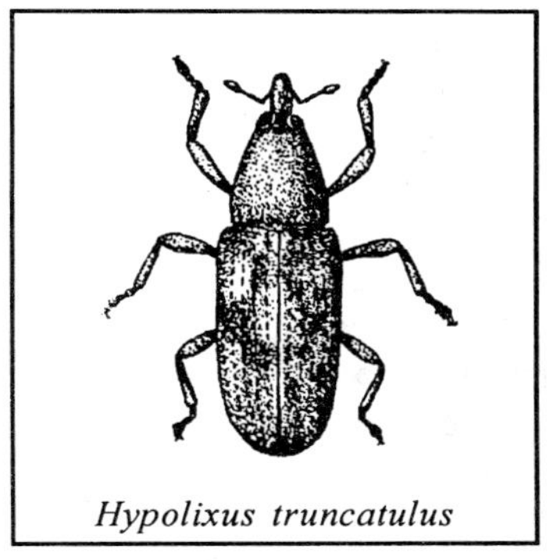
Hypolixus truncatulus

is 2 to 4 and 10 to 12 days during Summer.and Winter respectively. Grub stage lasts for 12 days in May and 20 to 24 days during October-December whereas adult longevity ranges between 12 and 66 days, average being 37 days (Ahmad, 1939; Gupta and Rawat, 1954).

For checking the damage by this weevil, remove and destroy all wild *Amaranthus* plants growing in the vicinity of cultivated crop, as the weevil breeds on these plants during the off-season and their destruction will prevent the carry over of the pest. As soon as infestation is observed, remove and destroy promptly all the affected plants with grubs inside. Spraying 0.05% dichlorvos or malathion is also effective.

In nature, *Poreuderus torymoides* Ferriere, *Telenomus javensis* Dodd and *Anastatus* spp. have been recorded as egg parasites; *Pareuderus torymoides*. *Dinarmus sauteri* Mani, *Eurytoma curculionum* Meyrick, *Aprostocetus krishnieri* Mani and *Xeridescopus* species have been recorded as larval parasites and *Eurytoma curculionum* as pupal parasite. Besides *Lastodiplosis* species has been reported to predate upon eggs of the weevil.

Leaf-Eating Caterpillars : *Eretmocera impactella* Walker (Heliodinidae), *Hymenia recurvalis* (Fabricius). *H. facialis* (Cramer) and *H. perspectalis* (Hubner) are the caterpillar pests which are commonly known as leaf webbers whereas related species *Nacoleia indicus* (Fabricius) has been found rolling the leaves and then webbing together the lop shoots to feed within on epidermal tissues including chlorophyll. The other lepidopterous pests include *Dichocrocis punctiferalis* (Guenee), *Psara basalis* (Moore), *Junonia orithya* (Linnaeus), *Helicoverpa armigera* (Hubner), *Othreis fullonia* (Clerck), *Plusia eriosoma* Doubleday. *Spodoptera exempta* (Walker). *S. exigua* (Hubner) and *S. littoralis* (Fabricius). All these are polyphagous pests causing severe damage to a number of crops of economic importance. On

Amaranthus, except the first two, the others have been reported as leaf defoliators of minor importance.

Hymenia recurvalis - *Amaranthus* leaf caterpillar - is a destructive pest often observed damaging *Amaranthus*. It is widely distributed in Tropical and subtropical regions including Africa. Asia, Australia and Hawaii Islands. In the Indian sub-continent it is found all the year round but is more active during warmer, rainy and early Winter months. Besides *Amaranthus*, it is a major pest of grasslands and pastures and also damages beans, Coleus, *Luffa* spp., melons and spinach. Eggs are generally laid singly but sometimes in batches of 2 to 5, usually in grooves of leaf veins. A female lays 50 to 80 eggs (maximum 150) during its lifetime. On hatching, the caterpillars feed on epidermis and pallisade tissues of leaves; later they web together the leaves with silvery silken threads secreted by them and feed within. Gradually these webbed leaves become completely devoid of chlorophyll and dry up. When full-fed-the caterpillars drop down to pupate in the soil.

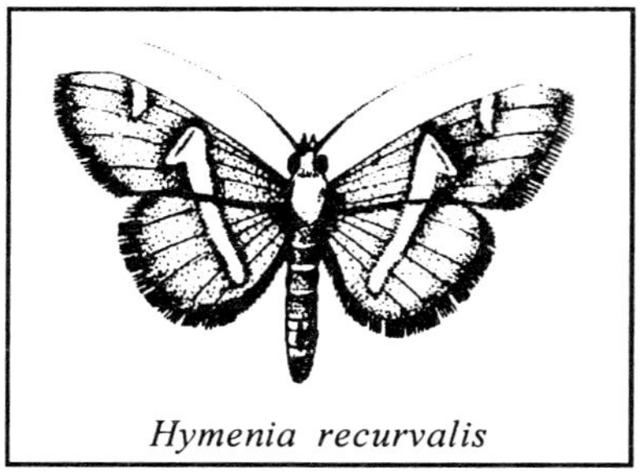
Hymenia recurvalis

Eggs are spherical in shape and snow-white in colour, Caterpillars are greenish in colour with white lines and black crescents on thorax below the lateral line; when full fed they measure 17 to 20 mm in length. Pupae are 10 to 14 mm long and brownish in colour. Adults are black coloured, slender bodied moths about 10 mm long and antennae. Both pairs of wings are dark fuscous in colour, having broad white fascia in the middle and outer margins are fringed with short hair. Wing spread is 15 to 20 mm. Incubation, caterpillar and pupal periods last for 3 to 4, 12 to 16 and 8 to 12 days respectively. A single life-cycle is completed in 3 to 4 weeks.

Eretmocera impactella is a sporadic pest of *Amaranthus* and is widely distributed in the Indian sub-continent. Eggs are laid on leaves preferably on top shoots. After hatching, the caterpillars web the leaves with white silken threads and remain hidden in the folds feeding from inside. Pupation takes place in white silken

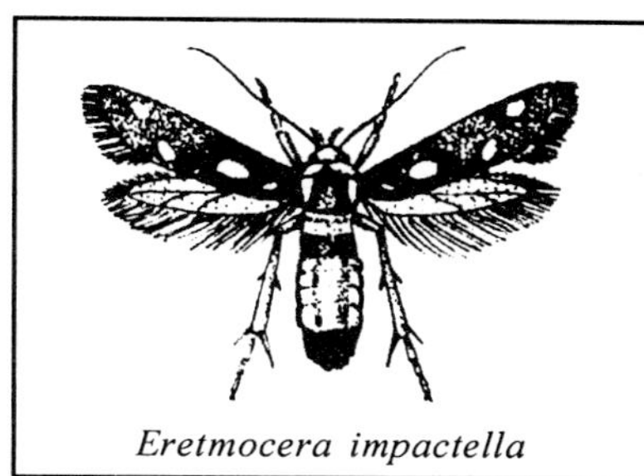
Eretmocera impactella

cocoons which remain attached to the leaves- Full-grown caterpillars are 9 to 12 mm long. cylindrical, brownish-yellow to brownish-gray in colour with a broad submedian dark stripe and black tubercles longitudinal hair, Pupae are about 6 mm long and uniformly brown in colour. Moths have cuperous head and thorax and yellow abdomen with second, third and terminal segments cuperous in colour. Fore-wings are also cuperous with yellow spots; hind wings are paler and the wing expanse is 14 to 18 mm. Life-cycle is completed in 3 to 4 weeks.

Psara basalis, though a minor pest of *Amaranthus*, is very common in Kerala. Its habits, symptoms of damage and life history are similar to that of *Hymenia recurvalis*. Full fed caterpillars are greenish in colour and 22 to 28 mm long. Head and thorax of adults are ochreous suffused with rufous -and abdomen is suffused with ruscous. Forewings are of various shades of rufous, costal and outer areas tinged with olive-green and basal area is irrorated with white. Hindwings are fuscous. Both wings have dark margin with white specks at the major veins. Wing expanse is 28 to 32 mm.

Dichocrocis punctiferalis is a major pest of castor. Its caterpillars are pale redish-brown in colour with numerous tubercle., on body and when full-fed, measure 15 to 25 mm in length. Moths are bright orange-yellow colured having a number of black dots on all four wings. Incubation, caterpillar and pupal periods last for 6 to 7. 12 to 16 and 7 to 10 days respectively and a life-cycle is completed in 25 to 33 days.

To control these caterpillars, spray 0.05% malathion or 0.03% endosulfan. The sprayed crop should not be harvested for at least a week. In nature. the caterpillars of *Hymenia recurvalis* are parasitsed by *Apanteles delhiensis* (Muesbeck and Rao, 1958) *A. ruidus* Walker and *Cardiochiles hymenise.*

Grasshopper : *Atmctomorpha crenulata* (Fabricius) is a highly polyphagous pest with a very wide range of host plants both cultivated as well as wild. This is recorded as a major pest of cabbage, cauliflower, carrot and radish and a minor pest of a number of other crops including amarathus. Nymphs and adults nibble the leaf lamina causing irregular holes. In ease of severe attack, dust the crop with 4% carbaryl or endosulfan.

Sap Sucking Insects : A few species of aphids - *Aphis craccivora* Koch, *Lipaphis erysimi* (Kaltenbach), *Myzus persicae* (Sulzer). *Hyadaphis indobrassicae* Das, as also mealybug *Ferrisia virgata* (Cockerell) and scale insects *Coccus hesperidum* (Linnaeus) and *Puluinoria durantae* Targioni have been reported damaging amaranthus by sucking the vital sap from leaves. Besides some species of thrips *Aeolothrips collaris* Priesner. *A. fulvicollis* Bagnall have been found feeding - on leaves and *Euroyaplthrips crassus* (Ramakrishna and Margabandu) and *Haplothrips ceylonicus* Schmutz infesting the inflorescences. These are also polyphagous pests and cause severe loss to a number of crops of economic importance but the loss caused to amarantbus is negligible and normally does not require any chemical control measures.

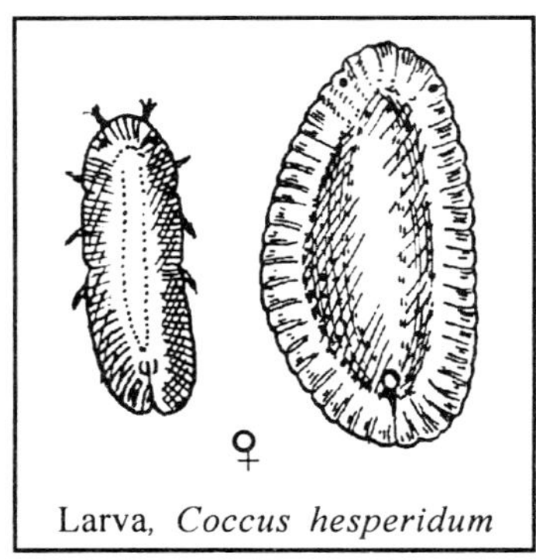

Larva, *Coccus hesperidum*

Thrips *Erankliniella intonsa* Trybom, coreid bug *Cletus pugnator* Fabricius and lygaeid bug *Nysius inconspicuous* Distant have been reported damaging *Amaranthus* plants in Bangladesh (Alam, 1970) but surprisingly there is no report of these insects damaging amaranthus in India.

Leaf Eating Beetle : *Aspidomorpha exilis* Boheman -Tortoise beetle - is a minor pest of *Amaranthus.* Eggs are laid singly on ventral surface of leaves. Grubs and adults feed by scrapping the outer tissues of leaves. Pupation takes place on leaf surface. A life-cycle is completed in 15 to 30 days.

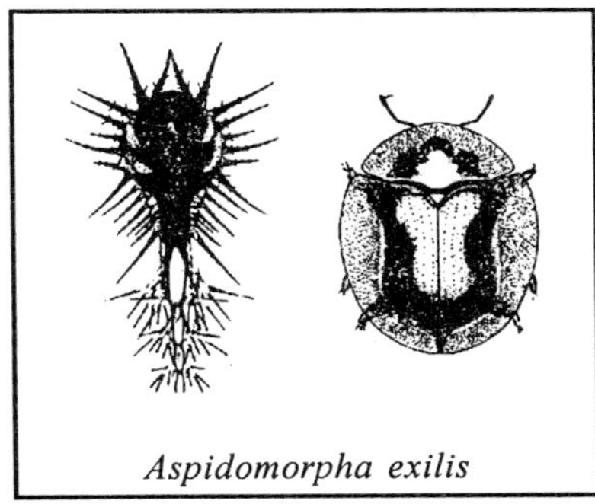
Aspidomorpha exilis

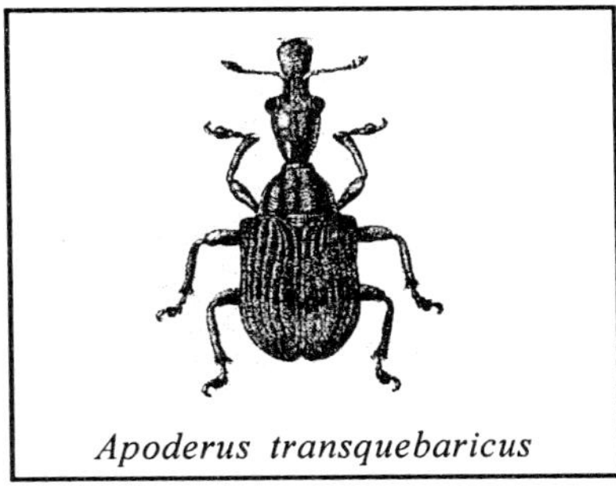
Apoderus transquebaricus

Leaf Twisting Weevil : *Apodems transquebaricus* (Fabricius) is another minor pest of *Amaranthus* recorded from South India. It is generally active from May to October. Its main host is country almond *(Terminalia catappa* Linnaeus*):* The oviposition and damage by this weevil are very peculiar. It cuts across a leaf from margin to midrib near the base; the leaf is then folded longitudinally from tip downwards. After a couple of folds, an egg is laid by the female within the outer fold and the folding or rolling of the leaf continues and a compact thimble-shaped structure is formed. The egg hatches inside and the grub feeds within the fold and also pupates there. The roll gradually starts drying and ultimately may fall down along with the pupa. The adult comes out by making a small hole in the dried, rolled mass of the leaf. Eggs are oval in shape and yellow in colour. Grubs are apodous and pale-yellow in colour while pupae are bright yellow. Adults are reddish-brown weevils with head drawn anteriorly into a long snout and posteriorly into neck.

Besides this weevil, *Cyratozernia cognata* Marshall and *Ptochus ovulum* Faust have also been found occasionally nibbling the leaves: both are minor pests.

Generally, no control measures are adopted against these weevils.

Termites : *Microtermes anandi* Holmgren (*obesi* Holmgren) and *Microtermes* species have been reported damaging amaranthus in Rajasthan (Kushwaha, 1960). The damaged plants wither and soon die away. The termites damage occurs generally in the amaranthus crop grown in sandy or sandy-loam soils. Soil application with 5%

HCH, chlorpyrifos dust @ 20 to 22 kg per hectare may be recommended in the endemic areas.

SPINACH

Spinach, (*Spinacia oleracea* Linnaeus) is commonly known as 'vilayati palak' native of South-west Asia while spinach-beet or palak (*Beta vulgaris bengalensis* Hortom) orginated from Indo-Chinese region, and the records show that it was known in China as early as 647 A.D. (Nath, 1976). Both the species are mainly Winter season crops and can withstand frost better than any other vegetable crop. There are, however, off-season varieties grown in cooler regions, practically throughout the year. Both are extensively grown in the plains of Northern India but in the Southern states, spinach is more common than spinach-beet. These are grown as herbaceous annuals for production of edible leaves whereas biennials are grown for seed production. Leaves of spinach have lobed margins while those of spinach-beet have unbroken margins. In spinach-beet, male and female parts are present in the same flower whereas in spinach, male and female flowers are usually separate. Beside providing necessary roughages in the diet these leafy vegetables are a cheap sorce of vitamins A. B_1 and C. an excellent source of iron and also contain proteins and carbohydrates. The spinach varieties recommended for cultivation in India include. All Green and Pusa Jyoti and those of spinach-beet are 'Virginia Savoy' and 'Early Smooth' Leaf.

Indian Spinach (Pol) *Basella alba* Linnaeus (Basellaceae), is a twining, succulent, undershrub grown on trellises and hedges. Its leaves and tender stems are used as vegetable; juice of the leaves can remove constipation in children and pregnant women (Peter and Devadas, 1989). Fleshy, spurious berries yield dye which is used for dying cheap confectionary (Singh *et al.*, 1983). Water spinach or swamp cabbage (Kalmisag), *Ipomoea reptans* (Linnaeus) (Convolvolaceae) is an aquatic or semi-aquatic, trailing herb found in Bihar, Orissa, Maharashtra

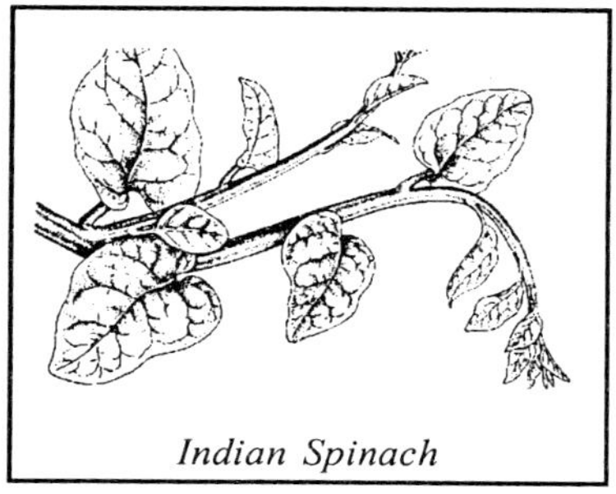

Indian Spinach

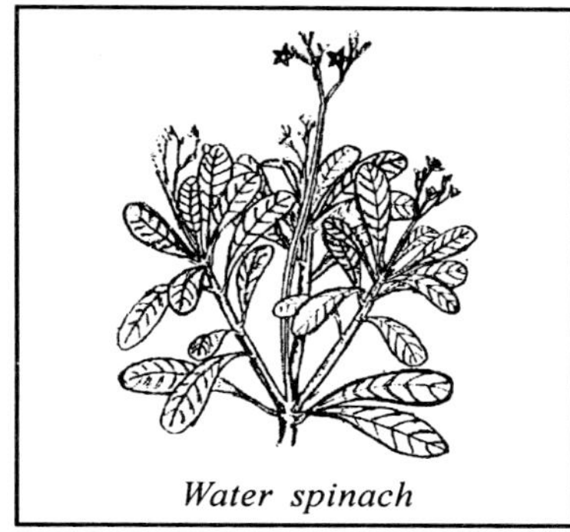
Water spinach

and South India.Young leaves and shoots are eaten as vegetable while the vines are used as fodder (Singh *et al.*, 1983)

Tree Spinach *(Chaya): Cnidoscolus chayamansa* and *C. aconitifolius*, are fast-growing ornamental and shade shrubs. Native of Southern Mexico and found upto Costa Rica, these are also grown in Florida (USA) and Cuba. A source of nutritious green leaves and shoots that are cooked and eaten like spinach. These must be properly cooked before eating since the fresh leaves contain toxic hydrocyanic glycosides that are inactivated during the process of cooking (Anonymous, 1979).

INSECT PESTS

Half a dozen insect species have been reported damaging spinach and spinach-beet leaves. Of these, those of regular occurrence which cause considerable damage are blue beetle and aphids. Grasshoppers appear occasionally and cause minor damage, leaf-eating caterpillars also appear sporadically, specially in the areas where spinach is grown in the vicinity of *Amaranthus*. Outside India, spinach flea beetle *Disonycha xanthomelaena* Dalman has been reported from Canada (Gibson, 1913) and green pentatomid bug *Nezara viridula* (Linnaeus) from Australia (Zeck, 1933). This bug is quite common in India also and among the vegetables potato is its preferred host. So far *Chaya* appears to be free of the pest and diseases that damage green leafy vegetables. This is of important economic and ecological advantage in favour of this crop. However, tomato hornworm *Manduca sexta* (Johannsen) has been reported from Central America causing extensive defoliation (Martin and Ruberte, 1978). but the plants recover quickly as soon as the pest is controlled.

Blue Beetle : *Altica caerulescens* (Baly)(Altlcidae) has been reported as a pest of cabbage and spinach. Besides, it has also been recorded as a minor pest on strawberry and plums. Eggs are laid in soil. On hatching, the freshly emerged grubs scrap and feed on chlorophyll containing tissues, later they mine inside the leaves; feed on the mesophyll tissues and pupate therein. Adults nibble the leaf margins causing very little damage.

Grubs are 8 to 10 mm long, dark brown in colour. Pupae are 12 to 15 mm long and brown in colour when freshly formed, then turn blackish-brown. Adults are 5 to 7mm long, steel-blue in colour with hind femora strongly thickened which enable the beetles in tumping movements.

To control their growth dust 4% carbaryl or spray with 0.1% carbaryl or 0.05% endosulfan or malathion. The treated crop should not be harvested for atleast ten days.

Aphids : *Lipaphis erysimi* (Kaltenbach), *Myzus persicae* (Sulzer) and *Hyadaphis* (*Siphocoryne*) *indobrassicae* (Das) have been recorded infesting leaves and causing damage to spinach: the last one being comparatively common. These aphids are polyphagous and preferred hosts of *L. erysimi* and *H. indobrassicae.* among vegetables are cole crops, whereas *M. persicae*, though found on cole crops, beet-root, lettuce, chillies, potato, radish etc., prefers eggplant. *Hyadaphis indobrassicae* was first described by Das (1918). Besides spinach, this aphid has been recorded on almost all cruciferous crops. Its adults are dull green in colour with two median rows of shining spots on dorsal side: antennae are dark coloured except at the base: cornicles arc longer than cauda and arc dark tipped. The pest appears early in October, the activity continues till the end of March though the peak activity is seen during February.

The damage caused by these aphids by sucking plant sap results in yellowing of leaves; the infested leaves become unfit for consumption.

To prevent spread of aphid infestation, cut and destiny promptly all the infested leaves along with colonies of aphids thereon during the early stage of attack. Spraying, suggested against the blue beetle, will control aphids and can be done when severe infestation is observed.

Grasshopper : *Atroctomorpha crenulata* Fabricius (Acrididae) has been recorded nibbling the spinach leaves, though it prefers fenugreek which is often grown as inter crop or mixed crop with spinach. This is a minor pest of leafy vegetables and eggplant. The other alternate hosts are cole crops on which this grasshopper is commonly found causing substantial damage. The affected leaves show presence of irregular holes. As it is a minor pest on leafy vegetables no control measures are adopted against this grasshopper. Dusting done against blue beetle etc., controls this pest as well.

Leaf Eating Caterpillar : *Hymenia recurvalis* (Fabricius) (Pyraustidae) sometimes appears in large number and feed voraciously on spinach leaves, especially when spinach is grown in the vicinity of amaranthus of which this is a major pest. Hand-picking of caterpillars in the initial stage of attack will prevent the population build-up of the pest. If necessary spray 0.03% endosulfan or 0.05% malathion but harvest the crop only after ten days of treatment.

LAMBS-QUARTERS

Chenopodium album Linnaeus also known as pigweed, is a small odourless herbaceous plant found all along the foothills of the Himalayas - wild as well as cultivated. The plant contains ethereal oil which closely resembles cholesterol oil. In addition to the oil, its other constituents include, ammonia and amines both in free and combined forms. *C. Lamaranticotar,* also known as Mexican tea, is an aromatic herb with a camphoraceous smell. It is found mostly wild in Eastern states of India. Anthelmintic and volatile oil extracted from it is used for medicinal purpose as it is effective against various intestinal parasites. Fruits of both the species are globular in shape, slightly compressed with thin pericarp. The seeds, small in size, orbicular in shape, brown in colour, smooth and shining having bitter pungent taste, are used as carminative and tonic. They yield diosgenin, precursor of steroids including sex hormones and oral contraceptives (Singh *et al.*, 1983).

Chenopodium blitum Hook, a wild species - occurs in Kashmir; reported to be used as pot-herb (Singh *et al.*, 1983). *C. murale* Linnaeus (khartua) - a herbaceous winter weed is found in Punjab,

Upper Gangetic plains, Kumaon (Uttarakhand) and Peninsular India. Leaves and tender twigs cooked and consumed as a vegetable; also used as fodder.

FENUGREEK

Trigonella foenumgraecum Linnaeus (Papilionaeecae), is an aromatic annual herb, found in Kashmir, Punjab and Gangetic plains. The full-grown plants are 30 to 60 cm tall with pinnate trifolialate leaves and white or yellowish-while flowers. Pods are 5 to 15 cm long and contain 10 to 20 greenish-brown seeds. The edible part is the leaves which are generally eaten mixed with spinach, or used for flavouring other vegetables like potato, carrot, turnip etc. Seeds are bitter in taste and are used medicinally,

Trigonella polycerata Linnaeus *(chini)* is the wild species of fenugreek occurring in the Himalayas and as Winter weed in plains of Northern India. Leaves are used as vegetable.

DILL

Anethum graveolens Linnaeus (Apiaceae) - a herbaceous plant is native to Eurasia, now cultivated on small scale in Jammu and Kashmir. Dried fruits are used as spice and condiment as also as carminative (Singh *et al.*, 1983).

Indian dill, A. *sowa* Kurz is cultivated all over Indian sub-continent. Leaves are used for flavouring other vegetables like amaranthus, spinach etc. Dried fruits are used as carminative, stomachi and stimulant. Freshly cut herb, on steam distillation yields oil of dill herb whereas mature crushed seeds yield oil of dill seeds. These oils are used as flavouring agents in food industry besides in pharmaceutical and perfumery industries.

AGATHI

Sesbania gfrandiflora (Linnaeus) (Papilionaceae) - a quick growing tree, native to Tropical Asia, is now grown in Assam, Bengal, Punjab, Gujarat, Andhra Pradesh and Tamil Nadu. Its tender leaves, flowers and fruits are used as vegetable. Plant is used as support and shelter for betel vine and black pepper: shade to coconut seedlings and also as wind-break in banana plantation in Tamil

Agathi

Nadu (Peter and Devadas, 1989). Leaves are chewed to disinfeet mouth and throat, while juice of roots mixed with honey is useful as an expectorant (Singh *et al.*, 1983).

INDIAN SALTWATER

Suada fruticosa (Linnaeus) and *B. maritima* (Linnaus) (Chenopodiaccae) also known as seepweed are herbs or undershrubs occurring in moist and swampy areas of Northern India, East of Punjab and along the Eastern and Western ghats. Green fleshy leaves are eaten as vegetable specially by the poor, in time of scarcity. The plants are reported to fix sand along the sea-shores and are recommended for treating water-logged and saline soils.

WATER-COCONUT

Water-coconut or nipa-palm, *Nypa fruticans* Wurmb. (Areaceae), a prostrate, coastal palm having a stout, branched, creeping rhizome, is found in the tidal swamps of Sundarbans in West Bengal and in the Andaman Islands. Its tender stern buds are eaten as vegetable, young peduncles and immature seeds are eaten raw or cooked. Leaves are used for making coarse mats, baskets as also thatch, while the midribs are used for making coarse brooms. The palm is much valued, particularly in Philippines, for the sweet sap obtained from its spadix. This sap is used for making palm-jaggery, alcohol, sugar and vinegar (Singh *et al.*, 1983).

No separate studies have been conducted on insect pests and diseases of Lambs-quarters, fenugreek, dill, *agathi* Indian saltwater and water-coconut.

13

SALAD CROPS

SALAD is a general term used for a number of uncooked herbs eaten raw with or without, dressing. The vegetables usually used for making salad include cabbage, carrot, radish, onion, beet-root, cucumber, tomato, celery, lettuce and parsley. Except the last three, others are also used in cooked form.

CELERY

Apium graveolens Linnaeus (Apiaceae) - a biennial marshy plant having strong smell, is native to Mediterranean region. It is a cold season crop and is usually confined to higher elevations. The petioles are 60 to 90 mm long, thick and grooved externally with compound leaflets attached to the apex. Flowers are very small, white and borne on compound umbels. Earlier, it was only grown as pot-herb for medicinal use and much later for flavouring soups and stews. In Southeast Asia, celery is grown throughout the year. It is generally taken raw, but is also served steamed and creamed along with various Chinese dishes or incorporated in soups and stews.

INSECT PESTS

Cutworms, *Agrotis* spp. and aphid *Acyrthosiphon malvae* (Mosley) are the only regular pests, besides mites and root-knot nematodes. Damage by cutworms is confined to young seedlings which are cut at the ground level and killed outright. Aphids suck sap from tender plant parts; the plants get devitalised and become unfit for use as salad. The mites may be seen on ventral surface of leaves, feeding under the protective covering of silken webs. To prevent the population build-up of aphids and mites, clip-off the affected plant parts in the initial stage of attack and destroy these promptly. For cutworms, dust the seedlings and soil around the seedling with 4% endosulfan or carbaryl or 5% malathion dust. The treated crop should not be used for at least 15 days after dusting.

LETTUCE

Lactuca sativa

Lactuca sativa Linnaeus (Asteraceae) is most common leafy herb that is used as salad. It has been grown for such a long time (over 2000 years) that no one seems to know when and where it originated - its place of origin can possibly be the Eastern Europe or Western Asia. It has descended from wild lettuce (*L. scariola* Linnaeus) which is one of the most common weeds found all over the World. It is said that lettuce was served at royal table of Persian Kings in about 550 BC and used as salad by Greeks and Romans. This is a herbaceous annual that grows as a basal rosette of leaves and later in the season produces a stalk bearing a bunch of flowers. The edible portions are the large-sized tender leaves and these contain vitamin A and minerals like calcium, phosphorus, sodium, sulphur, magnesium and potassium though in small quantities. Being low in calories and rich in nutritional properties, it serves as an excellent addition to our diet. Great Lakes, Slow Bolt and Chinese Yellow are some of the varieties grown in India.

INSECT PESTS

Aphids are the only destructive pests of lettuce in India. The species recorded so far include, *Aphis gossypii* Glover, *Myzus persicae* (Sulzer) and *Lipaphis erysimi* (Kaltenbach). All three are polyphagous pests causing severe damage to a number of economic crops including eggplant, cucurbit and cole crops. If and when the aphids attack in large number, plants get devitalised and leaves start fading and yellowing. The difference between other vegetables and lettuce in respect of damage caused by aphid, is that of the effect on quality which is important in the case of vegetables used as salad. Even a few spots caused by aphid, render the leaves unfit for table purpose. Collect and destroy the infested leaves in the early stage of attack. Cutworm *Agrotis segetum* (Dems and Schiffer-mmlcr) has also been reported from various parts of the Indian sub-continent, damaging the seedlings, but their occurrence is very sporadic and thus the damage caused is negligible.

In Europe, a few more species of aphids have been recorded damaging lettuee, namely *Nasonovia ribisnigri* (Mosley, 1841), *Myzus ascalonicus* (Doncaster, 1946), *Macrosiphum euphorbiae* (Thomas, 1878) and *Pemphigus bursarius* (Linnaeus, 1758). In addition to aphids, *Estigmene acrea* (Drury) and *Altica cyanea* (Weber) have been reported feeding on lettuce leaves in Latin America and Bangladesh respectively. *A. cyanea*- the blue beetle-occurs in India as a major pest of waternut (*Trapa bispionsa* Roxbergh) and a minor pest of sugar-beet.

Genetic manipulation in Lettuce : Wild types of lettuces are good source of donor genes, which has been utilized for manipulation of lettuces (Davey *et al.*, 2002). Aphids and whitefly cause significant damage to lettuce and act as viral vectors. Hence, breeding for insect resistance can also confer viral resistance in lettuce hybrids. Resistance to pests and diseases has been identified in some cultivars of *L. sativa* and in wild species, such as *L. scariola* and *L. saligna*. Resistance to *Nasonovia ribisnigri* (leaf aphid) has been identified in *L. virosa* and transferred into cultivated lettuce.

PALM-HEART

A Tropical delicacy, it is the tender growing up of several palm trees. It is the smooth portion at the apex of trunk of palm trees between fruit clusters and leaves, 150 to 300 mm in diameter and about a meter in length. After removal of palm-heart, the tree dies away, hence it is generally taken from unwanted or wild palmae. Some palmae like pejibave (*Guilielma gasipaes*) produce several stems from a common root cluster, so only excess stems and not the whole plant is sacrificed to harvest palm-hearts (Anonymous, 1979). The food value of palm-heart is more or less similar to that of cabbage; these may be added to salads, served as vegetable or used for enhancing the flavour of other vegetables. There is little awareness of its commercial potential outside Brazil and Costa Rica. In India there is sufficient scope of collecting and raising palm-hearts specially in Gujarat, North-west Rajasthan, Western Haryana and South-west Punjab.

INSECT PESTS

The only destructive pest of palm-heart is red palm weevil *Rhynchophorus ferrugineus* Olivier (Butani, 1974, 1975 c). This is a big, about 35 mm long, cylindrical, flattened, reddish-brown weevil with red spots on thorax and a long, slightly curved snout. Its incubation, caterpillar, pupal and whole life durations, on coconut-palm are, 2-5, 24-61. 18-34 and 50-90 days respectively (Menon and Pandulai, 1958).

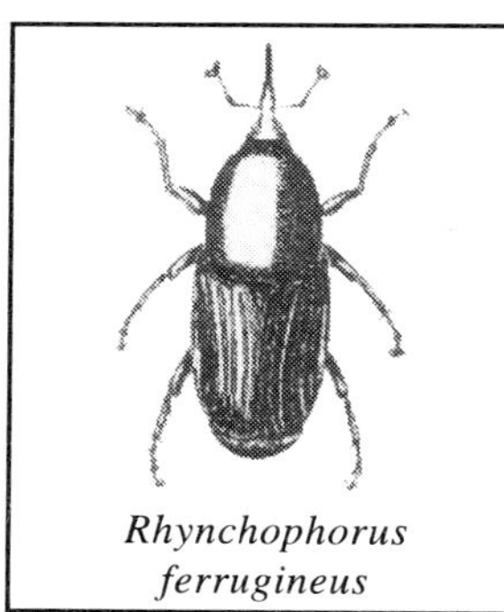

Rhynchophorus ferrugineus

PARSLEY

Petroselinum crispum (Miller) (- *sativum* Hoffm., *hortense* Hoffm.) (Apiaceae) has been cultivated for over 2000 years. It is said that during Pliny's era there was not a sauce or salad served without parsely. This is a biennial plant that forms a rosette of moderately long petioled leaves. It is rich in vitamins A and C. This is also a cold-season crop, grown at higher altitudes throughout the year. At low altitudes, the growth is slow and quality poor, unless grown as pot-herb with proper care. It is generally used in small quantities for garnishing the salads and other vegetables. Its demand is generally more in posh restaurants than in houses. Redolence of parsley can absorb the intoxicating fumes of alcohol and thereby can prevent the usual effect of over indulgence.

INSECT PESTS

There are no major insect pests of parsley reported from India. The minor ones include some leaf-eating caterpillars that occasionally defoliate the plants. However, nematodes can often become a serious problem in localised area.

14

ONION & GARLIC

ONION *Allium cepa* Linnaeus and garlic *A. sativum* Linnaeus (Alliaceae) are the two common commercial crops consumed not as a vegetable but as an essential supplement with various other vegetables, specially the tasty Indian and Mughalai, vegetarian and non-vegetarian gravies. Native to Central Asia, these probably originated from the region between North-west India and Palestine. Egyptian workers building the great pyramids (48000 years ago) are reported to have eaten onions and garlic. In India, onion and garlic were used in medicine in 600 BC (Verma, 1980). At present, onion is a widely cultivated vegetable crop found all over the World.

Onion is a bulbous herbaceous plant bearing linear hollow fleshy cylindrical green leaves and umbels of small white flowers, which mature into 3-celled capsular fruits containing small black seeds. The underground bulbs the edible portion vary in shape (round, flat or conical), colour (white, yellowish or reddish-brown), size, firmness, keeping quality, period of maturity and flavour. It is potentially a biennial crop producing large bulbs and hollow leaves during first year and it can flower and produce seeds only if kept for the next year. In India, onions occupy an area of over 100,000 hectares; of which nearly half the area is in Maharashtra; the other main onion producing states are Andhra Pradesh, Karnataka, Bihar and Assam. Both onion and

Onion plant

garlic are important foreign exchange earners and are exported to Malaysia, Japan, Myanmar, Pakistan, Sri Lanka, Iran, UAE and East African countries. Pakistan imported 2000 tonnes of onions from India in 1995 (Hindu, 7-12-1995). Allied species, *Allium rubellum* Bieberste (*Jangli piaz*) is found in North-western Himalayas (Singh and Arora, 1978), *A. cepa viviparum* (Metz.) (Egyptian onion) and *A. ascalonicum* Linnaeus (Shallot) in South India. *A. schoenoprasum* Linnaeus (*Chive*) in Western Himalayas and *A. tuberosum* Rottl. (fragrant onion or Indian leek) in Bengal (Singh *et al.*, 1983). Leek, *A. ampeloprasum porrum* (Linnaeus), a stout, non-bulb forming, tall herb is native to Mediterranean region.

Nutritionally, onion is rich in carbohydrates and minerals like phosphorus and calcium and also contains proteins as well as vitamin C. Primarily the bulb is the edible portion but in some places, green leaves are also eaten, raw as well as cooked. Bulbs are eaten either raw in salads or cooked with other vegetables and form an essential component of gravies and stews. The bulbs are also pickled and used in soups and sauses. Its various products like onion powder and onion salts are used for flavouring or seasoning different culinary preparations. The bulbs when cut emit strong pungent odour that brings tears in the eyes of one who is cutting or sitting nearby. Onions are also fed to cattle and poultry. It has some medicinal properties which are also mentioned in *Charaka Samhita* - one of the earliest medical treatise of India. It is believed that its consumption during Summer can protect from heat stroke. Fresh juice is used for curing flatulence, dysentry and cholera (Purewal, 1954). The improved varieties commonly grown include, 'Early Grano', 'Pusa Red', 'Pusa Ratnar' and 'Niphad 53'. In Maharashtra, No. '404' and '207-1' are grown during *kharif* (rainy)and *rabi* (Winter) seasons respectively while in Bihar, Patna Red and Patna White are more common.

INSECT PESTS

There are only two major insect pests reported damaging onion and garlic in India - onion thrips (erroneously called thrip) and onion fly. In addition, a few species of leaf eating caterpillars, groundnut earwig and some beetles damaging bulbs, may appear regularly but cause negligible damage.

Onion Thrips : *Thrips tabaci* Lindemann (Thripidae) was originally prevalent only in warmer parts of Palaearctic and Neoarctic regions, but it has now spread to all parts of the World through commerce (Ananthakrishnan, 1971) and has become cosmopolitan in distribution (C1E map No. A-20). It is highly polyphagous with a very wide range of host plants. In India, besides onion and garlic, it attacks cole crops, cotton, cucurbits, peas, pine apple, tobacco, tomato, turnip and ornamentals like carnation, lilies (*Agapanthus* spp.) *Osbeckia* spp., roses (*Rosa* spp.), *Verbena venosa* etc. (Ananthakrishnan, 1960).

The thrips is active throughout the year; it is found on onion and garlic during November to May, from these hosts it migrates to cotton and other summer crops in June and then to cold crops during September-October. A long spell of dry weather is favourable for its rapid multiplication whereas heavy rains and humid weather adversely affect its development and multiplication (Butani and Verma, 1976 e). Numerous nymphs and adults may be found between leaf sheaths and stems lacerating the epidermis of leaves and lapping the exuding sap. The affected leaves show silvery white blotches which later become brownish and get distorted from tips downwards, wilt and ultimately dry away. Heavy infestation at seedling stage results in retardation of growth and severe scarring of leaves which ultimately kills the seedlings outright. In case of heavy infestation at later stage the bulbs remain undersized and get distorted in shape.

Eggs are tiny and white in colour. Nymphs and adults are slender, fragile and yellowish in colour. Adults have four narrow and long wings fringed heavily with fine hair. Males are 0.8 to 1.0 mm long while the females are 1.0 to 1.2 mm long. The males are extremely rare; the ratio being 3000 females to one male. The reproduction is thelytokous parthenogenetic. A single female lays 40 to 60 eggs in 4 to 9 days in notches in the epidermis of leaves. Biology studied in Bihar showed that incubation period is 5 to 10 days; nymphal development takes 5 to 10 days; prepupal and pupal stages last for 2 to 3 and 4 to 7 days respectively and are passed in soil. Total life-cycle takes 14 days in Summer but extends up to 24 days during Winter (Lall and Singh, 1960, 1968}. There are more than 10 generations in a year and as *Thrips tabaci* the females have generally long oviposition periods the generations frequently overlap under natural conditions. In South India egg, nymphal, prepupal and pupal periods last for 8 to 9, 4 to 6, 2 and 3 days respectively (Nair, 1975). In Pakistan the entire life-cycle is completed in 11 to 21 days (Rahman and Batra, 1944) while in Japan, the life-cycle of this species varies from 17 to 38 days during Summer and is more than 38 days in Winter (Sakimura, 1937).

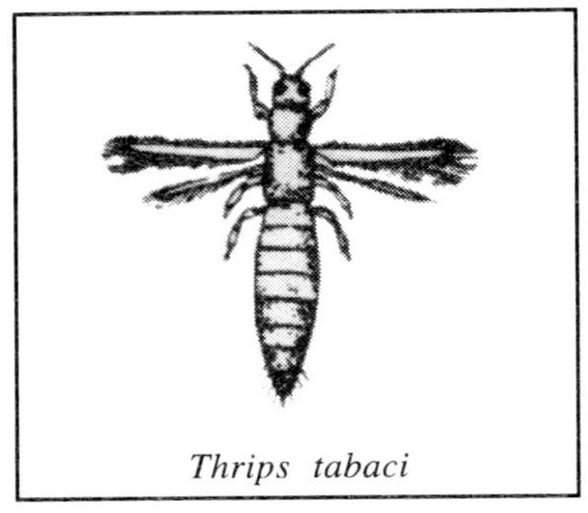

Thrips tabaci

Another thrips commonly found in association with *T. tabaci* on onion is groundnut thrips *Caliothrips indicus* (Bagnall). This is also a polyphagous pest though groundnut and *Dhencha* (*Sesbania bispinosa*, *S. cannabina* and *S. sesban*) are its preferred hosts. A heavy infestation results in white silvery sheens all over the surface and the leaves appear to be bleached totally. Both nymphs and adults are uniformly blackish-brown in colour with broad forewlngs having 4 dark and 3 pale bands. The life-cycle of this species does not show any apparent deviation from its allied species including *Thrips tabaci*. Egg to adult periods range between 13 and 33 days (Ananthakrishnan, 1973). On onion, Saxena (1971) observed four overlapping generations between February and April.

The seed crop suffers damage from blossom thrip *Aeolothrips collaris* Priesner, at the flowering stage, specially in Northern India. This is also a polyphagous pest - onion, mango and mustard are its main hosts in India. Adults of this species are brownish having broad forewings that are rounded at the apex.

To prevent the attack of thrips, grow resistant vaiieties of onion. Jones *et al.* (1934) have reported White Persian, Grano. Sweet Spanish and Crystal wax to be fairly resistant to *Thrips tabaci*. In India Spanish white variety is very tolerant to the attack of this thrips whereas 'Poona Red' is highly susceptible (Lal and Singh, 1968). Verma (1966) observed that 3 applications of gamma HCH (lindane) increased the yield of onion by 93%. Spraying 0.05% endosulfan, moncrotophos or dimethoate or 0.2%

carbaryl (Ananthakrishnan, 1971) or 0.03% quinalphos followed by second spraying of 0.05% monocrotophos or endosulfan (Reddy and Jagadish, 1980) have been found to be effective in controlling thrips and thereby increasing the yield. All these insecticides are reported to leave no residues beyond 10 days (Jotwani and Butani, 1980b). Outside India, Harding (1961) suggested spraying 0.05% endosulfan whereas Hole and Shorey (1966) obtained good control of this pest with granular application of fenthion, endosulfan or dimethoate. Mogal *et al.* (1982) recommended three applications with phorate 10 G @ 2 kg a.i. per hectare to be applied 2, 6 and 10 weeks after transplanting.

Onion Fly : *Delia antiqua* (Meigen) (Anthornyiidae) is another destructive pest of onion and garlic in India. Besides India, it has been reported from Canada, USA, Finland. England, Denmark, France, Germany, Russia. Pakistan, Bangladesh, Japan etc. Eggs are generally laid on soil near the base of host plants but sometimes even in the cracks and crevices in the soil. On hatching, small maggots move for sometime near basal portion of leaf sheath attached to stem and then burrow down into the underground portion of stem and often in the bulb. Large bulbs may he attacked by several maggots, each carving out a small cavity, which subsequently results in rotting of the bulbs in storage. The damage caused by this pest is generally followed by the attack of fungus -*Bacillus carotovorus*, causing soft rot of onions (Johnson, 1930). Pupation takes place in the soil.

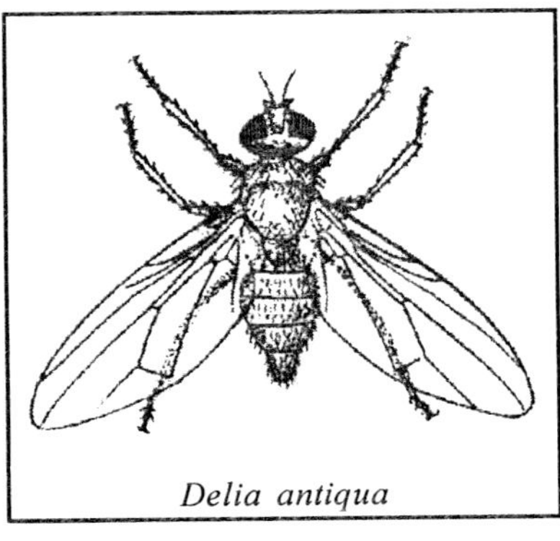

Delia antiqua

Eggs are elongate in shape and white in colour. Maggots are also while and about 18 mm long when full grown. Adult flies are slender, about 6 mm in length and grayish in colour having large wings. Eggs hatch in 2 to 7 days while maggot and pupal durations last for 2 to 3 weeks each (Singh, 1970).

To avoid infestation of this fly, grow resistant varieties of

15

TARO (ARUM)

ARUM was a common name given to various cultivated and some wild species of *Alocasia, Arum* and *Colocasia* (Araceae), which have now been scientifically separated out in different genera and species. *Arum jacquemontii* Blume is the only species of arum found in India (Kashmir valley) which grows mostly wild. Its corms are edible but only after repeated boiling. *Alocasia indica* (Roxburgh) and *A. macrorrhiza* Schott (giant taro) are tall aroids with an under-ground rhizome bearing a succulent swollen stem. These are cultivated in Assam, Bengal and Maharashtra and used as food that is easily digested.

Colocasia esculenta (Linnaeus) (*Antiquorum* Schott), arum or taro, is also known as elephant-ear because of characteristic resemblance of its leaves with elephant's ears. It is native of South-east Asia and grows wild on the banks of streams, ponds and marshes and is cultivated mostly in shady places. This is a perennial tuberous plant having large heart-shaped leaves borne on half to 1½ metres long petioles that arise from a burioh of underground farinaceous corms. The main edible part is the tuber, though leaves and young stalks can also be cooked and eaten in the same way as spinach. Low in protein and fat, taro is essentially a carbohydrate food. Nutritionally, it

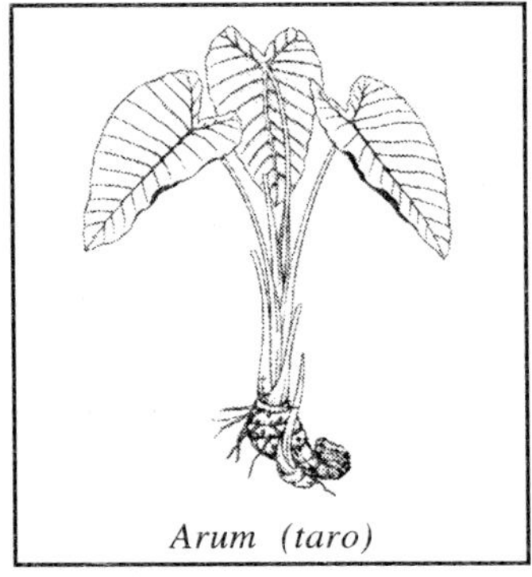

Arum (taro)

compares favourably with other root crops like sweet potato, tapioca and yam. It is a good source of vitamins A and B as also calcium and phosphorus. The tubers are also used for production of commercial alcohol. Their juice contains amylase. Starch extracted from these tubers is used for sizing textiles, where penetration and not coating property is the main consideration. Flour of raw or pre-cooked tubers is used in soups and gruels as also in making biscuits, cookies, bread and pan cakes. It is an excellent component for gravies and puddings as it is not glutinous like wheat flour. The juice of petioles can be used as an astringent.

INSECT PESTS

Like any other cultivated and wild plants, taro also suffers quantitative and qualitative losses due to ravages of insect pests. Fortunately, the tubers are not subjected to any direct attack by insect pests. Among the leaf feeders, flea beetle is of regular occurrence causing substantial loss. Leaf eating caterpillars are of sporadic occurrence, some species are causing severe loss but occasionally, while small grasshopper, aphids, thrips and lace-wing bug are the pests of minor importance.

Flea Beetle : *Monolepta signata* Olivier (Chrysomelidae), the white spotted flea beetle, is the most destructive pest of taro. This is a polyphagous pest having a wide range of host plants including beet-root, cabbage, cauliflower, chilli and radish. The pest is more active in South India than in the North. The beetles bite holes in leaves, and this damage to leaf lamina affects the development of tubers, resulting in low yield. In young plants, severe damage may result in its complete destruction. The adult beetle is 3 to 4 mm long having reddish-brown body and pale brown elytra with two big, white spots on each elytron. To control flea beetles, dust 5 to 10% HCH. but this insecticide should be applied on older root crops and tubers, as it is

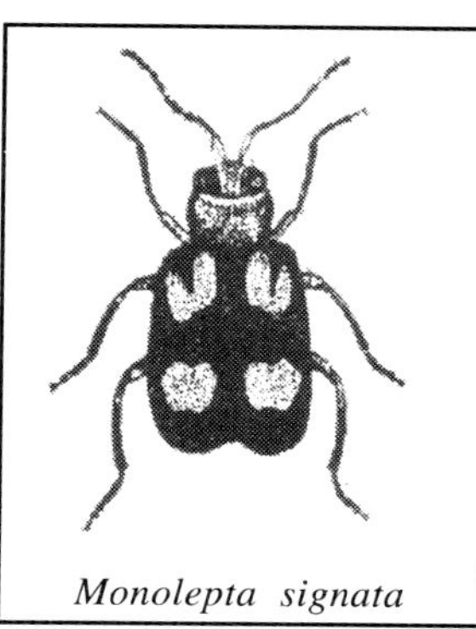

Monolepta signata

likely to impart off-flavour to these consumable plant parts. In such case, spray 0.05% endosulfan or 0.2% carbaryl or dust 4% endosulfan or 5% carbaryl.

Leaf-eating Caterpillars : *Pericallia ricini* Fabricius, *Agrius convolvuli* (Linnaeus), *Theretra gnoma* (Fabricius), *T. pinasfrina* (Marytn), *T. oldenlandiae* (Fabricius), *Panacra mydon* Walker. *Rhyncholaba acteus* (Cramer) and *Spodoptera littoralis* (Fabricius) have been reported feeding on leaves of taro and other *Colocasia* spp. These are polyphagous pests reported as serious pests on a number of economic crops but damage on taro and other *Colocasia* spp. is usually not serious.

Pericallia ricini (Arctiidae), also known as wooly bear or black hairy caterpillar, is a sporadic pest found all over the Indian sub-continent; banana is its preferred host. The caterpillars are nocturnal in habit and feed voraciously at night. Among the vegetables, this pest has also been recoded damaging drumstick trees.

Pericallia ricini

Agrius convolvuli (Siphingidae) is a polyphagous pest and besides taro, it has also been recorded as a pest of sweet potato and several leguminous crops. Caterpillars are voracious feeders and can defoliate the plants in a short time. Moths are crepuscular in habit and feed on nectar of flowers especially those having tubular calyx. Pupation takes place in soil at the depth of about 80 to 100 mm.

Theretragnoma (Fabricius), *T. oldenlandiae* (Fabricius) and *T. pinastrina* (Martyn) (Siphingidae) are minor pests of taro. The caterpillars are voracious feeders: *T. oldenlandiae* is comparatively more common, and it also attacks yams, sweet potato, elephant-foot etc. When these caterpillars appear in large numbers, the infested plants are completely defoliated affecting adversely the yield of the crop.

Hand-picking of the conspicuously coloured caterpillars and their mechanical destruction is the best method to combat these pests in early stage of attack. Spraying with 0.2% carbaryl or 0.05% endosulfan or 0.1% malathion can effectively check damage by these pests.

Grasshopper : *Gesonula punctifrons* (Stal) (Acrididae) is a minor pest of *Colocasia* sp. It is widely distributed in the Indian sub-continent and has also been reported from South China, Taiwan, Ryukyu Islands, Philippines, New Guinea and Australia. In India, it has been recorded from Tamil Nadu, Kerala, Andhra Pradesh, Karnataka, Orissa and Assam (Nayar *et al.*, 1976). The preferred host is water hyacinth *Eichhornis crossipes* – a noxious weed, due to which this grasshopper is often considered as beneficial insect rather than a pest. The females bore into the succulent stem with the help of sharp ovipositor and make 20 to 30 mm deep tunnel where it lays eggs in clusters of 8 to 12 eggs. The eggs are glued together and embedded in a frothy substance. Some gummy substance exudes from the entry holes which makes it easy to spot the infested plants. Nymphs and adults feed on leaves. Eggs are creamy-white in colour, elongated and slightly curved in shape and 2 to 3 mm long. Freshly hatched nymphs are 4 to 5 mm long and dark brown in colour; with each moult, the size increase is accompanied by fading of colour; the full-grown nymphs are 18 to 22 mm long and pale brown in colour while adults arc dusty brown, measuring 17 to 18 and 19 to 22 mm in case of males and females respectively (Butani and Verma, 198la). Eggs hatch in about 9 days; nymphal development takes 22 days and the adult longevity is one month. Nayyar *et al.*, (1976) reported egg period on water hyacinth to be 3 to 4 weeks and nymphal period 28 to 38 days.

These grasshoppers seldom attain the status of a major pest to warrant any control operations but if and when necessary dusting may be done with 2% phosalone or 4% endosulfan.

Aphids: small, delicate, soft-bodied, pear-shaped insects measuring about 1 to 2 mm in length. Most of the species are polyphagous having a wide range of host plants. Nymphs and adults suck the cell sap from tender shoots and leaves, secrete honeydew and some oven transmit vira! diseases. As a result, of desapping (he affected parts turn yellow, curl, become deformed and ultimately die away. The species reported damaging taro in India are *Pentalonia nigronervosa* Coquillett, *P. galadii* Distant and *Aphis gossypii* Glover; the first one is more common. Though these aphids are reported to be vectors, transmitting certain viral diseases, fortunately no such damage has been reported in the case of taro.

Pentalonia nigronervosa, or banana aphid - as the name suggests is a major pest of banana. In addition, it also attacks taro, *Alocasia sp., Colocasia sp.,* cardamom (large and small), tomato etc. This aphid was first recorded and described from Madagascar (Coquerell, 1959) and by now it has spread to Central and South America, West Indies. Hawaii, Pacific Islands, New Guinea, Australia, South-east Asia, South China, Indian sub-continent and some African countries (CIE map No. A-242). In India, it is more prevalent in Kerala. Tamil Nadu, Karnataka, Maharashtra, Bengal and Delhi. Warm and moist weather is quite favourable for rapid multiplication of this pest and under these conditions large colonies of this aphid, comprising of alate and apterous viviparous

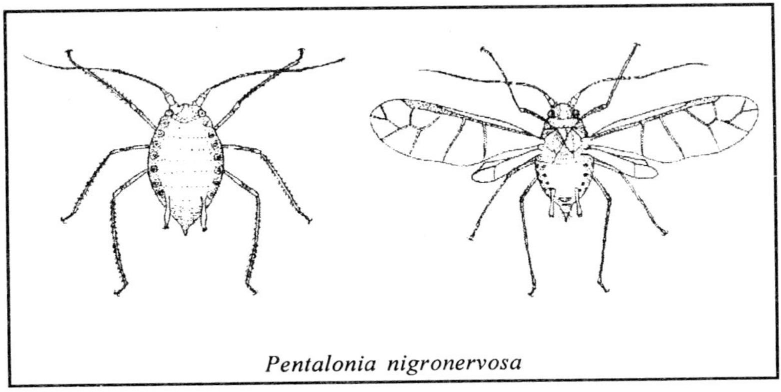

Pentalonia nigronervosa

females and nymphs, may be seen congregated on ventral surface of leaves, apterous females being predominant.

Males are rare. Thus these aphids mostly reproduce parthenogenetically. A single female produces 32 to 59 young ones in her life time @ 1 to 3 per day. Nymphal development takes 8 to 9 days and adults live for 23 to 27 days (Nayar *et al.*, 1976).

As a vector of virus diseases this aphid can transmit bunchy top disease of banana, katte disease of small cardamom and foorkey disease of large cardamom.

Aphis gossypii --a cotton or melon aphid, it is cosmopolitan in distribution. Though cotton and melons are its preferred hosts, it has also been recorded feeding on apple, bean, eggplant, chilli, *Citrus*, cole crops, cucurbits, guava, grapevine, okra, papaya, potato, tomato, many pulses etc. It is active throughout the year, especially during cloudy days when humidity is high and rainfall very low. The winged and wingless forms reproduce parthenogenetically and viviparously. Sexual reproduction also takes place mostly during cold season and the aphid then overwinters in egg stage.

To control these aphids, spray 40% nicotine sulphate (1: 600) or 0.5% dimethoate or methamidophos. One to two sprayings arc sufficient to check the pest effetively.

Lace-wing Bug : *Stephanitis typicus* Distant [Tingidae) is a polyphagous pest reported damaging banana, cardamom, coconut, *Colocasia* spp., turmeric etc. It is more common in South India than in the North. Besides the Indian subcontinent, the pest has also been recorded from South China, Philippines and Taiwan. Nymphs and adults are found in large number on ventral surface of leaves sucking the sap therefrom. As a result grayish-yellow spots appear on the leaves, and in case of severe infestation, plant's growth is stunted and ultimately, yield is adversely affected. Both nymphs and adults are small, dull coloured bugs: adults have transparent lace-like wings that make these bugs very conspicuous, Eggs are laid singly and are inserted in leaf tissues. A female lays on an average 30 eggs. The eggs hatch in about 12 days and nymphal development takes on an average 13 days (Nair, 1975). In China, this bug is reproted to have 6 generations in a year (Cheng, 1967).

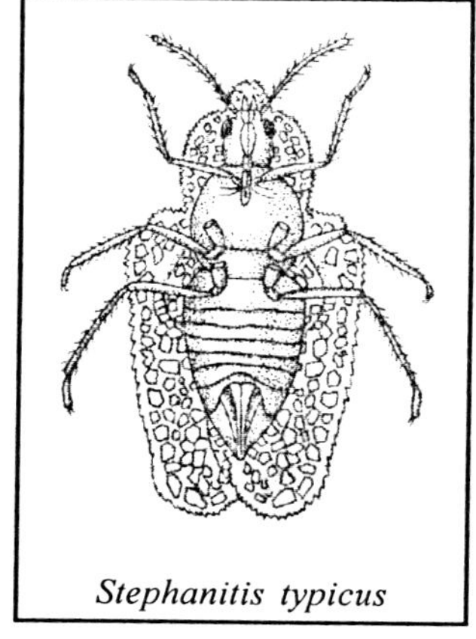

Stephanitis typicus

Normally no control measures are adopted against this bug on taro. Dusting recommended against grasshopper or spraying suggested for aphids can also control this bug.

Thrips : These are comparatively minute, slender and fragile insects, measuring 0.5 to 1.5 mm in length. Adults have long, narrow and heavily fringed wings. Females are larger in size and more abundant than males. Mostly phytophagous, though a few, are also predaceous. Nymphs and adults infest succulent parts of the plants, lacerate green tissues and lap the oozing sap. The affected parts are devitalised and often show white silvery sheens; ultimately these start drying. A severe infestation on leaves results in curling of leaf tips and the leaves gradually die away. Reproduction is sexual as well as parthenogenetic, the latter being more common as the males are rare. The species commonly found feeding on taro in India are, groundnut thrips. *Caliothrips indicus* (Bagnall), banana leaf thrips, *Helionothrips kadaliphilus* (Ramakrishna and Margabandhu), greenhouse thrips. *Heliothrips haemorrhoidalis* (Bouche) and cardamom thrips *Scirtothrips cardamomi* (Ramakrishna).

Caliothrips indicus: groundnut thrips is a polyphagous pest. Ananthakrishan (1971) has given a long list of its host plants which includes eggplant, chilli, cabbage, cauliflower, knol khol, onion and potato. The insects are comparatively small in size, being less than one mm in length; adults have comparatively broader forewings. A life cycle is completed in about 2 weeks.

Helionothrips kadaliphilus - banana leaf thrips, is a minor pest of taro. This species was first recorded by Ayyar and Margabandhu (1931) from South India. The pest is more common in Kerala and Karnataka. Maximum activity has been recorded during June to September. Both nymphs and adults are yellowish-brown in colour; adults being slightly darker with yellowish-gray wings. Its egg and nymphal durations last for 4 to 6 and 9 to 12 days, respectively (Murthy, 1958).

Heliothrips haemorrhoidalis: Greenhouse thrips, is cosmopolitan in distribution (C1E map No. A- 135). It has a very wide range of host plants, including taro and various ornamental plants. Outside India, it has also been recorded on mango leaves (Ananthakrishnan, 1971). Besides causing the usual damage, the nymphs also deposit faecal globules all over leaf lamina on which fungus develops causing brownish patches (Butani, I979b). Females make incisions in leaf tissues and insert eggs singly or in pairs. Eggs are bean-shaped and about 0.3 mm long. Freshly emerged nymphs are whitish in colour which gradually changes to pale green and then to greenish-brown. Adults arc uniformly dark brown with black head and thorax; antennae and legs are whitish, the latter having yellowish tinge. Wings are yellowish with a median longitudinal pale grey band, paler at the base; forewings are pointed at the apex.

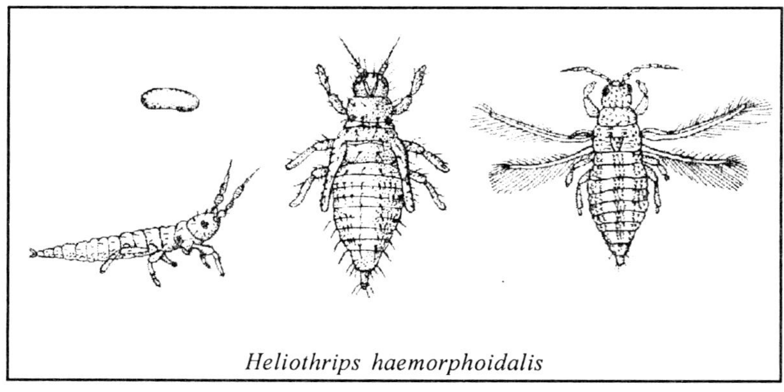

Heliothrips haemorphoidalis

Eggs hatch in 2 to 7 days and nymphal duration varies between one and two months depending upon the climatic conditions. Under controlled temperature condition a life cycle takes 30 to 32. 40 and 60 days at 26 to 28°, 23 to 25° and 20°C, respectively. Hill (1983) has reported that, the life cycle of this thrips was completed in 8 and 12 weeks at 19 and 15°C, respectively.

Scirtothrips cardamomi as the name suggests is a major pest of cardamom, especially in South India and a minor pest of taro. It is most active during May-June and heavy mortality occurs during Rainy season which brings down the pest population considerably (Butani and Verma, 1981). Nymphs are grayish-brown in colour and adults dark grayish-brown with pale yellow antennae and legs; upper portion of the wings is gray and basal more or less transparent. Eggs are laid singly and at random all over the plant parts. A single female lays 5 to 30 eggs during her life time.

These eggs hatch in 9 to 12 days and nymphal period occupies 18 to 21 days.

Any chemical control measures adopted against aphids will control thrips as well. In the absence of aphid infestation. thrips may be controlled by spraying 0.03% dimethoate, endosulfan, phosalone or phosphamidon.

16

ELEPHANT - FOOT

ELEPHANT-FOOT, *Amorphophallus companutatus* Blume (= *Arum campanulatum* Roxbergh) (Araceae) is native to Tropical Asia and is found throughout the plains of the Indian sub-continent as also some parts of South-east Asia and Africa, both growing wild as well as cultivated. It is a stout stemless, herbaceous plant having a big underground corm (200 to 400 mm in diameter), dull dark brown in colour and depressed hemispherical in shape. The corm bears a large solitary mottled leaf on a long petiole. The corms are either cooked and taken as vegetable or pickled for future use. Tender petioles are also edible. Corms are used in certain *Ayurvadic* preparations meant for the cure of piles and dysentery. In Philippines pigs and hogs are fed on boiled corms.

An allied species, *Amorphophallus commutatas* Englar (=*A. sylvaticus* Dalzell and Gibs.) is found growing wild in humid regions of South-western and South-eastern India. The underground corms after washing and prolonged boiling are eaten by the poor.

INSECT PESTS

No insect pest has been reported so far damaging the underground corm, but half-a-dozen species have been recorded as foliage feeders. The most destructive ones are, leaf-eating beetle and a hairy caterpillar while tobacco caterpillar and some sphingid caterpillars occasionally cause minor damage.

Leaf-eating Beetle : *Galerucida bicolor* (Hope) (Gallerucidae) – the yam beetle, is a pest of regular occurrence in South India.

Besides yam, it also causes considerable damage to elephant-toot. The grubs feed gregariously on leaves and after skeletonising the leaves they gnaw in to the petioles. Adults feed by nibbling the leaves and cause much less damage than the grubs.

Spraying 0.05% endosulfan or 0.2% carbaryl is recommended for the control of this beetle.

Leaf-eating Caterpillars : *Pericallia ricini* Fabricius *(Arctiidae)* – a hairy caterpillar is a polyphagous pest which in addition to elephant-foot has also been reported damaging eggplant, *Colocasia*, drumsticks, sweet, potato and yams. Tobacco caterpillar, *Spodoptera littoralis* Fabricius (Noctuidae), is yet another polyphagous pest of this crop that has been recorded on a large number of other crops of economic importance, including amaranthus, eggplant, chilli, cole crops, peas, sweet potato, tomato and yams. Both these caterpillars occur sporadically and can cause considerable damage to leaves. Besides these two pests, caterpillars of several species of sphingid moths including *Rhyncholoba acteus* (Cramer), *Hippotion celerio* (Linnaeus) and *Theretra gnoma* (Fabricius) have been found feeding on leaves of elephant-foot but the damage caused is seldom severe.

Eggs of *Rhyncholoba acteus* are smooth, shining, broadly ovoid in shape, 1.5 to 2.O mm long and bright green in colour. Full-grown caterpillars are stout, 68 to 76 mm long and have two different colour patterns - green and brown. The green form has rich grass-green coloured head; body segments 2 to 5 are pale bluish-green and rest of the body is rich pale grass-green above, pale bluish-green below with dark green longitudinal stripes. In the brown form the green is replaced by rich brown or ochraous-chocolate colour and the stripes are pale pinkish-brown. Horn is prominent and 6 to 8 mm long stout at the base, orange in colour and covered with minute tubercles. Pupae are 46 to 54 mm long having head and thoraxic region brown dorsally and pinkish ventrally; cremaster is elongate wedge-shaped and dull orange in colour. Adults have purplish-gray head, thorax and abdomen with a dorsolateral green stripe from vertex of head to anal end. Forewings are purplish-gray with a green oblique central area from just below the apex to inner margin. Hind wings are fuscous in colour with an anal patch and submarginal hand ochreous. Wing expanse is 64 to 76 and 70 to 80 mm in case of males and females, respectively (Bell and Scott, 1937).

Hippotion celerio resembles *Rhyncholoba acteus* but is slighty smaller in size. Full-grown caterpillars are 56 to 64 mm long with a small, 5 mm long, straight dark brown or blackish horn. In this species also, there are two larval forms - green and brown, the former is slightly bigger than the latter. Pupae are about 50 mm long and pale russet-brown in colour; legs, antennas and wing-case have golden lustre; eremaster is triagular in shape ending in minutely bifid needle-like shaft. Adults have brown head and thorax with white lateral stripe, abdomen is brown with a white darsolateral spot on each segment. Forewings arc pale brown and have silvery band from apex to inner margin with a median dark line all along its length and some ochreous and pale brown lines behind it. Hind wings have base and anal angle bright pink and outer area ochreous-brown with a black submarginal band. Wing expanse is 60 to 70 mm in males and 70 to 80 mm in females.

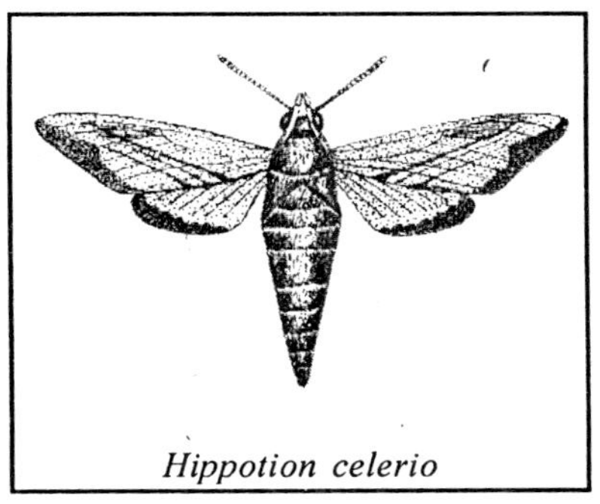

Hippotion celerio

Eggs of *Theretra gnoma* are broadly-ovoid in shape, about 2 mm long and shining pale green in colour. Full-grown caterpillars are about 80 to 85 mm long, having glaucous-green head and yellowish-green body speckled with dark green and a narrow broken dark green dorsal stripe: horn is smooth, stout and about 8 mm long. Pupae are 58 to 62 mm long and soil-brown in colour; cremaster is triangular, broad at the base with its basal half yellowish in colour. Adults have greenish-brown head and thorax with a white lateral stripe; abdomen is brown with a black dorsal patch. Forewings are brown with one discal line almost parallel to outer margin. Hind wings are black shading towards brown near the apex. Wing expanse is 80 to 96 and 84 to 109 mm in case of males and females respectively.

Generally, no chemical control measures are adopted against these caterpillars. Hand-picking and mechanical destruction of the caterpillars in the initial stage of attack can effectively check the pest population. Nevertheless, when necessary, spraying with 0.02% carbaryl or 0.05% endosulfan can successfully control the population of these caterpillars.

17

SWEET POTATO

SWEET POTATO, *Ipomoea batatas* (Linnaeus) (= *Batatas edulis* Choisy) (Convolvulaceae), a native of Latin America, is now grown all over the Tropics and subtropics including Japan, South China, South-east Asia, Africa and Tropical America extending from South USA to Brazil. In Japan, it ranks next to rice as staple food (Nuttonson, 1951). Sweet potato is a herbaceous creeping vine and though perennial, it is cultivated as annual. It produces fleshy roots which are commonly called tubers; these modified roots form the main edible portion and are also used for manufacture of starch and alcohol. The tubers contain 16% starch and 4% sugar which means 20% alcohol manufacturing material (Choudhury, 1967), Leaves of sweet potato are also highly nutritious and are used as leafy vegetables in some countries. In India the crop is grown in about 200,000 hectares and mainly serves as staple food for the poor, and is also eaten as delicacy after baking or making stew in syrup.

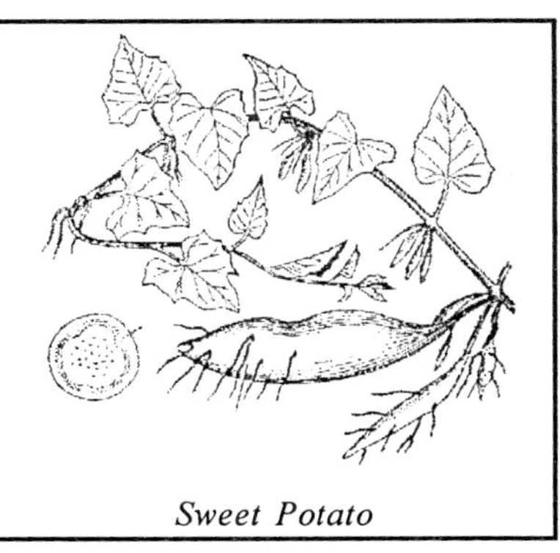

Sweet Potato

INSECT PESTS

The crop is attacked by a number of insect pests both in field as well as in godowns. The field pests reduce the production per unit area and also affect the quality of the produce. The major pests include, sweet potato weevil, sphinx caterpillar, Bihar hairy caterpillar and tortoise beetle while those of minor importance are, sap sucking bugs, vine-borer, leaf eating caterpillars, leaf-miners, melolonthid beetles, hispid beetles and root mealybugs.

Sweet potato Weevil : *Cylas formicarius* (Fabricius) (Curculionidae) is specific pest of sweet potato, extensively Pantropical in distribution (CIE map No. A-278). Grubs bore into stems, cause tunnelling inside and feed on soft tissues. Grubs as well as adults bore into the tubers both in field and in godowns. The affected tubers develop dark patches which may later start rotting; occasionally, adults feed on stems and leaves. Loss of tubers to the extent of 60 to 70 per cent has been reported by Nair (1975) and Nayar *et al.*, (1976). In Philippines, 20 to 50 per cent infestation has been reported at low elevation which occasionally results in total crop failure (Knott and Deanon, 1967). The pest is disseminated from field to field through infested vines and is carried over from season to season by breeding in damaged tubers left in the fields after harvest (Trehan and Bagal, 1957).

Grubs are fattish, about 8 to 10 mm long, legless, pale yellowish-white in colour. Adult weevils are ant-like, due to which they have been given specific name *formicarius* and are 6 to 8 mm long, slender bodied having elongated snout-like bluish-brown head with non-geniculate antennae, bright red thorax and legs and brownish-red abdomen. Adults are active and fly long distances during night in search of food. Females make small cavities on the tubers or sterns and lay eggs singly in these cavities. Each female lays 100 to 200 eggs in 50 to 100 days.

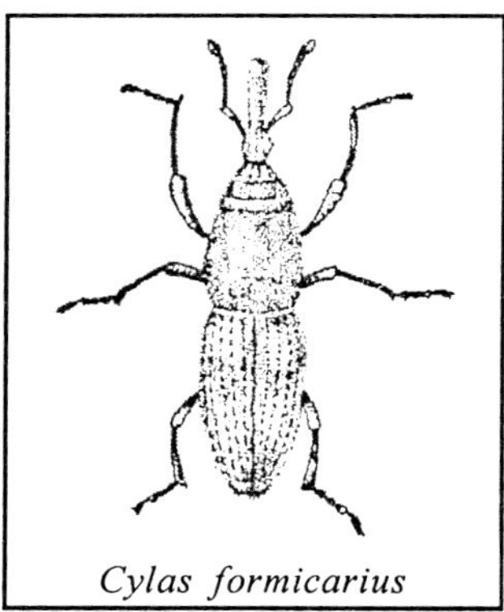

Cylas formicarius

Incubation, grub an pupal stages last for 5 to 10, 16 to 20 and 4 to 8 days respectively (Butani and Verma, 1976 b). Pupation takes place in larval burrows from where the freshly emerged weevils cut their way out. Life cycle is completed in. 4 to 5 weeks while male adults may live from 10 to 15 weeks and females from 13 to 16 weeks. The life history of the pest has also been studied in detail in Japan (Murakami, 1933), Taiwan (Fukuda, 1933) and Java (Franssen, 1934).

Other weevils found feeding on sweet potato leaves include, *Blosyrus asellus* Marshall, *Myllocerus sabulosus* Marshall and *Phytoscaphus triangularis* Olivier, whereas *Protocylas coimbatorensis* Subramaniam feeds on tubers. Fortunately. *Manophyes* species - a major pest in Bangaldesh (Alam *el at.*, 1964) - has not so far been reported from India.

In addition to these weevils, striped sweet potato weevil *Alcidodes fabricii* Fabricius (Curcullonidae) has been reported breeding inside the sweet potato (COPR, 1978). Eggs are laid at the base of tubers. On hatching,, the grubs bore into the tubers and feed on the seeds. Pupation takes place in the soil. Adult weevils are 12 to 15 mm long and reddish- brown in colour with four longitudinal stripes on each elytron. For preventing population build-up in the next crop season, collect all rubbish and trash and burn the same immediately after the harvest; care should be taken to see that no infested or rotting tuber is left in the field. Do not use the seed material from infested areas and plant only healthy cuttings. Spraying 0.2% carbaryl followed by 0.1 to 0.15% malathion ® 1000 litres per hectare at 10 days' interval is quite effective in controlling these weevils (Singh, 1970). In godowns, treat the outside of the bags containing the tubers with 5% malathion or carbaryl dust.

Leaf-eating Caterpillars : Sphinx caterpillars, *Agrius convolvuli* (Linnaeus) and *Theretra. oldenlandiae* Fabricius have been reported damaging sweet potato vines. The former, commonly called homed caterpillar or giant hawk moth, is widely distributed in the Old World and has been reported from Europe, Africa, Iran, Indian sub-continent. South-east Asia, South China, Australia and New Zealand. This is a polyphagous pest that attacks a number of crops, including fruit trees, some legumes, arum and sweet potato. The pest is active during monsoon season when moths can be seen hovering around light at night,. The females lay conspicuous seed-like shiny eggs, singly on the tender parts of plants. On hatching, the caterpillars feed voraciously on leaves often defoliating the vines completely. Another allied species *A. cingulata* Fabricius has been reported causing similar damage in Florida (Watson, 1944) and Jamaica (Edwards, 1937).

Eggs of *Agrius convolvuli* are subspherical in shape and about one mm in diameter. Full-grown caterpillars are robust, about 80 to 100 mm long, dark brown in colour with reddish patches on sides and a conspicuous curved hornlike process at the anal end. Pupae are reddish-brown in colour. Moths are stout, pale gray in colour having a well developed proboscis and frenulum and pale gray wings with transverse violet bands on abdomen. Wing expanse is 80 to 100 mm. Incubation, caterpillar and pupal stages last for 5 to 10, 14 to 21 and 7 to 1 1 days respectively. Pupation takes place in hard earthen cells in the soil and the pest hibernates in pupal stages. A complete life cycle occupies 4 to 5 weeks (Butani and Verma, 1976b).

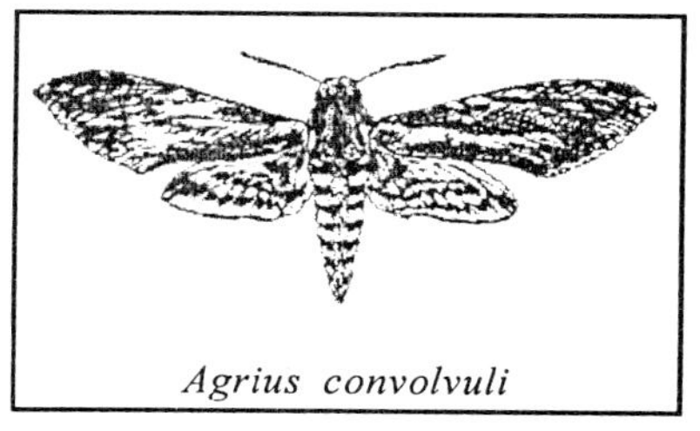
Agrius convolvuli

Spilarctia obliqua (Walker) (Arcticise) - the Bihar hairy caterpillar, is a serious pest of jute in Eastern India and Bangladesh. It is a polyphagous pest having a wide range of host plants. It has been reported damaging a number of fruit trees, tobacco, pulses and vegetables, including cabbage, cauliflower, cowpea, lettuce, potato, soybean and sweet potato. Lefroy and Howlett (1909) have called the caterpillars of this pest as omnivorous. According to Dethier (1970) the differential infestation involves host selection and is mainly due to the response of newly hatched caterpillars to specific odours of host plants.

Eggs are laid in clusters on ventral leaf surface in parallel rows on 3 to 4 consecutive nights. A single female may lay 400 to 1000 eggs (maximum 1600) in its life time. On hatching, the caterpillars feed gregariously on leaves, skeletonising the same. Later, the caterpillars segregate and feed voraciously on leaf lamina; often, the vines may be completely defoliated. The caterpillars move from plant to plant and from field to field; older leaves of older plants are preferred. Pupation takes place in the soil.

Full-grown caterpillars are stout, 25 to 40 mm long and have 7 broad orange coloured transverse bands with tufts of yellow hair, dark at both ends. Moths have crimson coloured body with black dots, antennae are black in colour, pectinate in males and filiform in females. Wings are pinkish-buff coloured with numerous black spots and wing expanse 40 to 50 mm. Incubation, caterpillar, prepupal and pupal stages last for 4 to 9, 28 to 43, 2 and 10 to 13 days respectively (Trehan and Bagal, 1957). The life cycle is completed in 5 weeks in Summer but is extended upto 8 weeks during Autumn and 10 weeks in Winter. There may be as many as 8 generations in a year; the caterpillars of Winter-brood enter into soil to pupate and hibernate.

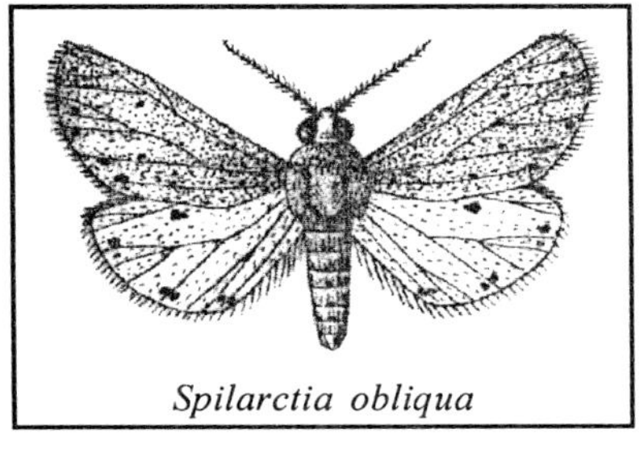

Spilarctia obliqua

Junonia orithya (Butler) (Nymphalidae)- blue pansy, is a beneficial insect as it normally feeds on *Striga euphrasioides* - a weed parasite, common on sugarcane roots. It is also a minor pest of sweet potato. Eggs are laid on leaves and the caterpillars feed voraciously on these leaves. Eggs are small, acorn-shaped and light green in colour with 15 longitudinal flanges or keels. Full-grown caterpillars are about 25 mm long, cylindrical in shape, sooty-black in colour with sharp and complex spines and bristles borne on 9 longitudinal rows of well-developed pointed scoli (Butani, 1958 a). Pupae are obtect and brown to dark brown in colour, suspended by cremaster. Moths are medium-sized with only two pairs of functional legs (forelegs being rudimentary). More than half of the forewings are velvety black, apical portion being dull fuliginous; hind wings are blue, shaped with velvety black towards the base - thus the beautiful wing pattern resembles the pansy flower from which the insect has acquired its common name. Wing expanse is 40 to 50 and 50 to 60 mm in case of males and females respectively. Incubation, caterpillar, pupal periods, adult longevity and entire life-cycle duration last for 3, 14 to 16, 4 to 8, 3 to 7 and 27 to 29 days respectively (Agarwala and Naqvi, 1954).

*Euchromia polymena (*Linnaeus) (Amatidae), is another minor pest. The moths are diurnal in habit; the females lay eggs on leaves during day. The caterpillars feed on leaves and may sometimes defoliate the entire plant. Caterpillars are *Euchromia polymena* reddish in colour with red tubercles; long anterior and posterior tufts of hair and shorter dense medial dorsal tufts. Pupation takes place in hairy cocoons. Adults are metallic blue coloured moths having transparent wings that are devoid of scales and have big metallic blue and orange spots on forewings and only orange spots on hind wings. Wing expanse is 40 to 45 and 50 to 55 mm in case of males and females, respectively.

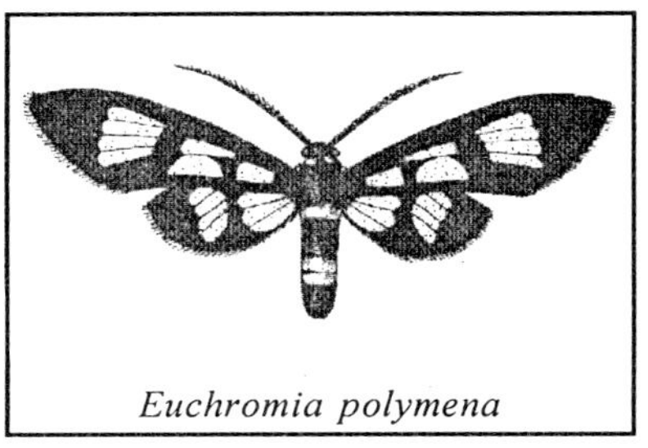

Euchromia polymena

Creatonotus gangis Linnaeus (Arctiidae), is a polyphagous pest which has also been observed damaging sweet potato; lucerne is its preferred host. It appears in July and its activity continues till November. Eggs are round in shape, 0.75 mm in diameter and shining creamy-yellow in colour. The caterpillars are cylindrical, slightly tapering posteriorly, 38 to 44 mm in length when full-grown and dark violet to black in colour. Meso- and metathorax are light golden-yellow and head black, hairy with characteristic yellow stripe dorsally. Pupae are obtect type 13 to 17 mm long and reddish-brown to dark red in colour. Adults have shiny black head, ochreous thorax with a black spindle-shaped mid-longitudinal band. Forewings are straw coloured with pinkish tinge and a characteristic set of black fascia medio-longitudinally. Hind wings are whitish with a few black dots at the margin. Wing span is 36 to 40 and 40 to 45 mm in case of males and females respectively. Pairing starts within 3 hours of the emergence of moths; pre-oviposition, oviposition and post-oviposition periods arc 10 to 24 hours, 2 to 3 days and 1 to 3 days respectively. A female lays 285 to 695 eggs in clusters of 160 to 317 eggs each, arranged in rows closely set together. Incubation period is 4 to 5 days, caterpillar 22 to 32 days and pupal 5 to 7 days. Adults are nocturnal in habit and are readily attracted to light in large numbers (Baser and Kushwaha, 1968).

Hyposidra successaria Walker, *Estigmene lactinea* (Cramer), *Catephia inquieta* Walker and *Spodoptera littoralis* (Fabricius) have also been occasionally found feeding on sweet potato leaves (Nair, 1975). Of these *H. successaria* is comparatively common. Its egg, caterpillar and pupal periods are 4 to 5, 18 to 29 and 8 to 9 days, respectively (Venugopal, 1958). Caterpillars of *E. lactinea*, commonly called black hairy caterpillars are black-in colour and thickly haired. These are voracious feeders and move rapidly from plant to plant. and field to field. Adults are large white moths with crimson markings on head, body and wings.

As the reproductive potential of these pests is rather high, it is important to take immediate steps to cheek their initial infestation and spread. After harvesting the crop, give deep ploughings and if possible flood the infested fields to kill the pupae present in the soil and thereby prevent the carry-over of the pest. Collect and destroy promptly the egg-clusters and leaves bearing the caterpillars to prevent the population build-up. Dusting 10% HCH @ 20 to 22 kg per hectare is quite effective especially against young caterpillars, and may be used to check the gregariously feeding caterpillars during the early stage of crop's growth. After formation of tubers, application of HCH may result in imparting off-flavour to them and therefore, its application should be avoided at this stage. Spraying 0.05% dichlorvos (DDVP) or endosulfan can effectively control the widespread infestation of caterpillars at advanced stage (Butani *et al.*, 1977).

In nature, *Apantetes obliqua* Walker and *Eurytome* species have been recorded as parasites of *Spilarctia obliqua* caterpillars.

Leaf-feeding Beetles : There are quite a few species of tortoise beetles reported damaging sweet potato vines. These include, *Aspidomorpha miliaris* (Fabricius), *A. furcata* Thunberg, *A. indica, A sanctaecrucis* (.Fabricius), *Cassida indicola* Duvivier, *Chiridia sexnotata* (Fabricius), *Glyphocassis trilineata* (Hope), *Metriona circumdata* (Herbest) and *M. varians* (Herbest). All are specific pests of sweet potato albeit *A. miliaris*, and *M. circumdata* are relatively more common, whereas *A. furcata* and *A. sanctaecrucis* have been reported regularly from Kerala and Karnataka states, respectively. *A. miliaris* is comparatively more destructive. These beetles are active during monsoon. Eggs are laid on ventral leaf surface. Grubs are nocturnal in habit and feed on epidermal tissues of leaves, skeletonising them. Later, the grubs as well as the adults gnaw holes in the leaf lamina. Pupation takes place in ventral surface of leaves.

Eggs of *Aspidomorpha miliaris* are laid in 5 to 10 rows. Grubs are flattened with spiny processes covering their bodies; dried excreta is invariably seen on their anal process. Adults are broad oval shaped (12x9 mm) beetles, brownish -red in colour with black dots. Egg, grub and pupal stages last for 9 to 11, 15 to 20 and 4 to 6 days, respectively. A single life cycle is completed in 28 to 36 days.

Aspidomorpha furcata lays eggs singly in rectangular trays (1.25 x 1.0 mm) of white papery material made by the female. Eggs are ovoid and faintly brownish in colour. Grubs are greenish-white and about 5 mm long. Pupae are green and about 6 mm long. Adults on emergence are white in colour but become deep brown within 2 to 3 hours and shiny golden after 7 to 9 days. These are also broad-oval in shape but much smaller than *A. miliaris* (6.5' 5.0 mm). Incubation grub, prepupal and pupal durations last for 3 to 4, 8 to 9, 3 and 4 days respectively (Visalakhi *et al.*, 1980). Mating takes place 7 to 9 days after emergence, pre-oviposition period is 6 to 8 days.

Eggs of *Metriona circumdata* are laid singly adjacent to each other and fastened on the leaf surface by means of short, brown filaments secreted by the female, Grubs are pale greenish in colour while pupae are greenish-yellow. Adult beetles are similar in shape and size to *A. furcata.* but greenish-yellow in colour. Incubation, grub and pupal periods last for 3 to 5, 10 to 15 and 6 to 8 days respectively and the total life cycle is completed in about 30 days.

Oncocephala tuberculata Olivier (Hispidae) has been recorded as a minor pest from South India. Eggs are thrust inside the leaf lamina. On hatching, the grubs mine the leaves and pupate within. while adults feed on leaves randomly gnawing small holes. Affected leaves wither and dry up. Adults are small hispid beetles, brown in colour with blunt spine-like projections all over their bodies. Egg, grub and pupal stages last for 7 to 10, 17 to 23 and 8 to 15 days respectively. Total life cycle occupies 35 to 45 days (Cheiran and Menon, 1940).

Leucophalis coneophora Burmeister (Melolonthidae) - a major pest of coconut (Nirula *et al.*, 1952) especially in Kerala where it is also a minor pest of other intercrops like banana, *Colocasia*, sweet potato, tapioca, yams etc. Adults are chestnut coloured beetles with glistening pubescence. Eggs are laid during June-July in soil, 80 to 150 mm deep. Grubs and pupae remain in soil while adults emerge out with the onset of monsoon and feed on leaves. These remain active for at least two months. Incubation, grub, pre-pupal and pupal stages last for 20 days, 13 months, 9 to 12 and 25 days respectively (Nayar *et al.*, 1976).

Anomala dussumieri Blanchard (Melolonthidae) and *Gonocephalum* species (Tenebrionidae) have also been found feeding on leaves and stems; their grubs gnaw the new shoots under soil, causing dead-hearts and outright death of that plant. *Trachy ipmeae* Theobold (Buprestidae) has been reported as a leaf-miner from Coimbatore (Nair, 1975).

If and when the high infestation of these beetles occurs, spray the crop with 0.2% carbaryl @ 1000 litres per hectare (Butani and Varma, 1976b). In nature, *A. miliaris* is parasitised by *Cassidocida aspidomorphae* Crawford and *Tetrastichus colemani* Crawford.

Sap Sucking Bugs : *Exitianus indicus* Distant (Cicadelhiae) -the sweet potato hopper, is a minor pest of sweet potato. Both nymphs and adults suck sap from leaves and tender shoots but the damage caused is negligible except under heavy infestation. Adults are active, slender, white leafhoppers, 6 to 8 mrn long, with head, thorax and scutellum greenish in colour.

Riptortus linearis Fabricius (Coreidae) - fig bug, is widely distributed in India and has been recorded infesting and damaging the tender shoots of sweet potato. A female lays 3 to 8 eggs singly on leaves, which are hatched in 6 to 7 days; nymphal development is completed in 14 to 18 days. Adults are elongated, dark brown bugs, about 14 to 18 mm long.

Halticus minutus Reuter (Miridae) has also been reported causing some damage in Bihar. Adult bug about 1.5 mm long, shiny black in colour, is broadly sub-ovate in shape; head is short, broad and laminately flattened posteriorly and pronotum obsoletely strigose.

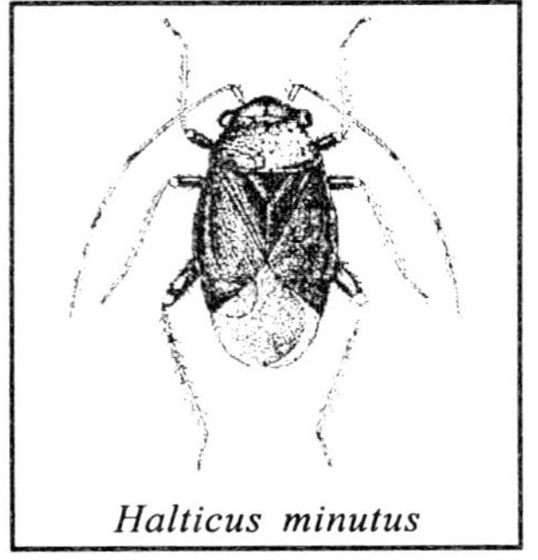
Halticus minutus

Graptosethus servus Fabricius (Lygaeidae) is another bug common in South India. These are grayish-black bugs, the adults being 8 to 10 mm long.

Nymphs and adults suck the cell sap from tender leaves devitalising the same. As these are all minor pests, no separate control measures are needed. Nevertheless, spraying 0.03% dimethoate, oxydemeton-methyl, phosphamidon or quinalphos is effective in checking these pests (Butani, 1979a).

Mealybug : *Geococcus coffeae* (Green) (Pseudococcidae). -coffee mealybug, was first described by Green (1933). It is Pantropical in distribution (CIE map No. A- 285). Though its main host is coffee *(Coffea* spp.*),* it has been reported from Coimbatore (Tamil Nadu) infesting tender roots and tubers of sweet potato (Muthukrishnan and Nagraja Rao, 1959). The damage is more pronounced when slightly infested tubers are stored. Under normal storage conditions the mealybug multiplies rapidly and the stored tubers get thickly coverd with these insects and become shrivelled due to loss of sap by sucking action of the pest. The problem is perpetuated by use of infested tubers as seed.

To avoid infestation by this mealybug, tubers from infested fields should not be used as seed; every tuber should be checked for being free from infestation. Dipping the tubers in 0.5% phenthoate solution, just before planting, can also help in checking the infestation and damage by this pest.

Leaf Rollers: *Brachmia arotraea* (Meyrick), *B. convolvuli* Walsingham, *B. effera* (Meyrick) and *B. engrapta* (Meyrick) (Gelechiidae), have been recorded damaging sweet potato vines in Coimbatore (Tamil Nadu), the last one and *B. macroscopa* Meyrick have been reported from Ludhiana (Punjab) and Hisar (Haryana) respectively. Fletcher

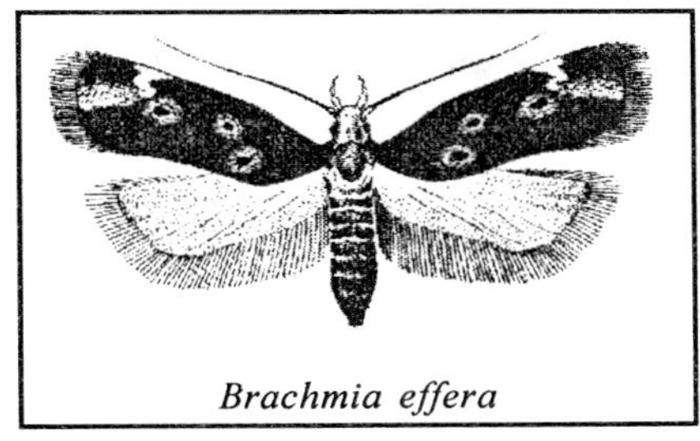
Brachmia effera

(1920) also reported *Helcystogramma lamprostoma* (Zeller) (Gelechidae) from Pusa (Bihar). Eggs are laid in clusters at the base of radiating veins on ventral surface of leaves. On hatching, the tiny caterpillars scrape the tender surface tissues of leaves and feed on them sheltered beneath the thin webbings. Later, the caterpillars fold single leaves longitudinally and feed on green tissues; the affected leaves subsequently dry up. Pupation takes place within the rolled leaves. These insects are, however, of minor importance.

Brachmia arotraea, originally reported from Myanmar (Meyrick, 1894), is now widely distributed in the Indian subcontinent. The caterpillars, when full-grown are 8 to 11 mm long, cylindrical in shape, tapering posteriorly; head is shiny black and abdomen has one dorsal light green stripe and several green broken lateral stripes - thus the caterpillars look greenish in colour. Pupae are also cylindrical in shape, tapering posteriorly, about 5 mm in length and brownish in colour. Adults have whitish-ochreous head and thorax with three fuscous stripes; forewings are long, and narrow with pointed apex; hind wings are pale gray. Wing expanse is 11 to 14mm (Ghai *et al.*, 1979).

Brachmia convolvuli is recorded as a pest of several *Ipomoea* species (Beeson, 1941). Its full-grown caterpillars are about 10 to 15 mm long, slightly flattened and tapering towards both the ends. Head is red dish-brown, glossy and flattened; mesothorax, metathorax and first two segments of abdomen are velvety black, other segments are yellowish-white with a velvety black band. Pupal case is clothed with scattered erect hair - a curious and unique feature, Adults are small-sized moths having light ochreous-brown head, purplish-fuscous thorax, grey abdomen and outstretched wings: forewings are elongate, apex obtuse, purplish- fuscous and hind wings are fuscous-whitish in males and gray in females; wing expanse is 12 to 14 mm. Incubation, caterpillar and pupal stages last for on an average 3, 14 and 7 days, respectively.

Brachmia engrapta moths are more or less of same size and pattern as those of *B. convolvuli*, head is whitish-fuscous, thorax brown and abdomen gray; forewings are narrow, elongate and brownish while hind wings are light gray.

Brachmia macroscopa caterpillars are slightly bigger in size (14 to 18 mm long) than the other species. Its moths have fuscous body; forewings are elongated, rather narrow, suffused with dark fuscous and having a marginal series of distinct blackish dots; hind wings are grayish-white (Meyrick, 1932). Wing expanse is on an average 18 mm.

Helcystogramma lamprostoma was originally described by Zeller (1847) from Sicily, and subsequently it has been reported from Spain, West Africa, Asia Minor, the Indian subcontinent and South-east Asia. In India, it has been recorded on sweet potato causing damage by rolling its leaves. It has also been reported on *Ipomoea reptans* (water spinach), causing minor damage.

Spraying 0.1% carbaryl or 0.025% phosalone or 0.075% tetrachlorvinphos is effective in controlling the pest population.

Leaf Miner : *Acrocercops prosacta* Meyrick (Gracilaridae) is a minor pest reported from Bihar. Caterpillars mine leaf lamina and cause blotches. The attacked leaves curl and become deformed. Adults are tiny moths having off-whitish head and thorax, pale abdomen; forewings are elongate-lanceolate in shape, brownish-fuscous in colour with white shining markings; hind wings are gray. Wing spread measures 5 to 8 mm.

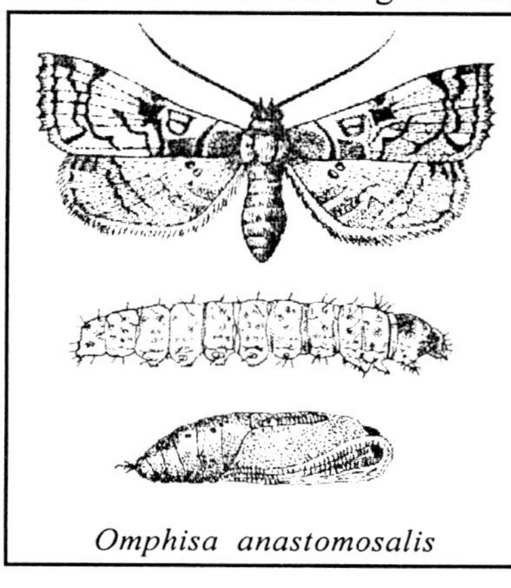
Omphisa anastomosalis

Vine Borer : *Omphisa anastomosalis* (Guenee) (Pyraustidae) is also a minor pest of sweet potato. The stout whitish caterpillars bore into stems, often killing the same outright. Pupation takes place within the tunnels made by caterpillars. Moths have head, thorax and abdomen suffused with ochreous and rufous. Both wings have

Omphisa anastomosalis rufous suffusion at basal area; forewings have hyaline patches at middle and end cell; hind wings have two irregularly waved postmedial lines. Wing expanse is 32 to 36 mm.

Tuber Borer : *Metosia coniotalis* Hampson - a pyralid tissue borer, is a minor pest of sweet potato. Caterpillars bore inside the tubers and feed within on the starchy material. Moths are grayish-brown in colour, having wing span of about 26 mm. Forewings are closely mottled with fine specks and have medial greyish-white lines and a series of black spots. Both wings have curved, brown post-medial line. As it is a minor pest no control measures are adopted against this borer.

18

YAMS

YAMS Dioscorea spp. (Dioscoreaceae) are widely grown throughout the Tropics and subtropics, especially in West Africa, part of Vietnam, Cambodia and Laos. The varieties commonly grown in India for obtaining edible tubers, are greater (white) yam, *D. alata* Linnaeus (=*globosa* Roxberg); potato yam, *D. bulbifero;* lesser yam *D. esculenta* Burkill, Linnaeus (=*sativa* Thunberg and *D. versicolor* Boch-Ham.) (= *D. aculeata* Linnaeus, *fasciculata* Roxberg, *spinosa* Roxberg and *sativa* Linnaeus). Besides, there are other dozen edible varieties that are found mostly growing wild in the Subtropical region of India extending upto Himalayas (Singh and Arora, 1978). The edible part is swollen tuber, which contains plenty of starch - over 20 per cent - but has little nutritional value.

Dioscorea alata, originated from South-east Asia and is now widely grown in Asia and the Pacific. This is a large climber, 16 to 18 metres high, having quadrangular winged stem and 5-nerved leaves; bulbils are globose, ovoid or obpyriform and tubers are stout, short, brown to black in colour, globose to pyriform in shape and lobed in various ways. These contain about 21% starch. The tubers are considered to be anthelmintic and are used for the cure of leprosy, piles and gonorrhoea (Chopra, 1933).

Dioscorea bulbifera is native to Tropics of Old World and at present grown throughout Asia and Africa. Tubers are solitary, globose to pyriform in shape and purplish-black in colour. In Kashmir the tubers are used for washing wool and as fish-bait.

Dried and powdered tubers are used for treating piles, dysentery and syphilis (CSIR, 1952).

Dioscorea esculenta originated from Indo-China and is now extensively grown in South-east Asia. It is a prickly climber; bulbils are absent, each plant produces 4 or more tubers in a bunch which are 100 to 140 mm long, globose or flattened and lobed in shape, fawn or reddish-tawny in colour. These are also very starchy but free from dioscorine.

INSECT PESTS

Yams do not suffer any severe loss in tuber yield due to ravages of insect pests. There are a few species of insects that infest the crop in South India regularly. These include yam beetle, some mealybugs and scale insects, sawfly and skipper butterfly. A few species of caterpillars as also a beetle have been reported as leaf defoliators from north India, Among the non-insect pests, nematodes cause substantial loss.

Leaf-eating Beetles : *Galerucida bicolor* (Hope) -Yam beetle and *Lemo locordairei* Baly are the pests of regular occurrence in South India whereas *Crioceris impressa* Fabricius is found in Northern India. Freshly emerged grubs of these beetles feed gregariously on the epidermal tissues of the ventral side of tender leaves; later they bite irregular holes in leaf lamina and often even cat away the entire lamina. The most conspicuous habit of the grubs is to carry exuviae and excreta on their dorsum. Adults generally scrap the surface tissues of leaves and cause only minor damage.

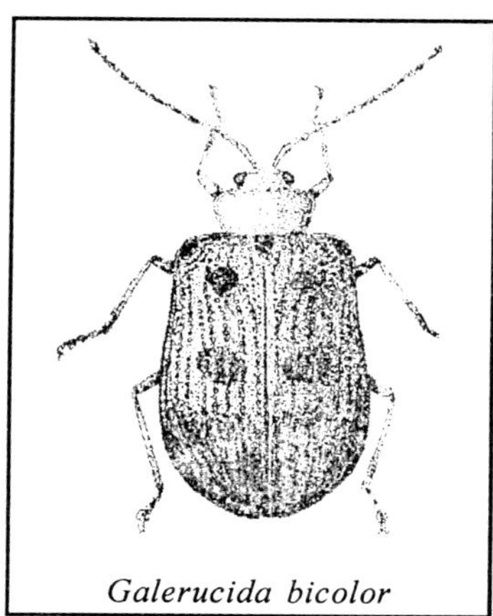
Galerucida bicolor

The yam beetles *Heteroligus metes* (Billb.) and *H. appius* (Burmeister) (Scarabaeidae) which are the major pests of yam in West Africa, have not been reported from India so far.

Galerucida bicolor has been reported as pest of elephant-foot and yams; it is also a minor pest of cucurbits and balsam. Red and black beetles appear in South India around May with the commencement of monsoon and lay eggs in clusters in the soil near the host plants. Grubs skeletonise the leaves and later gnaw into leaf stalks and may even enter the main stems. Pupation takes place in soil. Grub and pupal periods last for a month and fortnight respectively (Ayyar, 1940).

Lema locordairei is another important pest of greater yam in Kerala. Eggs are laid loosely on ventral surface of leaves. A single female lays 20 to 25 eggs in 30 to 40 days. Eggs are smooth, cylindrical, 0.5 to 0.7 mm long and yellowish in colour. Grubs are yellow in colour and have characteristic criocerine features of small head, narrow thorax and a disproportionately thick and fleshy abdomen. The full-grown grubs are 10 to 12 mm long. At the time of pupation, grub exudes from its mouth thick white frothy substance which dries to form a tough cocoon, within which the grub pupates. Pupae are 6 to 7 mm long and yellowish-brown in colour. Adults are 7 to 9 mm long beetles having yellow body and shining blue elytra. Pre-oviposition, incubation, grub and pupal periods last for 7 to 10, 2 to 3, 6 to 7 and 7 to 9 days respectively. Adult longevity is 70 to 75 and 90 to 95 days in case of males and females respectively (Visalakshi and Nair, 1978). The pest appears with the onset of monsoon in May and remains active till October-November.

Crioceris impressa (Chrysomelidae), reported from Kashmir valley as a major pest of potato yam, is also found along the. entire foot-hill region of Himalayas extending upto Assam. Besides, it has been reported from Myanmar, Sri Lanka and Malaysia, Lefroy and Howlett (1909) reported this pest on flowers and leaves of *Ficus elastica* Roxburgh (India rubber tree), *Dioscorea alatah. D. bulbifera, D. tomentosa* Roxburgh, *Callicarpa macropylla* Vahl. (ornamental shrub) and *Holarrhena antidysentrica* (Linnaeus) (Ivory tree). Eggs are laid singly or in batches of 3 to 5 on leaves and sometimes even on stems. Grubs feed on and around the tips of young tender leaves while the adults feed on mature leaves perforating the leaf lamina without damaging the veins (Sinha *et al.*, 1978). Eggs are oblong in shape and light reddish-brown in colour, becoming dark with age. Full-grown grubs are 5 to 6 mm long and dark gray in colour. Pupae are light red in colour and are found enclosed in round cocoons prepared by full-grown grubs. Adults are stout beetles, 7 to 9 mm long and uniformly fulvous in colour. Females are bigger and stouter than males. Mating takes place 3 to 4 days after emergence. Oviposition period lasts for 2 to 6 days during which a female lays 250 to 300 eggs. Incubation, grub and pupal periods last for 3 to 4, 5 to 8 and 10 to 13 days respectively (Srivastava and Bhagat, 1966). There are 5 overlapping generations during June-September, Adults hibernate during winter season.

These beetles can be effectively controlled by spraying with 0.05% endosulfan or 0.2% carbaryl.

Mealybugs : *Geococcus coffeae* Green and *Planococcus citri* (Risso) - the two coccid bugs have been reported as pests of yam in India. These bugs usually feed on roots and tubers hut are occasionally found on stems and leaves. The infestation is generally noticed only when yam crop is harvested as this operation coincides with the favourable period for the multiplication of these pests. As a result of the damage, the stored yams get shrivelled and if such affected yarns are planted, most of them do not sprout.

Geococcus coffeae (Pseudococcidae), as is evident from its name, it is a major pest of coffee. It is Pantropical in distribution and besides yam, it is also recorded on sweet potato.

Planococcus citri (Pseudococcidae) - the *Citrus* mealybug – is found in Tropical and subtropical regions (CIE map No. A-43) as also in glasshouses in Europe and America. Its main hosts are *Citrus* spp., though Ayyar (1930) has listed as many as 50 different host plants.

Eggs are ellipsoidal in shape and light creamy-yellow in colour. Freshly formed nymphs are yellowish in colour without waxy outer layer; the waxy coating appears gradually and full-grown nymphs are completely covered with a fluffy coat of wax. Adult females are slightly elongate-ovate in shape, 5 to 7 mm long, whitish-gray in colour, wingless, covered with a wax layer, and possesses 34 wax covered appendages round the entire periphery of the body (Butani, 1979 a). Male nymphs have wing-pads and the adults are winged.

Phenacoccus gossypii Townsend and Cockerell, and *Planococcus dioscoreae* Williams are the other mealy bugs reported

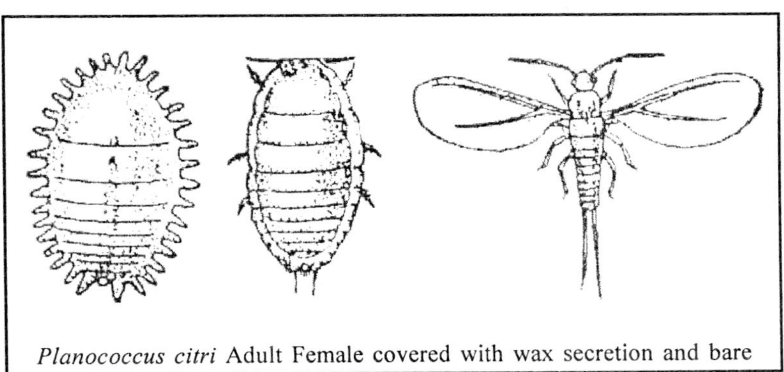

Planococcus citri Adult Female covered with wax secretion and bare

damaging yarn in French Guyana and Papua New Guinea but not in India.

To prevent damage by these mealybugs, plant only healthy tubers. Dip the seed in 0.3% chlorpyrifos or dimethoate just before planting. If and when the aerial parts get infested, spray 0.5% dimethoate or phenthoate.

Scale Insects : *Aspidiella hartii* (Cockerell) and *Quadraspidiaotus* (*Aspidiotus*) *destructor* (Signoret) are two common diaspid scale insects reported as pests of economic importance to yam plants.

Aspidiella hartii - the yam scale is widespread in the Tropics of the Old World (CIE map No. A-217). Both nymphs and adults are whitish-yelow in colour and are usually found clustered on tubers but occasionally on aerial parts of the plants as well. Even stored tubers may be found severely infested by these scales. Such tubers shrivel and become unfit as for sowing.

Quadraspidiatus destructor - the coconut scale is an armoured scale insect, Pantropical in distribution. It is a polyphagous pest found on a large variety of host plants including eggplant and chilli. Nymphs and adult females suck plant sap from ventral surface of leaves and tender shoots, resulting in loss of vitality, thereby affecting plant's growth. The pest is active during summer. Nymphs are about 2 mm long, oval in shape, transluscent, yellowish-brown in colour and are covered with waxy material. Females are circular in shape, semi-transparent and pale brown in colour (Butani, 1979 a); they reproduce oviparously. The eggs hatch under the body of the mother scale. A life cycle is completed in about a month.

To prevent infestation of *Aspidiella hartii*, plant only healthy tubers; tubers from infested fields should not be used for planting. Do not grow yam in the field immediately following harvest of ginger which is the main alternate host of *A. hartii*. Prune and destroy the affected plant parts in the initial stage of attack. Chemical control measures suggested for mealybugs can be effectively used against scale insects also.

Sawfly : *Senoclidia dioscoreae* Rohwer (Tenthredinidae) has been reported from South India occurring on yam. An allied species, *S. purpurata* F. Smith) recorded on yam from Papua New Guinea, has not yet been intercepted from India. The eggs are laid on young shoots and leaves. Larvae feed gregariously, often remaining in rows, beginning at the apex of leaf and moving backward towards the leaf stalk. In case of severe infestation, which is rare, the entire crop may be defoliated. The larvae closely resemble caterpillars in appearance but they have more than 5 pairs of prolegs. Pupation takes place in the soil. Adults are characterised by the broadly sessile abdomen - the 1st abodminal segment being only slightly fused with thorax - and have typical ovipositor which is adopted for sawing a hole for ovipostion. Parthenogenetic reproduction is common. Severe outbreaks of this pest are unusual and as such no specific control measures have been suggested. However, dusting the crop with 5% carbaryl or spraying 0.2% carbaryl will effectively control this pest.

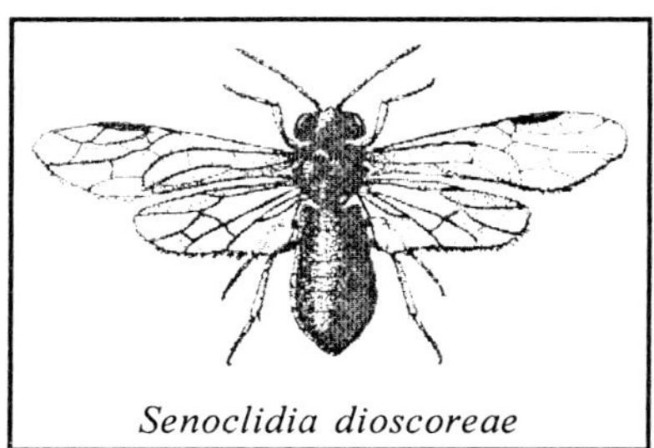
Senoclidia dioscoreae

Leaf-eating Caterpillars : *Theretra oldenlandiae* Fabricius, *Rhyncholaba acteus* (Cramer), *Pericallia ricini* Fabricius, *Spilarctia* *obliqua* (Walker), and *Spodoptera littoralis* (Fabricius)' are the common lepidopterous larvae recorded as minor peats. Of these the first one is comparatively common on yam and second one on elephant- foot; *P. ricini* is a major pest of drumsticks; *S. obliqua* prefers sweet potato and *S. littoralis* has several preferred hosts including tomato.

Eggs of *Theretra oldenlandiae* are broadly ovoid, smooth, shining and green in colour. Full-grown caterpillars are about 80 mm long and have black head, body segments 2 to 4 are velvety-blackish, rest blumbose with short black stripes; the terminal horn is about 8 mm long and black in colour with yellowish-white narrow tip. Pupae are on an average 40 mm long, pale yellowish-brown in colour, abdomen dotted with black and the cremaster is dark reddish-brown. Moths are stout, having brown coloured head and thorax and grayish-brown abdomen with a double silvery-white dorsal stripe. Due to their large size and colouration, they are popularly called as hawk moths. Forewings are grayish-brown with a dark brown discal band superimposed with a thin silvery-white streak; hind wings are dusky with a pale submarginal band not reaching apex. Wing span is on 54 to 74 mm in case of males and about 80 mm in females (Bell and Scott, 1937).

No control measures are generally adopted against these leaf eating caterpillars on yam. If and when these appear, hand-picking and mechanical destruction may be carried about to prevent their population build-up.

Skipper butterfly : *Parnara* species is another leaf eating caterpillar reported as a minor pest of yam from South India. Eggs are laid on leaves. Caterpillars feed on the foliage and in case of severe infestation, may completey defoliate the plant.Caterpillars are yellowish-green in colour with 4 white dorsal stripes. Hibernate in caterpillar stage. Adults are dark brown skipper butterflies - the name which they have acquired due to their peculiar skipping movement. They are distinguished by their antennae being widely separated at the base and dilated apically to form a gradual club.

19

DRUMSTICKS

DRUMSTICKS, *Moringa oleifera* Lamarck *(= pterygosperma,* Gaerin) (Moringaceae), is native of Northern Pakistan and North-west India, though now it is found all over the Indian sub-continent, Saudi Arabia, Turkey and Africa. In India it grows wild along entire sub-Himalayan range and is also cultivated in the plains of North, Central and South India. This is a quick growing tree and soon attains a height of 8 to 12 metres. The trunk is of soft wood and branches are delicate which break down easily while plucking the pods. The leaves are tripinnate and leaflets elliptical; flowers are white, fragrant and are borne on large panicles; pods are pendulous, 250 to 500 mm long, greenish in colour and angularly ribbed, containing trigonous seeds.The pods can be observed hanging loosely from different branches as if sticks have been tied with threads. The popular name has been given due to their stick-like appearance similar to the ones used for beating drums. The tender pods are cut into pieces and used in culinary preparations specially curries. Often the flowers and tender leaves are eaten as pot-herb. The seeds are often fried and eaten while the powdered roots are used as condiment. All the plant parts have some medicinal value. Leaves being rich in vitamins A and C are used in scurvy

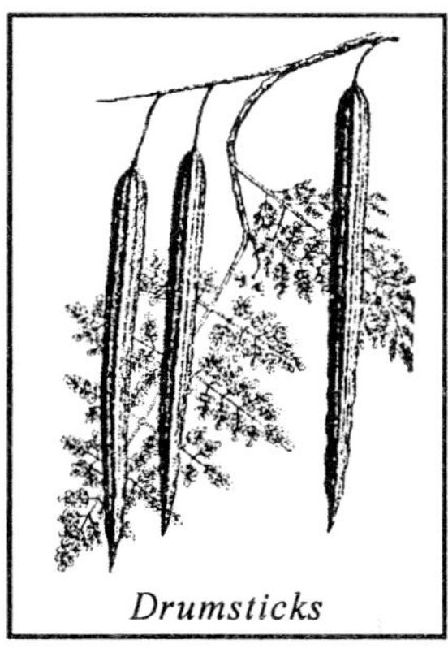

Drumsticks

bands. Wing expanse is 60 to 70 and 70 to 80 mm in case of males and females respectively. The bionomics of this species are similar to that of *E. mollifera.*

Noorda blitealis Walker (Pyraustidae) is an sporadically serious pest of drumstick trees especially in South India. Eggs are laid in batches usually on ventral surface of leaves. Caterpillars feed on leaf lamina and pupate in soil. Peak periods of infestation are March-April and December - January when even a medium infestation may defoliate the entire tree. Adults are medium-sized moths; forewings have rectangular apex with erect outer margin and are uniformly dark in colour with a small white streak at the inner area of base. Hind wings are hyaline with broad; black marginal hand narrowing towards anal side. Wing expanse is 22 to 26 mm. In Tamil Nadu, egg, caterpillar and pupal durations last for 3, 7 to 15 and 6 to 9 days respectively (Nair, 1975).

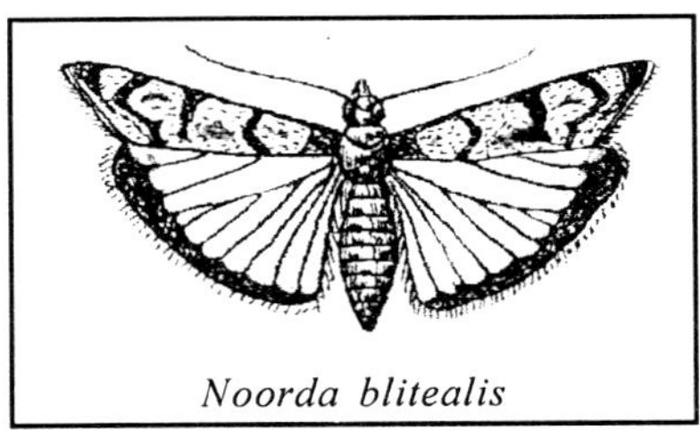
Noorda blitealis

Pericallia ricini (Fabricius), (Arctidae) - the black hairy caterpillar, is also a foliage pest, that occasionally appears in large numbers. This is a minor pest and besides drumstick trees, it has been reported on banana, black gram, cotton, cucurbits, castor, cowpea, soybean, tea, yam etc, Eggs are laid in clusters on ventral surface of leaves. On hatching, the caterpillars feed on leaf lamina; initially by scrapping epidermal layers and later by cutting the lamina. Full-grown caterpillars are 40 to 50 mm long, dark brown in colour, specked with white, and have dorsal and lateral tufts of long dark hair. Moths are stout with fuscous-brown forewings having numerous pale black ringed spots in the interspaces; hind wings and abdomen are crimson in colour having black bands and spots. Wing span in case of male and female is 38 to 50 and 60 to 68 mm respectively. Egg, caterpillar and pupal periods occupy

Pericallia ricini

4. 26 to 32 and 10 to 12 days, respectively; life cycle is completed in about 40 days during April-May (Nair, 1975).

Protrigonia zizanialis Swinhoe (Pyraustidae) is also a leaf eating caterpillar, recorded as a minor pest of drumstick trees. The caterpillars mine the leaves and feed within the tunnels. Advance stage caterpillars web together a few leaves and feed inside the same. Activity of this pest is more pronounced during February to June(Aiyar, 1944). Full-grown caterpillars are 8 to 12 mm long and light, yellowish in colour. Adults are pale brown moths with dark bands on abdomen; forewings are gray, thickly mottled with brown while the hind wings are pale suffused with fuscous on outer area. Wing spread is 14 to 18 mm. Egg, caterpillar and pupal periods last for 3, 11 to 13 and 7 days, respectively.

Actias selene Hubner (Saturniidae) - the moon moth, is widely distributed all over India. Caterpillars are voracious feeders and defoliate the trees in no time. Full grown caterpillars arc smooth, a|pple-green in colour and 100 to150 mm long. Pupation takes place in dense, firm silken cocoons. Adults are big, pinkish moths. Wings are whitish at the base, pale greenish all over and yellowish near the margins with a conspicuous cross band and a white eye-spot in the centre; hind wings have a long tail, pinkish in colour. Wing expanse is 130 to 160 mm and 140 to 180 mm.

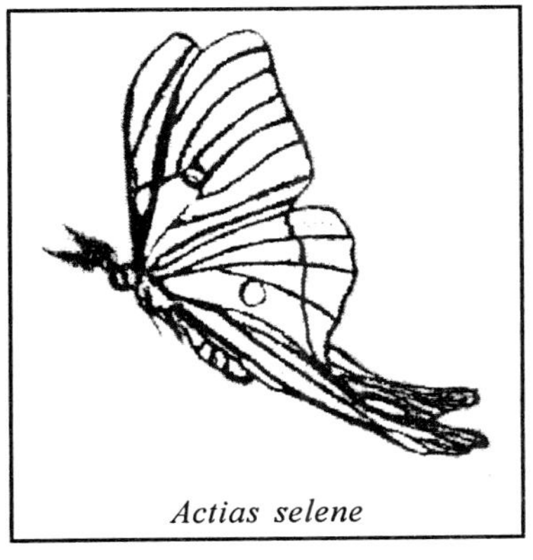

Actias selene

Metanastria hyrtaca (Cramer) (Lasiocampidae) - gristly citrus caterpillar, is found all over the Indian sub-continent. It is polyphagous and its preferred hosts are several *Citrus* species. Caterpillars are nocturnal in habit and feed gregariously and voraciously; during day, they remain crowded on shady side of the tree trunks. Eggs are spherical in shape and pale white in colour. Full-grown caterpillars are cylindrical in shape, grayish-brown in colour, stout, hairy and about 70 mm long. Adults are stout, grayish-brown moths exihibiting sexual dimorphism. Male and catarrhal infections. Flowers are used as tonic, diuretic and cholagogne. Seeds are acrid and antipyretic; whereas massage with seed-oil is beneficial against gout and rheumatism (Butani and Verma, 1981b).

INSECT PESTS

The most desructive insect, pests of drumstick trees are a few species of leaf-eating caterpillars and beetles. Besides, whifefly, scale insects, sap sucking bugs, thrips, midges, trunk borers and leaf-eating weevils have been recorded as minor pests,

Leaf Eating Caterpillars : *Eupterote mollifera* Walker (Eupterotidae) - The common hairy caterpillar, is the most destructive and specific pest of drumstick trees. Eggs are laid in clusters on leaves and tender stems. On hatching, the caterpillars feed gregariously by scrapping the bark and gnawing the foliage. A severe infestation may result in complete defoliation of the tree.

Full-grown caterpillars measure 40 to 50 mm in length; are brownish in colour and densely hairy. Adults are large-sized moths, uniformly light yellowish-brown in colour; forewings suffused with brownish-rufous. Wing expanse is 60 to 70 mm and 75 to 85 mm in case of males and females respectively. Moths appear with the onset of monsoon and lay eggs which hatch in 6 days, caterpillar and pupal periods last for 12 to 14 and 8 to 10 weeks respectively; there in only one generation in a year.

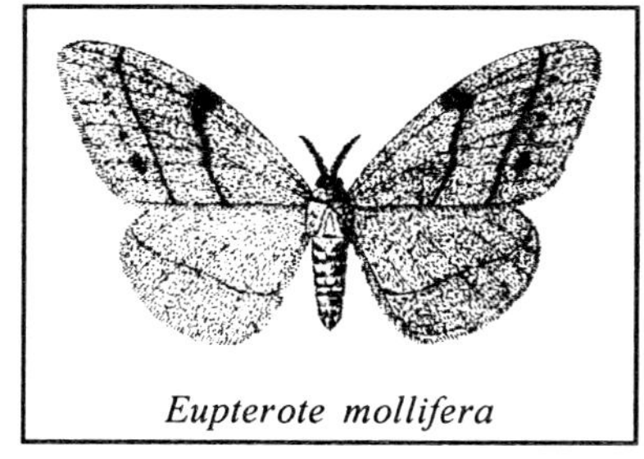

Eupterote mollifera

Eupterote geminata Walker (Eupterotidae), another hairy caterpillar, which has been recrorded damaging drumstick trees in Tamil Nadu, though it is a minor pest. The damage symptoms are same as those caused by *E. mollifera.* Full-grown caterpillars are 35 to 40 mm long, covered with tufts of grayish-white hair having black tips. Moths are yellowish-orange in colour with dark transverse

moths have pectinate antennae and chocolate-brown patch in the middle of (brewings, wing expanse being 50 to 60 mm.

Females are bigger in size, have longer and broader wings

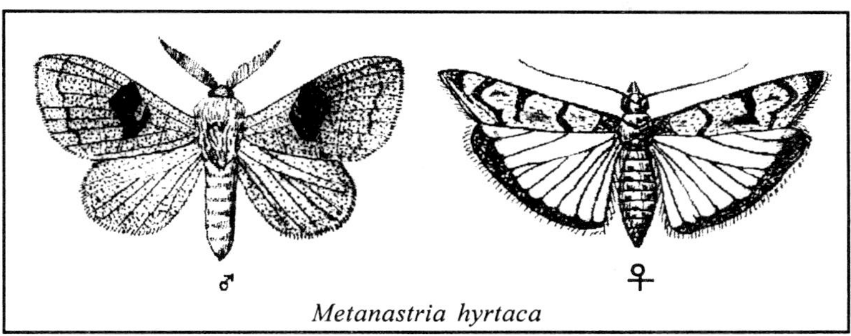

Metanastria hyrtaca

with wavy transverse bands of light and dark brown colour; the wing expanse is 70 to 85 mm. Incubation, caterpillar and pupal periods last for 9 to 12, 45 to 100 and 9 to 18 days respectively and a life cycle is completed in 75 to 110 days (Beeson, 1941).

Streblote siva (Lefevre) is another hairy caterpillar occasionally found feeding on drumstick trees. This is also a polyphagous pest, rose being its preferred host. Full grown caterpillars are 40 to 50 mm long, pale ochreous-brown in colour with small, black spots and long lateral tufts of ochreous hair. Moths have grayish-white head and thorax and whitish abdomen; forewings are beautifully coloured with reddish-brown spot ringed with white; hind wings

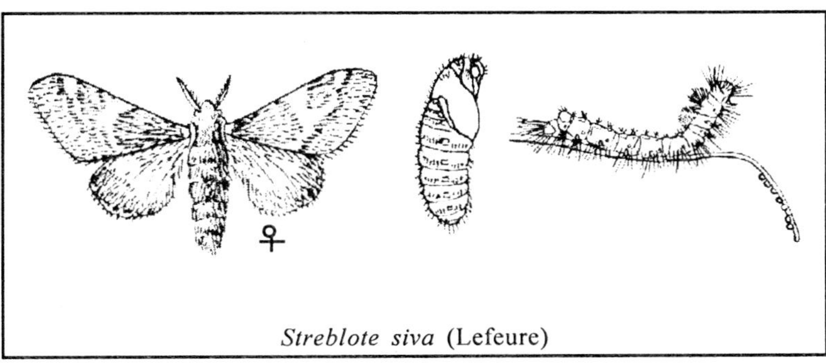

Streblote siva (Lefeure)

are white with slight fuscous on outer margin. Wing expanse is 42 to 52 and 70 to 80 mm in males and females, respectively.

If and when a serious infestation is observed, check the damage caused by these lepidopterous pests by spraying the trees with 0.2% carbaryl or 0.1% malathion or 0.05% lindane or endosulfan.

In nature, caterpillars of *Eupterote geminata* are parasitised by *Henicospilus tufas* Tosq and *Sturmia* species, while those of *Pericallia ricini* are parasitised by *Apariteles ricini* Bhatnagar *Meteorus* species, *Sturmia* species, *Thelaria nigripes* Fabricius, *Eulectrus* species and *Henicospilus rufus.*

Bark Borer : *Indarbela tetraonis* (Moore) (Metarbelidae) has a number of host plants including drumstick trees, of which it is generally a minor pest except in South India where it often causes severe damage. The attack is more pronounced on neglected trees and at places where crop sanitation is poor. Eggs are laid in cuts and cracks in the bark of trunk or main branches of the tree. On hatching, the caterpillars feed superficially below the bark, making zigzag galleries and later bore inside the bark or main stem, remain within these burrows during day but come out at night and feed on the bark. The conspicuous symptom of attack is presence of huge silken webbed masses comprising chewed wooden particles and excreta of the caterpillars, hanging loosely on tree trunk covering the entry holes. Pupation takes place inside the tunnels, and generally, there is only one caterpillar or pupa in each burrow, but there may be several such holes and burrows on a single tree.

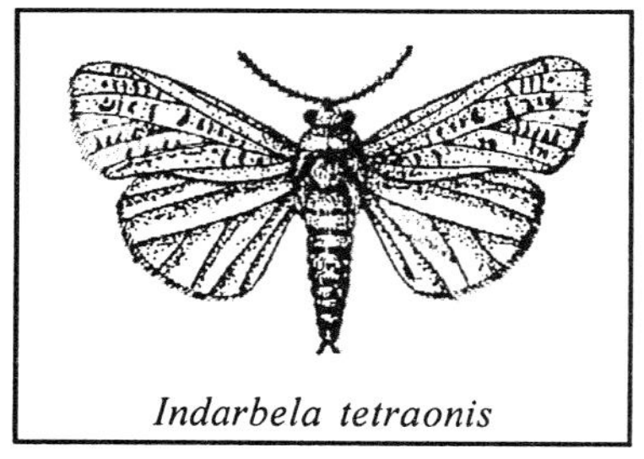

Indarbela tetraonis

Full-grown caterpillars are 40 to 45 mm long, stout and dirty brown in colour. Pupae are also stout, reddish-brown in colour and 14 to 18 mm long. Moths are pale brown; forewings have prominent brown spots and streaks; hind wings-are whitish. Wing expanse is 35 to 38 in males and 42 to 50 mm in females (Butani,

1977 b). Egg period is 6 to 8 days, while caterpillar stage is extremely long as it takes about 10 to 11 months: pupal period is completed in 14 to 20 days and there is only one generation in a year.

Longhorn Beetles : *Batocera rubus* (Linnaeus), *Coptops aedifactor* Fabricius *and Monochamus (-Monohammus) versteegi* Ritsema have also been reported boring the trunks of drumstick trees.

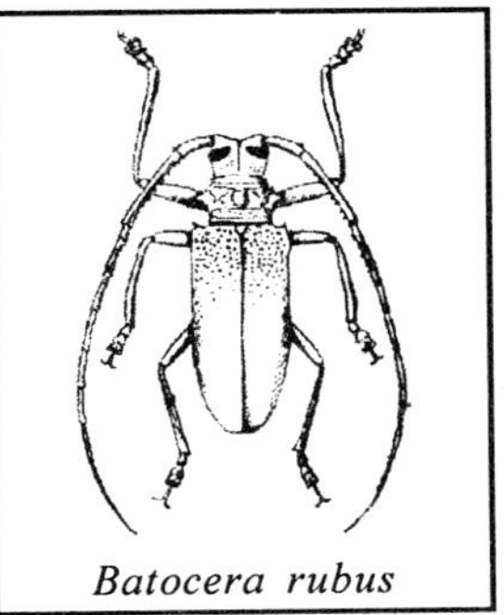
Batocera rubus

Batocera rubus is widely distributed all over the Indian subcontinent. Eggs are laid singly in cracks and crevices in the bark of the tree. On hatching, grubs make zigzag burrows beneath the bark, feed on internal tissues and may reach sapwood, causing thereby the death of the affected branch or stem. Pupation takes place within these tunnels. Adults come out and feed on the bark of young twigs and petioles. Grubs are stout, about 100 mm long, yellowish in colour with well defined segmentation. Adults are large-sized beetles about 50 mm long and yellowish-brown with white spots on elytra. Egg, grub and pupal periods last for 1 to 2, 24 to 28 and 12 to 24 weeks respectively; there is only one generation in a year.

Monochamus versteegi is found all over the Indian subcontinent mainly on *Citrus* spp. It is a minor pest of drumstick trees and is more common in South India. For oviposition, the female selects a suitable soft spot and makes a small depression or cavity in which one or two eggs are deposited. The grubs, immediately after hatching, bore inside the stem, plugging the entry hole with their excreta. Feeding inside the tissues, results in withering of growing point; leaves fall off and stems give an appearance of dry wood. Eggs are cigar-

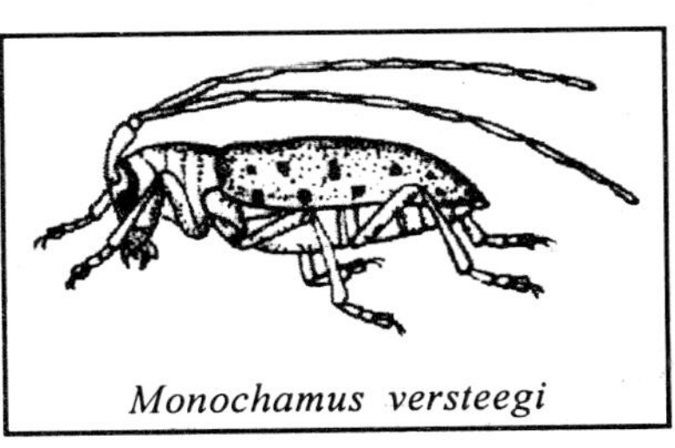
Monochamus versteegi

shaped, about one mm long and whitish in colour. Full-grown grubs are characterised by a well developed thoracic hump; their bodies measuring 30 to 35 mm, have shiny appearance and arc covered by short hairs, yellow on dorsal side and brown on ventral side. The pupae are yellow in colour and measure 20 to 25 mm in length. Adults are 14 to 18 mm long with heads bent downwards under the prothorax. Head, prothorax and elytra are brown with dark grey patches on elytra. Antennae are long, 12-segmented and filamentous. Incubation, grub and pupal stages have been reported to be 1 to 3, 65 to 117 and 7 to 10 days respectively (Subramaniam, 1920).

To kill these trunk and stem boring caterpillars and beetles, clean the affected portion of the tree by removing all the webbed material, excreta etc. and insert in each hole cotton-wool soaked in any good fumigant like carbon bisulphide, carbon tetrachloride, chloroform or even petrol and seal the treated hole with mud.

Aphid : *Aphis craccivora* Koch (Aphididae) - the bean aphid, is a polyphagous pest, beans being its preferred host. Nymphs and adults suck the vital sap from the twigs. As reproduction is mostly parthenogenic, the population build-up is very fast. Therefore, to avoid serious damage, spray promptly 0.03% dimethoate, phosphamidon or 0.1% malathion as soon as the infestation is observed. All the pods should be removed before spraying as it is not desirable to eat the pods for at least 15 days after spraying.

Whitefly : *Trialeurodes raja* Shumsher (Aleyrodidae) is a minor pest of drumstick trees. All stages of this whitefly are found on ventral surface of leaflets where eggs are laid in small batches. The nymphs, which are sluggish and more or less stationary, suck the vital sap; but. the loss caused is usually negligible. Adults are cream coloured, tiny insects with white wings having wing expanse of 4 to 5 mm. A life-cycle is completed in about 3 weeks - 16 to 20 days during July to September and a little longer during the end of December to mid-February. David *et al.* (1973) observed that temperature above 32°C favours increase in the activity of this whitefly whereas below 32°C there is decrease in population at various stages of the whitefly. Generally, no control measures are needed to check the incidence of this pest. Nevertheless, spraying 0.03% dimethoate or 0.1% malathion can effectively control this pest, if and when necessary.

Scale Insects : *Ceroplostodes cajani* Maskell (Coccidae) and *Diaspidiotus* species (Diaspidae) are the scale insects, reported so far, as pests of drumstick trees in India. *C. cajani* – the soft scale – is a serious pest of various varieties of gram and bean especially in South India. Though each insect takes only a few drops of sap during its life time, presence of enormous number of insects sucking the sap continuously may at times weaken the trees and ultimately affect the size of pods. If and when necessary, spray the trees with 0.1% monocrotophos to check growth of these scale insects. Spraying should be given before fruiting starts. In nature, *C. cajani* is parasitised by *Krishnieriella ceroplostodis* Mani, *Aphycus fusidorsum* Gahan and *Anicetus ceylonensis* Howard.

Thrips : *Scirtothrips dorsalis* Hood (Thrtpidae) - chilli thrips attacks a large number of plants including drumstick trees. Both nymphs and adults are very small in size (1 to 2 mm long) and therefore, their presence often goes unnoticed till the symptoms of damage become conspicuous. The attack is more pronounced during dry weather and confined to tender leaves only, and the loss caused is usually not of much economic importance.

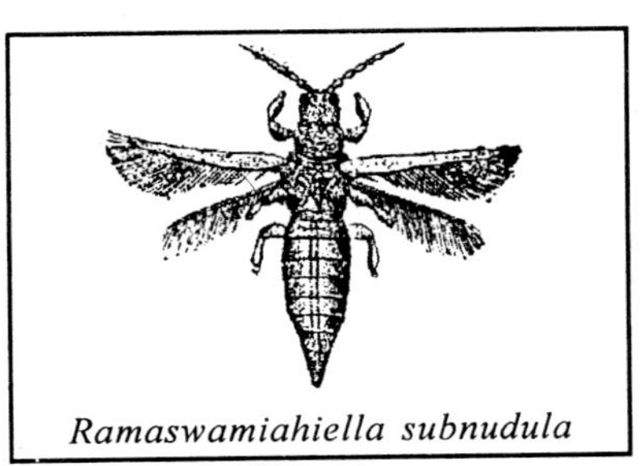

Ramaswamiahiella subnudula

Ramaswamiahiella subnudula Karny (Thripidae) - the blossom thrip - is also polyphagous. It feeds and breeds on buds and flowers of various host plants. Generally the loss caused to drumstick trees is negligible. In case of severe infestation, flowers fade and drop prematurely. Both nymphs and adults are very delicate, slender insects of orange-yellow colour; adults have heavily fringed grayish-yellow wings.

Control measures suggested against aphids and whitefly, will control thrips as well or else spray 0.1% malathion or 0.2% carbaryl to cheek the population of thrips.

Termite : *Neotermes fletcheri* Holmgren and Holmgren (Katotermitidae) is a subterranean pest causing economic damage to drumstick trees grown in sandy or sandy-loam soils. These insects feed on roots and tunnel into the main trunk. In ease of severe infestation, the lower most portion of the trunk may be completely hollowed and often filled with earth. To save the trees from damage of this pest, mix thoroughly with the soil around the trees, 5% HCH, or chlorpyrifos dust @ 500 g per tree.

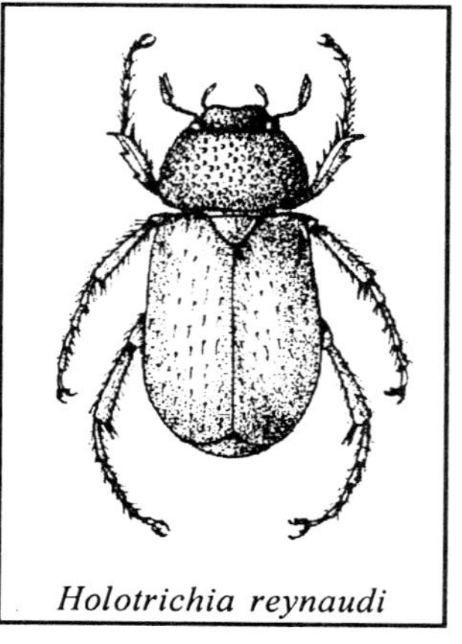
Holotrichia reynaudi

Beetle Grubs : *Holotrichia reynaudi* Blanchard (*insularis* Brenske) (Melolonthidae) has been reported as a pest from Rajasthan (Srivastava and Khan, 1963) whereas occurrence of *H. rustica* Burmeister has been recorded from Karnataka (Veeresh, 1977). Both the species are similar in appearance and have same habits and symptoms of damage. *H. rustica* is smaller in size (16 to 17 mm long) than *H. reynaudi* (18 to 20 mm long), Eggs are laid in moist soil near about the host plants. On hatching, the grubs feed on roots of different plants as well as organic matter present in the soil. When full-fed, they go deep down in the soil and pupate. On emergence, the adult beetles remain in the soil till the first shower of monsoon, after which they come out of the soil and feed voraciously on foliage of the available plants and trees. Mating also takes place during this period. The adults feed during night and hide in the soil during day. Their feeding activity continues till the commencement of heavy showers, when they lay eggs and die away. Incubation, grub and pupal durations last for 1 to 2, 8 to 12 and 2 to 3 weeks respectively. There is only one generation in a year; overwintering is usually in pupal stage but often the adults too are seen overwintering in the soil.

For controlling these beetles, plough around the trees during Winter to expose and kill the hibernating pupae and adults. Soil treatment with chlordane or heptachlor is effective against grubs. To kill adults, spray the tree and tree trunk with isofenphos or chlorpyrifos especially during the breeding season.

Bud Borers : *Stictodiplosis moringae* Mani (Cecidomylidse) -a midge - is a minor pest of drumstick trees reported from Tamil Nadu. Eggs are laid in clusters on anthers within the flower buds. Maggots feed on internal tissues of buds especially on ovaries. The pest is comparatively more active during August to January. The infested buds soon fall down and the full-fed maggots come out to pupate in the soil. Egg, maggot and pupal periods last for 1 to 2, 6 to 9 and 5 to 8 days, respectively: a single life cycle is completed in 12 to 19 days (Cherian and Basheer, 1938).

Noorda moringae

Gitona species - a drosophil - is a small yellowish fly with red eyes, that has been reported infesting the pods (Nayar *et al.*, 1976).

Noorda moringae Tams - a pyralid - is another bud border reported from South India. This is also a minor pest. Eggs are laid usually singly but occasionally in batches of 2 and 3 on flower buds. After hatching, the young caterpillars bore inside the buds and feed on Lender tissues. Generally, the infested buds contain only one caterpillar. The damaged buds seldom blossom; these fall down prematurely and the full-fed caterpillars come out to pupate in minute brownish cocoons, either in the soil or on the ground itself, below the dried leaves and debris.

Egg, larval and pupal durations occupy 3 to 4. 8 to 16 and 6 to 10 days, respectively, during June-July; adult longevity has been found to be 24 to 30 days.

Ploughing around the trees to expose and kill the pupae has proved to he effective in reducing the population of these bud borers.

Larval parasites of *Noorda moringae*, recorded from field, include *Microbracon brevicornis* Wesmael. *Elasmus hybleae* Ferriere, *Pristomerus* species, *Chelonus* species, *Perilampus* species, *Systasis* species and *Apanteles* species (Cherian and Basheer, 1939).

Leaf-eating Weevils : A number of weevils have been recorded nibbling leaflets of drumstick trees including *Myllocerus discolor variegatus* Boheman, *M. teniclavis inferior* Marshall, *M. undecimpustulatus maculosus* Desbrochers, *M. viridans* Fabricius and *Ptochus ovulum* Faust (Curculionidae).

Myllocerus spp. feed on a variety of crops. Eggs are laid in soil. Grubs feed on roots of cultivated crops, grasses etc. and pupate in soil. Adults come out of the soil and nibble the leaves causing some minor damage.

No control measures are generally adopted against these weevils on drumstick trees.

20

JACK - FRUIT

JACK-FRUIT, *Artocarpus heterophyllus* Lamarck (Moraceae), is a large evergreen tree, native of India. It is also grown in Sri Lanka, Bangladesh, Myanmar, Malaysia, Indonesia, Philippines, Brazil, etc. In India, estimated area under this fruit is 26000 hectares. Of this, a large chunk is in Karnataka and Kerala where it is grown in mixed plantation with arecanut and coconut along the Western coast. There are about 8500 hectares under its cultivation in Assam, 6200 hectares in Tripura, 4500 hectares in Bihar and smaller areas in Uttar Pradesh, Maharashtra, Andhra Pradesh and Tamil Nadu. In spite of so much area under jack-fruit, it is not classed as a commercial vegetable or fruit and is seldom grown in regular plantations. The tree flourishes in humid climate especially on hill-slopes and is sensitive to cold, frost and drought. As such these trees are grown from sea level to about 1500 metres elevation in the South and at the foothills of the Himalayas in the North. The fruits contain some minerals as also vitamin A and B. Beside being used occasionally as table fruit, it is widely used as a vegetable, as also for making pickles and other forms of culinary preparations. Timber is used for high class furniture, musical instruments, toys etc., while leaves are eaten by cattle.

Allied species, bread-fruit, *A. altitis* (Park.), native to Pacific Islands is grown in Maharashtra and in South India along the Western and Eastern ghats. Monkey-jack (*Barhal*) *A. lakoocha* Roxberg is cultivated in Bengal, Khasi Hills and Uttar Pradesh. Fruits of both these varieties are eaten as vegetable.

INSECT PESTS

Jack-fruit is attacked by over 35 species of insects. Among these, those of major importance are, shoot borer, bark borer. bud weevil, spittle bug and mealybug, while the minor pests include, scale insects, aphids, whiteflies, thrips, leaf webbers, and longicorn beetles.

Shoot Borer : *Diaphania caesalis* (Walker) Pyraustidse) was originally described by Walker (1859) as *Glyphodes caesaiis*. sp. *nova*; subsequently, it was transferred to genus *Margaronia*, then to genus *Palpita* and finally Wang (1963) placed it under genus *Diaphania*. Fletcher (1914) reported *G. caesalis* as a pest of jack-fruit in Karnataka and Maharashtra; Chowdhury and Majid (1954) recorded it as a major pest in Assam. It is also recorded from Sikkim, Bihar, Uttar Pradesh, Andhra Pradesh and Tamil Nadu, Outside India, it has been reported from Borneo, Sri Lanka (Hampson, 1896) and Bangladesh (Alam *et al.*, 1964).

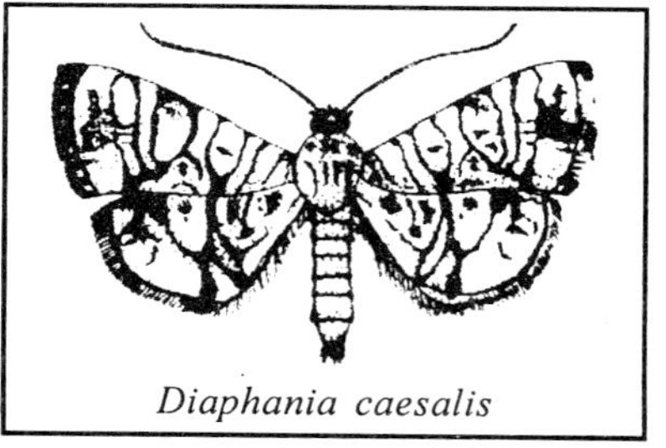
Diaphania caesalis

The pest is active from May to October. Eggs are laid on tender shoots and flower-buds. On hatching, the caterpillars bore into tender shoots, flowering buds arid developing fruits and tunnel through the same. As a result, the shoots wilt and droop, buds dry and drop down and the fruits start rotting (Butani, 1978). Pupation takes place inside the tunnels.

The caterpillars have yellowish head and prothorax with reddish-brown body having numerous black flattened horny warts each bearing one short bristle-like hair. Pupae are reddish-brown. Adults are whitish-brown; moths have wings with grayish elliptical patterns and a marginal series of black specks; wing expanse is 26 to 34 mm; females are slightly bigger than males.

Madhava Rao (1965) suggested covering of fruits with alkathene bags, but this is neither practicable nor economical. Remove and destroy the affected shoots, flower buds and fruits in the initial stage of attack to prevent the spread of infiltration. Spraying

0.03% phosphamidon or 0,2% HCH is effective in checking the pest population (Alam, 1962).

Bark Borers : *Indarbela tetraonis* (Moore) (Metarbelidae) is the only lepidopterous borer found boring into trunk and main stems of jack-fruit trees in India. Besides, a number of coleopterous beetles, namely, *Apriona rugicollis* Chevrolat, *Batocera rufomaculata* (de Geer), *B. rubus* (Linnaeus), *Epepeotes luscus* (Fabricius), *Glenea belli* Gahan, *Sthenias grisator* Fabricius and *Platypus indicus* Strohmeyer have been recorded boring the jack-fruit, trees, These are all polyphagous pests and in case of jack-fruit, *I. tetraonis* and *Batocera* spp. are comparatively more destructive. These are widely distributed all over India and their presence is conspicuously felt in areas where surroundings are not kept clean or where trees are not well looked after.

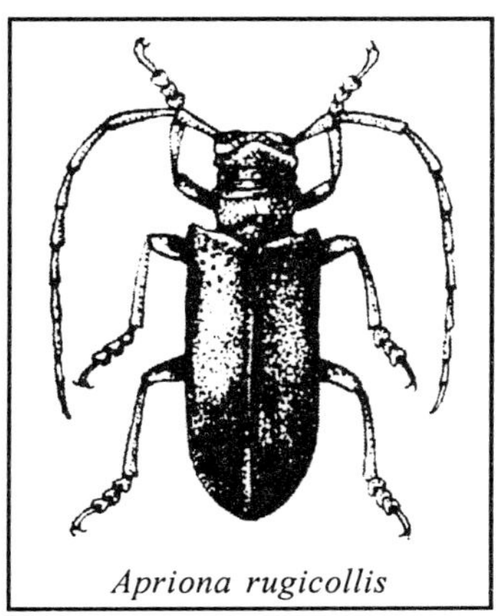

Apriona rugicollis

Indarbela tetraonis has been recorded damaging various fruit trees. The young caterpillars nibble the bark of the trees for a couple of days and then bore inside the trunk or main stems, making a short tunnel downwards. Often, more than one borer attacks the same tree, each making its independent hole. A severe infestation arrests the growth and adversely affects the fruiting capacity of the tree.

Apriona rugicollis (Cerambycidae) is a main pest of forest trees (Beeson, 1941). Adult beetles are about 50 mm long, yellowish - gray with elytral structure and margins bluish-gray. These are active during March to October and feed on bark of living shoots; the affected shoots are girdled and gradually die away.

Batocera spp. (Cerambycidae) are widely distributed in Oriental regions and, besides jack-fruit, they also attack a number of other Tropical and Temperate fruit trees. A female lays about 200 eggs, placed one by one in incisions cut in the bark with its mandibles.

On hatching, the grubs tunnel into the main stem or trunk and feed on meristem and then penetrate deeper; the excavations made in the early stage of infestation are extensive, irregular and deep in sapwood. In case of severe infestation, translocation of cell sap is interrupted, affecting adversely the growth and fruiting capacity of the trees; often, the bark may also split open exposing the inner tunnels. Adult beetles are nocturnal in habit and feed by gnawing the bark of living twigs.

Epepeotes luscos (Cerambycidae) has been reported as minor pest of jack-fruit and mango trees (Nair, 1975). It has also been reported on various forest trees including *Ficus* spp. The adult beetles are 17 to 27 mm long, reddish-brown with vague gray markings and two black shoulder spots on elytra. It has three generations in a year; shortest life cycle being 2½ to 3 months and prolonged one, 8 to 10 months. Females live for over 4 months and each lays as many as 1400 eggs (Beeson, 1941).

Glenea belli (Cerambycidae) is a longicorn beetle reported attacking jack-fruits in South India. Adult beetles are shining black and 12 to 15 mm long. Life cycle is annual, with emergence starting early in May and continuing up to July.

Sthenias grisator (Cerambycidae) is commonly known as girdler beetle as it girdles the stems of the host, trees and the affected stems are often killed. The main host of this beetle is grapevine but occasionally, citrus, jack-fruit and mulberry trees are also attacked. The beetles have only one generation in a year and adults are commonly found during August to October in South India.

Platypus indicus (Platypodidae) - the pinhole borer, is a main pest of forest trees (Beeson, 1941). It has also been reported as a minor pest of jack-fruit (Nair, 1975). These are tiny bettles, 3.5 to 3.8 mm long and have a very peculiar habit of boring into the bark of wood. The main and branch tunnels curve sinuously through the soft wood in a horizontal plane and the pupal chambers are offset vertically on both sides, spaced slightly apart.

To control these borers, clean the affected portion of the trunk or main stem and insert into the holes a wick of cottonwool soaked in carbon bisulphide, chloroform, petrol or even kerosene oil and seal the holes with mud. For minor pests, especially *Apriona rugicollis* and *Sthenics grisator*, hand-picking and mechanical destruction of the beetles is suggested, which is more economical.

Bud Weevil : *Ochyromera artocarpi* Marshall (Curculionidae) is an specific pest of jack-fruit. It is found all over India and has been recorded as a major pest from Assam and Western coast of Karnataka. The small whitish grubs bore into tender flower buds and fruits, with the result the affected buds and fruits fall prematurely. Adults are grayish-brown weevils found nibbling the leaves.

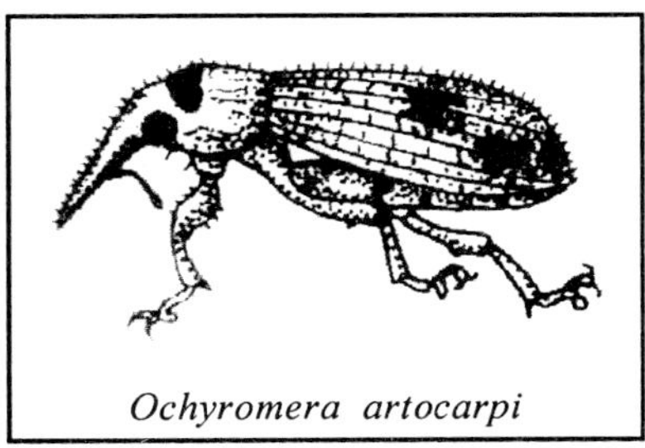

Ochyromera artocarpi

Two more curculionid weevils, namely, *Onychocnemis coreyae* Marshall and *Teluropus ballardi* Marshall have been reported feeding on leaves of jack-fruit trees in South India (Nair, 1975), but not causing any appreciable loss.

To control growth of these weevils, collect and kill mechanically the grubs and adults in the initial stage of attack. Later, remove and destroy the affected shoots as also infested and fallen flower-buds and fruits. In case of severe attack, spray 0.2% lindane or carbaryl.

Spittle Bugs : *Cosmoscarta relata* Distant (Cercopidae), is a serious pest of jack-fruit trees, specially in South India (Beeson, 1941). These bugs appear in swarms and feed on young shoots and leaves. The affected leaves often become curled. Nymphs are wingless, dark purple with yellowish head, pronotum and legs. These are found living inside common mass of froth generally on stalks and young shoots and ripening fruits. The frothy mass consists of the fluid voided from the anus, and mucilaginous secretion from the epidermal glands. After the last moult, nymphs leave this frothy mass and become adults. Adults are large-sized bugs, about 15 mm long with reddish head and pronotum, and forewings with reddish markings on the tagmina. The bugs are very active and difficult to catch.

Two more spittle bugs, namely, *Clovia lineaticollis* (Motschulsky) (Cereopidae) and *Ptyleus* species (Cercopidas) have been reported from South India (Nair, 1975). These bugs are smaller than *C. relata*. Adults of *C. lineaticollis* are about 9 mm long, chestnut-brown in colour with yellowish stripes while those of *Ptyleus* species are 10 to 12 mm long and grayish in colour.

To control these bugs, collect the masses of froth and destroy mechanically the young nymphs therein.

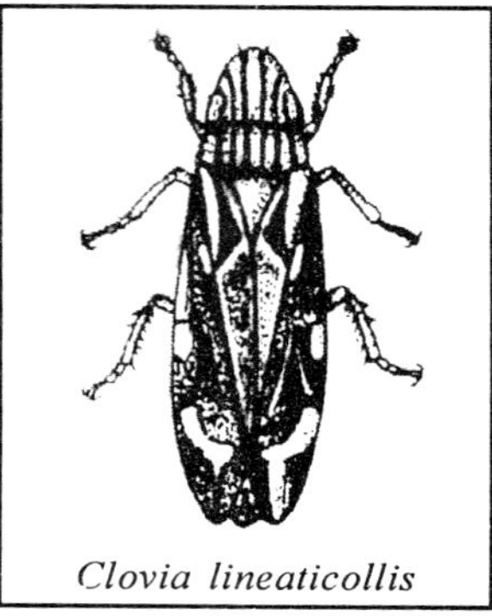

Clovia lineaticollis

Mealybugs : *Nipaecoccus viridis* (Newstead) [*corymbatus* (Green, *flamentosus* (Cockerell), *vastator* (Maskell)] (Pseudococcidae), is a polyphagous pest recorded on aonla, ber, *Citrus*, fig, grapevine, guava, mango, mulberry, tamarind etc. It is found all over India. Peak period of activity is around November when clusters of these mealybugs may be seen on leaves and tender shoots, sucking the cell sap. Eggs are roundish cylindrical, flattened at both ends, chestnut-brown in colour and fully covered with fine cretaceous material forming the ovisac. Nymphs are deep chocolate coloured, having their dorsum covered thinly with a whitish mealy material. Adult females are dark castaneous, covered completely with sticky cretaceous white ovisac. If removed from this ovisac, the females are seen covered with whitish mealy substance. Eggs are laid in ovisacs and the females die soon after egg-laying. A single female lays 400 to 700 eggs. The eggs hatch in 7 to 10 days and the nymphal development takes about 15 and 20 days in case of males and females, respectively.

Ferrisia virgata (Cockerell), the white tailed mealybug, is also a polyphagous pest and is widely distributed all over India (Ali, 1962, 1968). It feeds on leaves and tender shoots and during dry weather it moves down and inhabits the roots. A prolonged period of drought may result in a severe outbreak of this pest. The peak period of activity is August to November, when the affected parts turn yellow, wither and ultimately die away.

To control these mealybugs, spraying may be done with 0,05% dichlorvos or monocrotophos. The spraying will also check to some extent the infestation of mango mealybugs and scale insects.

Mango Mealybugs : *Drosicha mangiferae* (Green) and *D. stebbingi* (Green) (Monophlebidae) are the two allied species often confused with one another. Both have been reported as minor pests of jack-fruit. Apparently, these species look alike and show same symptoms of damage. The gravid females descend down the trees and enter into the soil for egg-laying. On hatching, the nymphs come out from the soil and climb up the trees to suck sap from succulent and tender leaves and twigs.

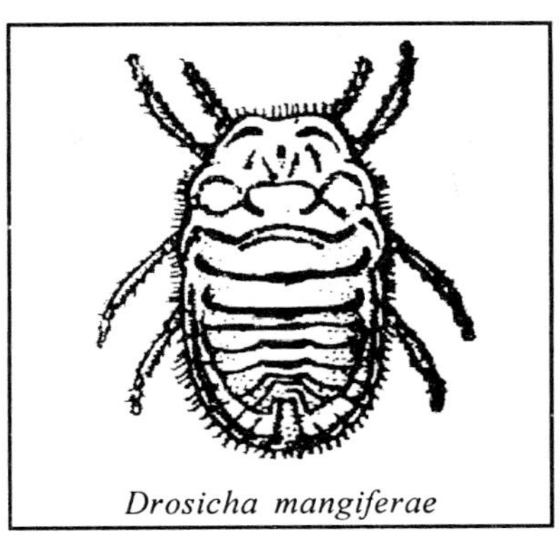

Drosicha mangiferae

Raking the soil around the trees during Summer, exposes and destroys the eggs; soil application of 5% HCH, or chlorpyrifos dust @ 300 to 500 g per tree, just before winter sets in, checks the nymphal population. Also, tying, the alkathene band around the tree trunk is effective in preventing the nymphs from climbing up the tree (Srivastava and Butani, 1972).

Leaf Webbers : Leaf webbers or leaf eating caterpillars, commonly found damaging jack-fruit trees in India, include *Perina nuda* Fabricius (Lymentridae) and *Diaphania bivitralis* (Guenee) (Pyraustidae). Both arc sporadic pests and of minor importance. The caterpillars roll, fold or web together the leaves and feed within.

Perina nuda has been reported from China, India and Sri Lanka. Besides jack-fruit, it has also been recorded as a minor pest of fig and mango trees in India. Eggs are about 0.7 mm long, cylindrically round tapering towards the end attached to the loaf. These are light pink in colour when freshly laid, later becoming red and finally brick-red. Full grown caterpillars are 22 to 25 mm long, having short erect lulls of dusky gray to brownish hairs. Pupae are hairy and brightly coloured, brownish-green dorsally and pale yellowish ventrally. Male pupae are on an average 16 mm long and female pupae 18 mm long. In moths, there is extreme sexual dimorphism. Male moths are smaller than female moths. A single female lays 60 to 400 eggs which hatch in 4 to 6 days. Larval and pupal periods are 16 to 20 and 5 to 9 days, respectively. The adult longevity is 3 to 11 days and the total life cycle is completed in 27 to 39 days (Cherian and Israel, 1939).

Diaphania bivitralis cause considerable damage to jack-fruit foliage especially in Tamil Nadu (Muthukrishnan *et al.*, 1958). It has also been reported, from Sikkim and Assam, and outside India, from Taiwan, Borneo, Myanmar and Sri Lanka, Full-grown caterpillars are about 30 mm long, olive-brown in colour with conspicuous white markings. They pupate within the leaf folds in thin white silken cocoons. pupae are red and 20 to 24 mm long. Moths are chestnut-brown having two semi-hyaline white blotches with a small black discocellar spot between the two. Hind wings are iridescent hyaline white having a broad chestnut marginal band with a black line on inner edge. The eggs hatch in 5 to 6 days; caterpillars pupate, after 14 to 16 days and the pupae become adults in 8 to 10 days (Nagaraja Rao and Subramaniam, 1958).

To control these webbers, hand-picking and mechanical destruction of caterpillars in the initial stage of attack is suggested. In case of severe infestation, dust 5 to 10% HCH, Spray 0.1% lindane or 0.1% carbaryl to check the pest population.

Castor Capsule Borer : *Dichocrocis punctiferalis* (Guenee) (Pyralidae) is a major pest, of guava and a minor pest of various fruit trees including jack-fruit. Eggs are usually laid on young fruits and occasionally on flower-buds. On hatching, the pinkish-brown caterpillars bore inside the flower buds or young fruits and feed within. The bored holes are plugged with excreta and in case of severe infestation, the

Dichocrocis punctiferalis

excreta falls on the ground below and can be seen there in small heaps. The affected buds do not open, wither and die away while the, fruits fall prematurely.

To check the populatin of this borer, remove and destroy all the infested flower-buds and fruits.

Scale Insects : The scale insects found damaging jack-fruit trees include, fluted scale, *Icerya (Crossotosomo) aegyptica* (Douglas): hard scales, *Hemiberlesia* (*Aspidiotus*) *lataniae* (Signoret), *Parlaspis papillosa* (Green), *Pinnapis* (*Chionaspis*) *aspidistrae* (Signoret) and *Sernelaspidus artocarpi* (Green) *Aspidiotus cistuloides* Green, *trigtandilosus* (Green) as also soft scales, *Ceroplastes rubens* Maskell, *Chloropulvinaria psidii* (Maskell) and *Coccus* (*Lecanium*) *ocutissimus* (Green). Except *S. artocarpi* and *P. papillosa*, the others are all polyphagous pests. Loss caused to jack-fruit trees by these insects is usually negligible. It is only *C, rubens* that occur regularly, the rest of the species occur sporadically and that too only in certain areas.

Ceroplastes rubens commonly known as waxy scale is a main pest of *Citrus* spp. and is often found on fig, jack-fruit, rnango, pear, etc. Hill (1975) also mentions coffee and tea is its alternate host plant. The pest has limited distribution CC1E map No. A-118) and has been reported from East Africa, India, Sri Lanka, parts of China, Japan. Malaysia, Philippines, East Australia, Pacific Islands, New Zealand, USA (Hawaii) etc. It is widely distributed in India. Colonies of these scales may be found on jack-fruit trees, covering the shoots and fruit stalks and sucking the cell sap therefrom. These insects also excrete copious amount of honeydew which keeps dripping on lower leaves and fruits; the honeydew attracts the ants and also favours the development of sooty mould which covers the affected parts with a black superficial coating. Adult females are convex in shape, 3 to 4 mm long and are covered with a pink waxy shell often with white vertical stripes. Males are rare. There is only one generation in a year.

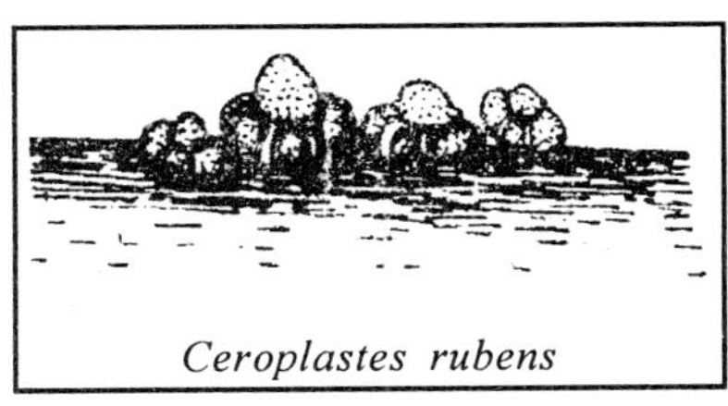

Ceroplastes rubens

Generally, chemical control measures are not required against scale insects on jack-fruit. If and when necessary, spray 0.05% dichlorvos or dimethoate.

Aphids : Jack-fruit aphid, *Greenidia artocarpi* (Westwood) and black citrus aphid, *Toxoptera aurantii*(Fonscolmbe) have been recorded feeding on jack-fruit trees; the former is an specific pest of jack-fruit whereas the latter is a polyphagous pest, *Citrus* spp. being the main hosts. Clusters of nymphs and adults may be seen on tender shoots and leaves sucking the cell sap and giving out honeydew. As a result, the entire tree is devitalised and the affected parts are distorted and covered with sooty mould. A dry spell of weather followed by rain favours rapid multiplication of the aphids resulting in severe infestation.

Greenidia artocarpi is a pale green aphid having long and hairy cornicles. It has been reported from Tamil Nadu where the pest is active from January-February to September-October (David, 1956). Reproduction is usually parthenogenetical, hence, population build-up is very rapid.

To control these aphids, spray 0.03% dimethoate, monocrotophos or phosphamidon. Two sprayings given at an interval of 10 to 12 days are quite sufficient to keep the pest population under check. These sprayings will also control whiteflies and thrips.

Whiteflies : *Aleurotrachelus caerulescens* Singh. *A. rachipora* Singh and *Pealius schimae* Takahashi (Aleyrodidae) have been recorded on jack-fruit but not as pests; the first two are polyphagous while the last one is specific to jack-fruit. Occasionally the nymphs of these whiteflies may be seen sucking the sap from leaves and giving out honeydew but the loss caused by them is negligible.

Thrips : *Pseudodendrothrips dwivarna* (Ramakrishna and Margabandhu) (Thripidae) was originally described by the authors (1931) as *Dendothrips dwivarna* but subsequently, Singh (1942) placed it under genus *Pseudodendrothrips*. This is typically a monophagous species infesting tender leaves of jack-fruit (Ananthakrishan, 1971). It is confined to Peninsular India and has been recorded from Karnataka, Kerala and Tamil Nadu. Each affected leaf may have 15 to 20 adults and numerous eggs and nymphs on it (Ananthakrishnan, 1955). Concentrated feeding of nymphs and adults produces a number of whitish spots and patches on the leaves and the affected leaves ultimately fade and die away.

Eggs are minute, elongated oval, slightly curved and 0.2 mm long. Freshly hatched nymphs are very small, less than 0.5 mm long and orange yellow with dark gray antennae and pale gray legs. Adults are also small, females being 0.9 to 1.1 mm long and males 0.6 to 0.7 mm, having head and thorax reddish-brown and abdomen yellow with heavily fringed wings. Females are always more in number; sex ratio is five females to one male.

Red Ant : *Oecophylla smaragdina* (Fabricius) (Formicidae) is found in most orchards, plantations and forests all over India. Innumerable workers of these ants may be seen going up and down the trees especially on sunny days. They cause no direct injury to the trees except webbing together a few apical leaves for constructing their nests. On the contrary, being carnivorous, they prey upon a number of insect pests. But they also relish honeydew for which they carry the nymphs of aphids, mealybugs and scale insects from tree to tree – thus helping the dissemination of these harmful insects. Besides, they are a real nuisance to persons who have to climb the trees.

Oecophylla smaragdina (Ants' nests)

To control these ants, their nests should be removed and destroyed mechanically, or these nests may be sprayed to run off point with 0.1% lindane (Butani and Tahiliani, 1974).

21

PAPAYA

PAPAYA (PAPAW), *Carica papaya* Linnaeus (Caricaceae), is native of Mexico. It was introduced into India in 16th century and is by now grown all over the country in frost-free areas. It had occupied about 10,000 hectares (Singh, 1969), which is now 12,000 hectares, of which 4000 hectares are in Bihar. It is essentially a Tropical crop and is highly susceptible to frost and needs well-drained soil, sun and high temperatures. The fruit is an important source of vitamin A and B and is rich in sugar and digestive enzymes. Ripe fruits are eaten as dessert while unripe fruits are cooked and taken as vegetable. Milky latex obtained from green, immature fruits contains papain - a proteolytic enzyme, used for tenderising meat. It is also used in preparing chewing-gum and in textile industry for reducing shrinkage of certain types of wool. In beer industry it is used for clarifying beer and in tanning industry for bating hides. The seeds and pulp of unripe fruits have anti-fertility property while use of ripe fruits is a cure for various stomach disorders. Variety 'Coorg honeydew' is high yielding and requires no cross pollination.

INSECT PESTS

Hitherto, no plant protection measures were being carried out in case of papaya as these were considered to be uneconomical. The trees usually do not suffer any substantial loss from insect pests but mites and virus diseases (mostly transmitted by insects) do take a heavy toll. With consumers becoming quality conscious, and the high price that the papaya fruits now fetch, it has become imperative to protect this crop, against the ravages of pests and diseases. The major pests are, *ak* grasshopper and mites. However, aphids and whitefly, though sporadic in occurrence, may cause serious loss by transmiting various viral diseases. Other insect pests recorded on papaya trees include, fruit-fly, coccoid, stem borer and gray weevil. In addition to these nematodes and also monkeys, bats, crows and other birds cause damage to ripe fruits.

***Ak* Grasshopper :** *Poekilocerus pictus* Fabricius (Acrididae) has been reported from India, Bangladesh, Pakistan and Africa. In India, it is widely distributed throughout the plains. It is primarily a defoliator of *ak* plants (*Calotropis* spp.), but has also been reported defoliating ber, banana, citrus, fig, grapevine, guava, melon, papaya, peach, various ornamental plants, vegetables etc. In case of severe infestation even the bark of papaya trees is not spared. The attack may start as early as April and continues till the onset of winter, maximum damage is done during July-August.

Pruthi and Nigam (1939) have described its immature stages. Eggs are elongate, curved and yellowish-orange in colour. Nymphs are yellowish-white with orange and black stripes and dots all over their bodies. Adults are stout, yellowish with broad bluish-green stripes on head and thorax; antennae are bluish-black with yellow rings; abdomen yellowish with transverse blue-black bands. Forewings are bluish-green with yellow veins and reticulations, hind wings hyaline.

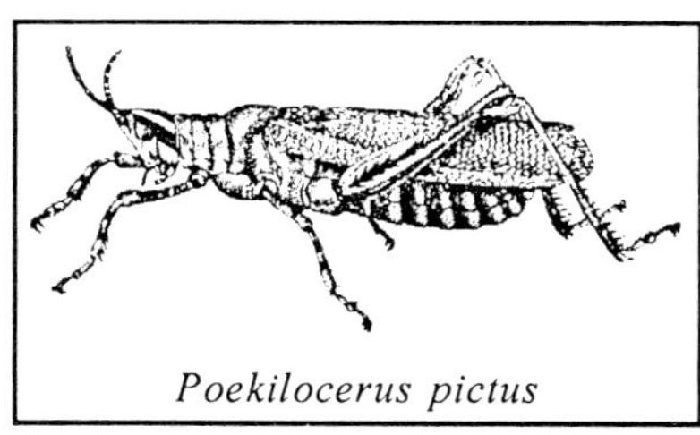
Poekilocerus pictus

Mating lasts for 5 to 7 hours and pre-oviposition period is 3 to 4 weeks. The female thrusts its abdomen into the soil and lays about 150 to 180 eggs at a depth of 120 to 200 mm, depending upon the texture of the soil. There are at least two broods in a year - a short one with incubation period of one month (June-July) and nymphal period of two months and a long generation when eggs laid during September to November overwinter and hatch around end of March or early April and become adults after another 2½ months.

To conrol these grasshoppers, Batra (1955) suggested dusting the plants trees with 5% HCH. Spraying 0.03% lindane, 0.05% chlorpyrifos or 0.1% malathion is also effecive (Verma *et al.*, 1970).

Aphids : *Aphis gossypii* Glover and *Myzus persicae* (Sulzer) (Aphididae) have been found feeding on papaya leaves (Singh, 1972); the former is a major pest of melon and the latter of peach. Nymphs and adults inject their saliva in the plant tissues and suck the cell sap, but do not cause any serious damage. In additon, these aphids act as vectors by transmitting various viral diseases and the loss thus caused is often phenomenal. The affected trees seldom recover from the disease and gradually die away. Reddy (1968) and Singh (1972) listed, *Aphis gossypii*, *A. medicaginis* Koch, *Myzus persicae* and *Uroleucom sonchi*(Linnaeus) as vectors of papaya mosaic virus.

The early symptoms of the disease are appearance of necrotic dots on the leaves; later blistered patches of green tissue may be seen distributed indiscriminately over the yellowing leaf lamina. Affected fruits become smaller in size, elongate and develop circular water-soaked lesions with solid spot in the centre.

It is quite easy to control the aphid as pest of papaya leaves, but not as virus vector. By the time the insects are noticed and dusting or spraying is done to check their population, the function as vector, by transmitting the virus disease has already been accomplished, and then, killing the vector is like locking the stable after the horse has been stolen. Therefore, only prophylactic sprayings can save the trees from attack of aphids and subsequent appearance of the disease. It is suggested to spray 0.03% dimethoate, fenitrothion, monocrotophos or phosphamidon etc, If and when any viral disease is noticed, cut and destroy immediately the affected leaves, to prevent the disease from spreading.

Whitefly : *Bemisia tabaci* (Gennadius) (= *gossypiperda* Misra and Lamba) (Aleyrodidae) is found in most parts of Tropics and subtropics (CIE map No. A-284). Its main hosts are tobacco, cotton and tomato on which the infestation is sporadically severe.

Tiny, white, scale-like objects may be seen clustering in between the veins on ventral surface of leaves. Shake the leaf slightly and a herd of tiny whiteflies flutter out for a few seconds and resettle rapidly on the leaf. Pest activity is more common during dry season and declines rapidly with the onset, of rains. As a result of their sucking sap from leaves, the affected leaves become yellowish, curl down-wards, wrinkle and there is early shedding of such leaves. In addition, this whitefly also acts as vector by transmitting the virus, causing leaf curl disease.

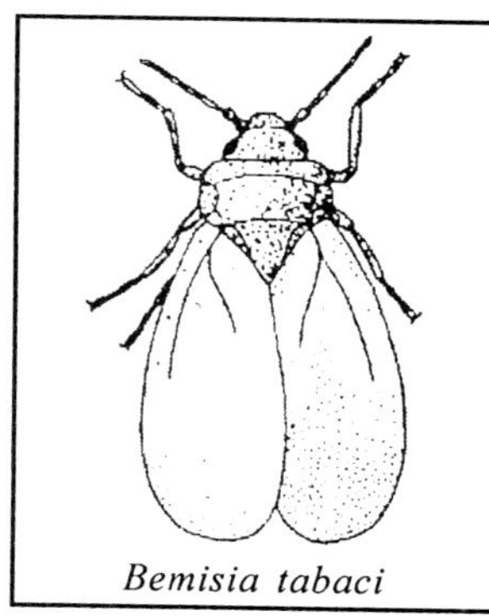
Bemisia tabaci

The control measures suggested for aphids are good enough for whitetly also.

Fruit-flies : Male flies of *Bactrocera diversus* (Coquillett) (Trypetidae) have been observed in large number visiting female flowers of papaya (Pruthi and Batra, 1960) but these cause practically no damage. Most species of fruit-flies do not have that long ovipositor with which they can thrust their eggs inside papaya fruits. However, *B. cucurbitae* (Coquillett) - a major pest of melons has been recorded on papaya (Narayanan and Batra, 1960) so also *Toxotrypma curvicauda* Gerstaecker (Trypetidae) (Dutta, 1966) and *B. pedestris* Bezzi (Hill, 1975). But all these are of minor importance. Unripe or ripening fruits are seldom attacked by any species of fruit-fly whereas ripe and over-ripe fruits may be occasionally attacked, but the infestation is never severe.

Outside India, Mediterranean fruit-fly, *Ceratitis capitata* (Wieldernann) has been recorded as major pest of papaya.

Coccoids : Among mealybugs and scale insects *Drosicha mangiferae* (Green) (Monophlebidae), *Planococcus citri* (Risso) (Pseudococcidae) *Aspidiotus destructor* (Signoret) (Diaspididae) and *Pseudoporlatoria* *ostriate* Cockerell (Diaspididae) have been occasionally reported as minor pests sucking cell sap from leaves of papaya trees. But none of these cause any economical loss. Clipping off the affected leaves with scales on, checks the spread of these coccoids.

Stem Borer : *Cerostoma rugosella* (Stainton) (Tinaeidae) has been recorded as a minor pest of papaya from Bihar, Maharashtra and Tamil Nadu (Fletcher, 1921; Hayes, 1966). The pest has earlier been reported feeding on bark of mango and *gular Ficus glomerata* Roxb. trees (Lefroy and Howlett 1909). The caterpillars bore into the stems. feed within and pupate in silken cocoons cither under the bark of the trees or in the soil.

To check the infestation of this borer, hand-picking of caterpillars and pupae and their destruction is suggested.

Grey weevil : *Myllocerus viridanus* Fabricius (Curculoinidae), has been recorded from South India by Jayaraj *et al*., (1960). skeletonising the leaves and sometimes causing premature death of the tree. However, this is a minor pest and can be controlled by dusting 5% HCH or carbaryl.

MITES

Red Spider Mite, *Tetranychus cinnabarinus* (Boisduval) - (Tetranychidae) has Worldwide distribution and has been reported from Central and South America. Africa, Middle East, Pakistan, India, Sri Lanka. S.E. Asia, Japan, Indonesia. Philippines and Australasia. It is polyphagous in habit having a very wide range of host plants including almond, apple, cherry, *Citrus*, grapevine, melons, papaya, peach, pear and strawberry. On papaya, it has been occasionally reported as serious. Clusters of yellow spots may be seen on dorsal surface of leaves. On turning these leaves innumerable silken webs, spun by adult males, are visible sheltering minute reddish creatures - the mites. The larvae, nymphs and adults suck the cell sap from plant tissues directly.

Eggs are spherical, about 0.1 mm in diameter and whitish in colour. Larvae are pinkish, about 0.2 mm long and have 3 pairs of legs. Nyrnphs are greenish-red and have 4 pairs of legs like adults.

Adults are ovate and reddish-green, males being 0.3 to 0.4 mm long and females 0.4 to 0.5 mrn.

A female lays on an average 200 eggs singly on ventral leaf surface stuck on the strands of silken webs. These hatch in 4 to 7 days. Larval stage is 3 to 5 days, nymphal 6 to 10 days and adult females live for about 3 weeks (Hill, 1975).

Another mite *Bravipalpus phoenicis* (Geijskes) (Tenuipalpidae) has also been reported damaging papaya and other fruit trees though tea and *Citrus* spp. are its main hosts. This species is widely distributed and occurs throughout the Tropical and subtropical regions (CIE map No. A-106). It is a minor pest of papaya.

To control these mites, remove the infested leaves and burn the same. Chemical control measures are not usually required but if and when there is a heavy infestation the trees may be dusted with fine sulphur dust or sprayed with wettable sulphur or 0.05% chlorphenamidine or dicofol.

BIBLIOGRAPHY

Abate, T., 1991. Seed dressing insecticides for beanfly *Ophiomya phaseoli* (Tryon) (Diptera: Agromyzidae) control in Ethiopia. *Tropical Pest Management,* **37** (4): 334-007.

Abraham, C.C. and K.S. Remamony, 1978. New records of Myrmecine ants as pests of bhendi, *Abelmoschus esculentus* Moench. *Journal of Bombay Natural History Society,* **75**(1): 242-243.

Adachi, T. and K. Futai, 1992. Changes in insecticide suceptibility of the diamondback/moth in Hyogo, Japan, *Japan Agricultural Research Quarterly*, **26**: 144-151.

Agarwal, Ashok and R.S. Mehrotra, 1988. effect of systemic and non-systemic fungicides on mycelial growth and respiration of *Phytophthora colocasiae. Indian Phytopathology*, **41**(4): 590-593.

Agarwal, M.l., D.D. Sharma and O. Rahman, 1987. Melon fruit fly and its control. *Indian Horticulture,* **32** (2): 10-11.

Agarwal, N.S., 1955. Bionomics of *Atractomorpha crenalata* Fabricius (Orthoptera: Acrididae). *Indian Journal of Entomology*, **17** (2): 230-240.

Agarwal, N.S. and N.D. Pandey, 1961. Bionomics of *Melonogromyza phaseoli* Coq. (Diptera: Agromyzidae). *Indian Journal of Entomology*, **23** (4): 293-298.

Agarwal, S.B.D. and S.Z.H. Naqvi, 1954. *Precis orithya* Swinhoe (The Blue Pansy) as a controlling agent for *Striga euphrasiodes*: A root Parasitic weed On Sugarcane. *Proceedings Bihar Academy of Agricultural Sciences*, **2-3**: 120-125.

Agnihothrudu, V. and N.S. Mithyantha, 1978. *Pesticide residue - a review of Indian work*, 173 pp. Rallis India Limited, Bangalore.

Agnihotri, N.P., R.S. Dewan and A.K. Dixit, 1974. Residues of insecticides in food commodities from Delhi. Part 1: Vegetables. *Indian Journal of Entomology,* **36** (2): 160-164.

Agnihotri, N.P., V.T. Gajbhiye, H.K. Jain, K.P. Srivastava and S.N. Sinha, 1983. Residues of synthetic pyrethroids on *Capsicum frutescens* Linn, (Chilli), *Abelmoschus esculentus* Moench. (okra) and *Allium cepa* Linn, (onion). *Journal of entomological research*, **7**(1): 10-14.

Agnihotri, N.P., H.K. Jain, S. Rai and V.T. Gajbhiye, 1980. Relative Toxicity and residues of synthetic pyrethroids on cabbage and cauliflower. *Indian Journal of Entomology*, **42**: 717-722.

Agnihotri, N.P., S. Rai, H.K. Jain and V.T. Gajbhiye, 1985. Persistence of some synthetic pyrethroids on cucurbits. *Pestology*, **9** (4): 20-23.

Agnihotri, N.P., S.N. Sinha, H.K. Jain and A.K. Chakarbarti, 1990. Bioefficacy of some synthetic pyrethroid insecticides against *Leucinodes orbonolis* Guen. and their residues on brinjal fruit. *Indian Journal of Entomology,* **52**: 373-37.

Ahmad, Taskhir, 1939. *Aulocophora foveicollis* (Lucas). *Indian Journal of Entomology*, **1** (1-2): 108.

Ahmad, Taskhir, 1939. The Amaranthus borer, *Lixus truncatulus* (F.) and its parasites. *Indian Journal of Agricultural Sciences,* **9** (4): 609-627.

Ahmad, Taskhir and R.L. Gupta, 1941. The pea leaf-miner, *Phytomyza atricornis* (Meign) in India. *Indian Journal of Entomology*, **3** (1): 37-39.

Ahmed, H. and S.M.A. Rizvi, 1993. Residues of some synthetic pyrethroids and monocrotophos in or on okra fruits. *Indian Journal of Plant Protection*, **21** (1): 44-46.

Ahmed, R. and Shashi Verma, 1984. Resistance of dimethoate and phosphamidon on brinjal fruits and determination of waiting perioid. *Indian Journal of Entomology,* **46** (4): 398-40.

Aiyar, K.S. Padmanabha, 1944. *Protrigonis zizanialis* Swinhoe, a new pest of *Moringa pterygosperma. Indian Journal of Entomology*, **6** (1-2): 165-167.

Alam, M. Zahurul, 1962. *Insect and Mite pests of Fruits and Fruit Trees in East Pakistan and their control.* 115 pp. Department of Agriculture, East Pakistan, Dacca.

Alam, M. Zahurul, 1970: *Insect Pests of Vegetables and their control in East Pakistan,* 146 pp. Agricultural information service and East Pakistan Agricultural research institute, Dacca.

Alam, M. Zahurul, Alauddin Ahmed, Shamsul Alam and Md. Ameerul Islam, 1964. *A review of research, Division of Entomology* (1947-1964), 272 pp. Agricultural information service and East Pakistan Agricultural research institute, Dacca.

Alam, M.Z. and D.L. Sana, 1964. Biology of *Leucinodes orbonalis* Guenee in East Pakistan. In: *A review of research, Division of Entomology* (1947-1964): 192-200. Agricultural information service and East Pakistan Agricultural research institute, Dacca.

Ali, M.H. and P.M. Sanghi, 1962. Observations on oviposition, longevity and sex ratio of brinjal shoot and fruit borer, *Leucinodes orbonalis. Madras Agricultural Journal,* **49** (8): 267-268, Madras.

Ali, S. Mohammad, 1962. Coccids affecting sugarcane in Bihar. *Indian Journal Sugarcane research,* **6** (2): 72-75, New Delhi.

Ali, S. Mohammad, 1968. Coccids (Coccidae = Homoptera: Insecta) affecting fruit plants in Bihar (India). *Journal of Bombay Natural History Society,* **65** (1): 120-137.

Anand, R.K., S. Rai and V.K. Sharma, 1967: Notes on hover flies (Diptera: Syrphidae) from Delhi and adjoining areas. *Indian Journal of Entomology,* **29** (3): 307-308.

Ananthakrishnan, T.N., 1955: Notes on *Pseudodendrothrips dwivorna* (R & M) from India, *Indian Journal of Entomology,* **17** (2): 213-216.

Ananthakrishnan, T.N., 1960. Thysanoptera from the Nilgiris and Kodaikanal Hills (S. India). *Journal of Bombay Natural History Society,* **57** (3): 557-578.

Ananthakrishnan, T.N., 1969: *Indian Thysanoptera. Zoological Monograph* No. 1, 171 pp. Publication and Information Directorate, CSIR, New Delhi.

Ananthakrishnan, T.N., 1971: Thrips (Thysanoptera) in Agriculture, Horticulture & Forestry - Diagnosis, Bionomics and control. *Journal of Scientific & Industrial* Research, **30**(3): 113-146, New Delhi.

Ananthakrishnan, T.N., 1973. *Thrips: Biology and Control,* 120 pp. The Macmillan company of India Ltd., New Delhi.

Ananthanarayanan, K. and T.R. Subramanian, 1958. Recent observations on the control of sweet potato weevil *Cylas formicorius* (Fb.). *Madras Agricultural Journal,* **45** (2): 74-75.

Anonymous, 1955. Control of white ant pest in vegetable gardens. Indian *Farm Mechanisation,* **6** (8): 14, New Delhi.

Anonymous, 1955. Onion thrip - method of controlling the pest. *Indian Farm Mechanisation,* **6** (8): 30, New Delhi.

Anonymous, 1955. Tomato fruit borer. *Indian Farm Mechanisation,* **6** (11): 4, New Delhi.

Anonymous, 1956. A brinjal pest - Steps for control. *Indian Farm Mechanisation,* **6** (13): 8, New Delhi.

Anonymous, 1956. Chilli thrips, dusting for control. *Allahabad Farmer,* **3** (4): 133, Allahabad.

Anonymous, 1957. Treating soil with DDT, measure against potato cutworm. *Indian Farm Mechanisation,* **8** (10-11): 4, New Delhi.

Anonymous, 1958. Studies on *Leucinodes orbonolis* Guenee (Lepidoptera, Heterocera, Pyralidae). *Kanpur Agricultural College Journal,* **17** (1-2): 72, Kanpur.

Anonymous, 1959. What is new in farming - potato tuber moth, *Indian Farming,* **9** (9): 25, New Delhi.

Anonymous, 1979. *Underexploited Tropical Plants with promising Economic value*, 189 pp. National Academy of Sciences, Washington, D.C.

Anonymous, 1980. Chilli can prevent blood cancer. *Sunday Standard, August* 3, 1980. 5, New Delhi.

Anonymous, 1987. Annual report. Indian Agricultural research Institute: New Delhi, 54-55, New Delhi.

Anonymous, 1988. Annual report. Indian Agricultural research Institute: New Delhi, 70, New Delhi.

Anonymous, 1989. Annual report (1988-89) Indian Agricultural research Institute: New Delhi.

Anonymous, 1990. Annual report (1989-90). Indian Agricultural research Institute: New Delhi, 76-77, New Delhi.

Anonymous, 2007. ICRISAT cure for Bollworm. *Pestology*. ***XXXI***(9): 74.

Arya, Meenakshi and B. Tiagi, 1982. Changes in total proteins in three carrot cultivars infested with *Meloidogyne incognita. Indian Journal of Nematology,* **12** (2): 398-400, New Delhi.

Asokan, G. and M.S. Venugopal, 1992. Bio-efficacy of certain synthetic pyrethroids to chilli thrips, *Scirtothrips dorsalis* (Hood), *Pestology,* **16** (1): 11-13, Bombay.

Atwal, A.S., 1986. *Agricultural Pests of India and South-east Asia,* 529 pp., Kalyani Publishers, New Delhi.

Atwal, A.S. and S.l. Sethi, 1963: Predation by *Coccinella septempunctata L.* on the cabbage Aphid, *Lipaphis erysimi* (Kalt.) in India. *Journal of Animal Ecology,* **31** (3): 481-488, Oxford.

Atwal, A.S. and K. Singh, 1969. Comparative efficacy of *Bacillus thuringiensis* Berliner and some contact insecticides against *Pieris brassicae* L. and *Plutella xylostella* L. on cauliflower. *Indian Journal of Entomology,* **31**: 361-363.

Atwal, A.S. and Kuljit Singh, 1969. Chemical control of cotton jassid (*Empoasca devastans* Distant) and whitefly (*Bemisia tabaci* Genn.). *Research Bulletin, Punjab Agricultural University, New Series,* **10** (3-4): 247-255, Ludhiana.

Atwal, A.S. and N.D. Verma, 1972. Development of *Leucinodes orbonolis* Guen. (Lepidoptera: Pyraustidae) in relation to different levels of temperature and humidity. *Indian Journal of Agricultural Sciences,* **42** (9): 849-854.

Avtar Singh and Dhamo K. Butani, 1963. Control of cotton jassids. *Indian Farming,* **13** (3): 9 & 12.

Avtar Singh and Dhamo K. Butani, 1964. Insect pests of cotton: III - A note on control of Jassid by dusting. *Indian Cotton Growing Review,* 18 (6): 361 -365.

Awasthi, M.D., R.S. Dewan, S.S. Misra, V.K. Chandla and S.M.A. Rizvi, 1977. Oxydemeton-methyl and phorate residues on/in potatoes. *Indian Journal of Plant Protection,* **5** (1): 41-46.

Awasthi, M.D., A.K. Dixit, Shashi Verma, S.K. Handa and R.S. Dewan, 1974: Dissipation of endosulfan residues from 'cauliflower crops. *Indian Journal of Plant Protection,* **2** (1): 24-29.

Awasthi. M.D., A.K. Dixit, Shashi Verma, S.K. Handa and R.S. Dewan, 1977: Fate of monocrotophos and carbaryl on cowpea. *Indian Journal of Plant Protection,* **5** (1): 55-61.

Awasthi, M.D., A.K. Dixit, Shashi Verma, S.K. Handa and R.S. Dewan, 1978: Residues following Application of monocrotophos and endosulfan on pea crop. *Indian Journal of Plant Protection,* 6 (1): 1-6.

Awate, E.G., D.N. Gandhale, A.S. Patil and L.M. Naik, 1982. Control of diamondback moth, *Plutella xylostella* L. with synthetic pyrothroids. *Pestology,* **6** (3); 1 L-12.

Ayyar, G.R., 1954: Unrecorded host of *Pentalonia nigronervosa* Coq. *current Science,* **23** (12): 408-409.

Ayyar, P.N.K., 1943. Biology of new Ichneumonid parasite of Amaranthus stem weevil of South India. *Proceedings of Indian Academy of Sciences,* **17-b** (2): 27-36.

Ayyar, T.V. Ramakrishna, 1922. The weevil Fauna of South India with special reference to species of economic importance. *Imperial Agricultural Research Institute, Pusa (Bihar), Bulletin No. 125,* 21 PP., Calcutta.

Ayyar, T.V. Ramakrishna, 1928. A contribution to our knowledge of the Thysanoptera of India. *Memoirs of Department of Agriculture in India, Entomological Series,* **10**: 217-316, Calcutta.

Ayyar, T.V. Ramakrishna, 1929. The economic status of Indian Thysanoptera. *Bulletin of Entomological Research,* **20**(1): 77-79.

Ayyar, T.V. Ramakrishna, 1930. Contribution to our knowledge of South Indian Coccidae (Scales and Mealy-bugs). *Imperial Council of Agricultural Research, India, Miscellaneous Bulletin No. 107,* 73 PP., Calcutta.

Ayyar, T.V. Ramakrishna, 1940. The Bionomics of the yam beetle, *Galerucida bicolor* (hope), a pest of cultivated yam in South India. *Journal of Bombay Natural History Society,* **47** (4): 874-876.

Ayyar, T.V. Ramakrishna, 1943. Notes on some fruit sucking moths of the Deccan. *Indian Journal of entomology,* **5** (1-2): 29-33.

Ayyar, T.V. Ramakrishna, 1963: *Handbook of Economic Entomology for South India.* 576 pp., Government of Madras, Madras.

Ayyar, T.V. Ramakrishna and V. Margabandhu, 1931: Notes on Indian Thysanoptera with brief description of new species. *Journal of Bombay Natural History Society,* **34**(4): 1029-1040.

Ayyar, T.V. Ramakrishna, M.S. Subbiah And P.S. Krishnamurti, 1935: The leaf curl disease of chillies caused by thrips in the Guntur and Madura tracts. *Madras Agricultural Journal,* **23** (10): 403-410, Madras, Coimbatore.

Azam, K.M. and T. Bala Bhaskar, 1983: Efficacy of certain insecticides in the control of leaf hoppers, *Amrasca devastans* Dist. on Okra. *Pestology,* 7(7): Ll-13.

Azariah, M.D., RP. Rai and K.K. Nirula, 1963. Survey of aphid infesting potatoes in Nilgiris. *Indian Potato Journal,* **5** (2): 80-85, New Delhi.

Babu, K.H. and K.M. Azam, 1985. Relative efficacy of synthetic pyrethroids against pest complex of cabbage. *Pestology,* **11** (12): 11-13, Bombay.

Babu, P.C.S. and B. Rajasekaran, 1981. A note on the control of stemfly, *Ophiomyia phoseoli* Coq. on cowpea, *Vigna unguiculata*(l.). *Madras Agricultural Journal,* **68:** 205-206, Madras, Coimbatore.

Bagle, B.G. and Shashi Verma, 1990. Decontamination of cauliflower and okra treated with methamidophos: *Plant Protection Bulletin,* **42** (1 & 2): 17-19.

Bagle, B.G. and Shashi Verma, 1991. Persistance and residual toxicity of monocrotophos against jassids on okra leaves. *Journal of Maharashtra Agricultural Universities,* **16** (1): 135- 136, Pune.

Bahl, N. and N.P. Agnihotri, 1989. Translocation of Nematicide residues In mushroom following their application at composting. *Mushroom Science.* **12**: 883-888. Braunschweig.

Bakhavatsalam, N., 1994. Chemical control of Pod borer, *Lampides boeticus* L. in Rainfed pea in Nagaland, *Pestology,* **18** (2): 18-19, Bombay.

Ballard, E. and Y. Ramachandra Rao, 1924. A Preliminary note on the life-history of cèrtain Anthomyid flies, *Atherigona* Spp. and *Acritochaeta excisa* Thomson. *Report of Proceedings of 5th entomological Meeting, February 1923,* Pusa (Bihar): 275-277, Calcutta.

Banerjee, S.N. and A.N. Basu, 1955: On the control of brinjal stem and fruit borer, *Leucinodes orboncdis* Guen. in West Bengal. *Science and Culture,* **20** (7): 350-351.

Banerjee, S.N. and A.N. Basu, 1956: Preliminary studies on the epidemiology of the potato aphids in West Bengal. *Beitrage Zur Entomologie,* **6**: 510-516, Berlin.

Banerjee, S.N. and A.N. Basu, 1956. Evaluation of insecticides against the brinjal stem and shoot borer in India. *FAO Plant Protection Bulletin,* **5** (1): 7-8, Rome.

Banerjee, S.N. and A.N. Basu, 1956. Experiments on the control of the brinjal mealy bug, *Centrococcus insulitus* (Green), in West Bengal. *Proceedings of Zoological Society of Bengal,* **9** (1): 45-51.

Banerjee, S.N. and B.K. Chatterjee, 1955: Studies on the phytotoxic action of BHC and DDT on the common cucurbitaceous Plants in West Bengal. *Proceedings of Indian Academy of Sciences,* **41-b** (5): 227-239, Bangalore.

Banerjee, Sourendra Nath and Benoy Krishna Chatterjee, 1955. Studies on *Gryllotcdpa africana* Palisot De Beauvois, a new major pest of potato crop *(Solanum tuberosum)* and its control. *Indian Journal of Entomology,* **17** (2): 219-229, New Delhi.

Barlett, B.R., 1956. Natural Predators. Can selective insecticides help into preserve biotic control. *Agricultural Chemistry,* **11:** 42-44 & 47.

Barlow, E., 1900. Notes on insect-pests from the entomological section, Indian Museum. *Indian Museum Notes,* **4** (4): 198, Calcutta.

Baser, S.L. and K.S. Kushwaha, 1968. Biology and external morphology of Forage Pests IV - Lucerne hairy caterpillar, *Creotonotus gangis*

Linnaeus (Arctiidae: Lepidoptera). *University of Udaipur, Research Studies,* **6**: 124-164.

Basha, A. Azeez and M. Balasubramanian, 1980. Newer chemicals for the control of green peach aphid, *Myzus persicae* Sulz. on chillies. Pesticides, **14** (*5*): 12-13 & 20.

Basha, A. Azeez, S. Chelliah and M. Gopalan, 1982. Effect of synthetic pyrethroids in the control of brinjal fruit borer (*Leucinodes orbonalis* Gen.). *Pesticides,* **16**(9): 10-11.

Basu, P.K. and V.R. Wallen, 1966. Influence of temperature on the viability, virulence and physiologic characteristic of *Xanthomonas phaseoli* var. *fuscans in Vivo* and in *Vitro. Canadian Journal of Botany,* **44**: 1239-1245.

Batra, H.N., 1953: Aphids infesting peach and their control. *Indian Journal of Entomology,* **15** (1): 45-51.

Batra, H.N., 1955. *AK* grasshopper, *Poekilocerus pictus* (fabr.) (Acrididae) as a pest of papaya and some other plants in Delhi. *Indian Journal of Entomology,* **17**(1): 132.

Batra, H.N., 1955. Danger of *Dacus* (*Leptoxyda*) *longistylus* vied. becoming a pest of cucurbitaceous plants in India. *Indian Journal of Entomology,* **17** (2): 278-279.

Batra, H.N., 1958. Bionomics of *Bagrada cruciferarum* Kirkaldy (Heteroptera: Pentatomidae) and its occurrence as a pest of mustard Seeds. *Indian Journal of Entomology,* **20** (2): 130-135.

Batra, H.N., 1960. Practical hints for controlling the red Pumpkin beetle. *Indian Horticulture,* **4** (2): 10-11.

Batra, H.N., 1960. The cabbage aphid *(Brevicoryne brassicae L.)* and its control schedule. *Indian Journal of Horticulture,* **17** (1): 74-80.

Batra, H.N., 1961. First record of *Haltica cyanea* Webber And *Bagous* species on singhara crop. *Indian Journal of Entomology,* **23** (1): 66-68.

Batra, H.N. and N.S. Bhattacharjee, 1960. Occurrence of *Hymenia recurvalis* (Fabricius) (Lepidoptera: Pyralidae) as a bad pest of some leafy vegetables. *Indian Journal of Entomology,* **22** (2): 128-130.

Batka, H.N. and S. Sarup, 1962: Technique of Mass Breeding of the painted bug, *Bagrada cruciferarum* Kirk. (Heteroptera: Pentatomidae*). Indian Oilseeds Journal,* **6**(2): 135-143.

Batra, H.N. and G.R. Sharma, 1959. A preliminary note on the varietal susceptibility of *Amaranthus* Spp. To the attack of Amaranthus borer,

Hypolixus truncatidus (Boheman) (Coleoptera: Curculionidae). *Indian Journal of Entomology*, **21** (2): 141-142.

Battu, G.S., V.N. Dilawari and O.S. Bindra, 1976. Studies on the larval susceptibility of lepidopterous pests of vegetables and field crops to infection by *Bacillus thuringiensis* Berliner. *Punjab Vegetable Grower*, **11** (1-4): 1-8.

Beeson, C.F.G., 1941. *The Ecology and Control of Forest Insects*, 1007 pp., C.F.G. Beeson, Forest Research Institute, Dehradun.

Bell, T.D.R. and F.B. Scott, 1937: *The Fauna of British India including Ceylon and Burma. Moths*, 5 (Sphingidae), 357 pp. Taylor and Francis, London.

Beri, Y.P., 1958. Field experiments on the cabbage pest control. *Indian Journal of Entomology*, **20** (1): 63-66.

Beri, Y.P., 1958: Preliminary studies on the effect of DDT Spray on the oviposition of *Plutella macidipennis* Curtis. *Indian Journal of Entomology*, **20** (3): 235-236.

Beri, Y.P., 1961. Studies on the application of Dyar's Law to the larval stages of *Plutella maculipennis* Curtis. *Indian Journal of Entomology*, **23** (1): 69-70.

Betala. S.R. and K.S. Kushwaha, 1965. Biology and external morphology of the cabbage Semilooper, *Plusia nignsigna* Walker (Lepidoptera: Noctuidae). *University of Udaipur, research studies*, 3: 159-161.

Bhandari, K.G. and G.S. Soni, 1962. Bionomics and control of *Urentius sentis* Distant (Hemiptera: Heteroptera: Tingidae). *Punjab Horticultural Journal*, **2** (1): 44-6l

Bharathan, N. and S.P.S. Beniwal, 1984 Transmission characteristic of urd bean leaf crinkle virus by the *Epilachna* Beetle *Henosepilachna dodecastigma*. *Indian Phytopathology*, **37**: 660-664.

Bhattacharjee, N.S. and M.G. Ramdas Menon, 1964. Bionomics, biology and control of *Hymenia recurvalis* (Fabricius) (Pyralidae: Lepidoptera). *Indian Journal of Entomology*, **26** (2): 176-183.

Bhatia, Ranjeet, Divender Gupta and A.K. Sharma, 1995. Field Screening of brinjal germplasm for resistance to shoot and fruit borer *(Leucinodes orboncdis* Guen.*). Indian Journal of Plant Protection*, **23** (1): 78.

Bhatia, S.L. and M. Shaffi, 1932. Life-histories of some Indian Syrphids. *Indian Journal of Agricultural Sciences*, **2** *(4)*: 543-570.

Bhatnagar, S.P. 1980. The distribution, abundance and status of chilli white

Grub. *Holotrichia insularis* Brenske in Rajasthan State. *Rajasthan Journal of Pesticides,* **7** (2): 100-110.

Bheemanna M., B.V. Patil, S.G. Hanchinal, A.C. Hosamani and N. Kengego wda. 2005. Bio-efficacy of emamection benzoate (Proclaim) 5% SG against okra fruit borers. *Pestology, XXIX* (2): 14-16.

Bijaya, P,S. Subharani and T.K. Singh. 2005. Bioefficacy of certain botanical insecticides against green peach aphid, *Myzus persicae* (Sulzer) on cabbage *Pestology,* XXIX (3): 28-31.

Bindra O.S. and Harcharan Singh, 1979. Phycitid pod borer, *Etiella zinckenella* Treitschke (Lepidoptera: Phycitidae). *Pesticides,* **13** *(5):* 14-17.

Bindra, O.S. and R.L. Kalra (Eds.), 1973. *Progress and Problems in Pesticide Residue Analysis,* 336 pp. The Punjab Agricultural University and Indian council of Agricultural research, New Delhi.

Bindra, O.S. and M.S. Mahal, 1981. Varietal resistance in Eggplant (*Solanum melongena L.*) to the cotton jassid (*Amrasca biguttula biguttula* Lshida). *Phytoparastica,* **9** (2): 119-131.

Bingham, C.T., 1907. *The Fauna of British India Including Ceylon and Burma. Butterflies,* 2. 480 pp. Taylor and Francis, London.

Borle, M.N. and S.A. Nimbalkar, 1980. Residual Toxicity of different insecticides against larvae of *Prodenia litura* Fabr. on cauliflower. *Pesticides,* **14**(2): 31-32.

Brar, K.S., A.S. Sidhu and M. l. Chadha, 1993. Screening onion vrieties for resistance to *Thrips tabaci* Lind. And *Helicoverpa armigera* (Hubner). *Journal of Insect Science,* **6** (1): 123-124.

Burkill, I.H., 1953. Habits of man and the origins of the cultivated plants of the old world. *Proceedings of Linnean Society of London,* **164**: 12-41, London.

Butani, Dhamo K., 1958 A. Note on *Hymenia fascialis* C., A Pest of sugarbeet. *Current Science,* **27** (5): 182-183.

Butani, Dhamo K., 1958 B. A note on the external Anatomy of *Precis orithya* Swinhoe (Nymphalidae: Lepidoptera). *Current Science,* **27** (11): 453-454.

Butani, Dhamo K., 1960. External Anatomy of *Leucama (Cirphis) unipuncta* Haworth (Noctuidae: Lepidoptera). *Indian Journal of Sugarcane Research and Development,* **4**, (4): 213-219, New Delhi.

Butani, Dhamo K., 1974. Les Insectes Parasites Du Palmier - Dattier En Inde Et Leur controle. *Fruits,* **29** *(10)*: 689-691, Paris.

Butani, Dhamo K., 1975 A. Crop pests and their control - 1: Cotton, *Pesticides,* **9** (3): 21-27 & 29.

Butani, Dhamo K., 1975 B: Insect pests of fruit crops and their control - 15: Date Palm. *Pesticides,* **9** (3): 40-42.

Butani, Dhamo K., 1976 A. Spotted bollworms of cotton, *Earias* Spp. (Noctuidae: Lepidoptera). *Cotton development,* **6** (3): 17-22.

Butani, Dhamo K., 1976 B. Pest`s and diseases of chillies and their control. *Pesticides,* **10** (8): 38-41.

Butani, Dhamo K., 1977 A. Insect pests of vegetables -Tomato. *Pesticides,* **11** (1): 33-36.

Butani, Dhamo K., 1977 B. *Indarbella quadrinotata* (Wlk). In: *Diseases, Pests and Weeds in Tropical Crops*: 438-439, Verlag Paul Parey, Berlin.

Butani, Dhamo K., 1978: Pests and diseases of jackfruit in India and their control. *Fruits,* **33** (5): 351-367. Paris.

Butani, Dhamo K., 1979 A: *Insects and Fruits*, 415 pp., Periodical Expert Book Agency, New Delhi.

Butani, Dhamo K., 1979 B. Insect pests of citrus and their control, *Pesticides,* **13** (4): 15-21.

Butani, Dhamo K., 1980. Insect Pests of vegetables and their control: Cluster Bean - *Pesticides,* **14** (10): 33-35.

Butani, Dhamo K., 1987. Key Pests of vegetables and their management, in: *Recent Advance in entomology,* **1**: 433-454 (Eds. Y.k. Mathur, A.K. Bhattacharya, N.D. Pandey, K.D. Upadhyay and J.P. Srivastava). Gopal Prakashan, Kanpur.

Butani, Dhamo K. and M.G. Jotwani, 1975. Trends in the control of insect pests of fruit crops in India. *Pesticides Annual,* 1975: 139-149.

Butani, Dhamo K. and M.G. Jotwani, 1983: Insects as a limiting factor in vegetable production. *Pesticides,* **17** (9): 6-13.

Butani, Dhamo K. and M.G. Jotwani, 1984. *Insects in vegetables,* 356 pp., Periodical Expert Book Agency, New Delhi.

Butani, Dhamo K., M.G. Jotwani and Shashi Verma, 1977. Insect pests of vegetables and their control - Cole crops. *Pesticides,* **11** (5): 19-24.

Butani, Dhamo K. and Shashi Shekhar Juneja, 1984. Pests of radish and their control. *Pestology,* **18**(5): 10-12.

Butani, Dhamo K. and Shashi Shekhar Juneja, 1985. Insects pests of elephants - foot and their control. *Pestology,* **9** (4): 37-39.

Butani, Dhamo K. and Shashi Verma, 1976 A. Pests of vegetables and their control - Brinjal. *Pesticides,* **10** (2): 33-35.

Butani, Dhamo K. and Shashi Verma, 1976 B. Pests of vegetables and their control - Sweet potato. *Pesticides,* **10** (2): 36-38.

Butani, Dhamo K. and Shashi Verma, 1976 C: Pests of vegetables and their control - 3: Potato. *Pesticides,* **10** (4): 46-51.

Butani, Dhamo K. and Shashi Verma, 1976 D: Insect pests of vegetables and their control - Lady's Finger. *Pesticides,* **10** (7): 31-37.

Butani, Dhamo K. and Shashi Verma, 1976 E: Insect Pests of vegetables and their control - Onion and Garlic. *Pesticides,* **10** (11): 33-35.

Butani, Dhamo K. and Shashi Verma, 1977 A. Insect pests of vegetables and their control - Cucurbits. *Pesticides,* **11** (3): 37-41.

Butani, Dhamo K. and Shashi Verma, 1977 B: Pests of vegetables and their control - Leafy vegetables. *Pesticides,* **11** (4): 34-36.

Butani, Dhamo K. and Shashi Verma, 1981 A. Insect pests of vegetables and their control - Colocasia. *Pesticides,* **15** (9): 33-37.

Butani, Dhamo K. and Shashi Verma, 1981 B. Insect Pests of vegetables and their control - Drumsticks. *Pesticides,* **15** (10): 29-32.

Butani, Dhamo K. and B.D. Tahiliani, 1974. Red ants on mango in South Gujarat. *Entomologists' Newsletter,* **4**(7): 37-38, New Delhi.

Butler, G.D. Jr. and T.J. Henneberry, 1991. Sweet potato whitefly control: effect of tomato cultures and plant derived oils. *Southwestern Entomologist,* **16** (1): 37-43, College Station, U.S.A.

Campbell, A.S., 1925. Studies on some borers and miners. *Lingnaam Agricultural Review,* 3: 16-17, Canton.

Carson, Rachel, 1962. *Silent Spring,* 304 pp., Eawcett Publications, Inc. Greenwich, Conn., USA.

Chahal, B.S., Jagir Singh and S.S. Sekkon, 1975. Control of *Myzus persicae* (Sulzer) a vector of potato viruses on autumn seed crop with phorate. *Indian Journal of Horticulture,* **32** (3-4): 184-186.

Chandele, A.G. and Anita Mane, 1994. Laboratory bioassay of *Bacillus thuringiensis* Ber. Against *Plutella xylostella* L. *Petology,* **18** (8): 27-28.

Chandla, V.K. and S.M.A. Rizvi, 1977. Chemical control of *Heliothis armigera* Hb. (Noctuidae: Lepidoptera) in Potato. *Pesticides,* **11** (9), 60-61.

Chandra, J. and O.P. Lal, 1973. Record of thrips on some vegetables and ornamental plants from Kulu Valley, Himachal Pradesh. *Indian Journal of Entomology,* **35** (2): 164-166.

Chandra, J. and O.P. Lal,, 1976. Development and survival of caterpillar of cabbage butterfly, *Pieris brassicae* Linn. on some varieties of cabbage. *Indian Journal of Entomology*, **38** (2): 187-188.

Chandramohan, N. and K. Kanjan, 1994. Control of white grub, *Anomola communis* (B) on cabbage. *Pestology*, **18** (4): 12.

Chandrasekaran, J., A. Ragupathy, K. Bhuvanesvar and V. Balasubramani, 1994. Biological evaluation of carbosulfan and Biobit (B.t.k.) for the control of diamondback moth *Plutella xylostella* (L.) on cabbage and cauliflower. *Pestology*, **18** (5): 5-12.

Chandrasekharan, S.N. and T.S. Ramakrishnan, 1929. Botany of some useful plants - VI: Ternstroemiaceae. *Madras Agricultural Journal*, **17** (1): 7-16.

Chaudhary, O.P. and R.K. Kashyap, 1992. Effect of some management practices on the incidence of insect pests and yield of eggplant in India. *Tropical Pest Management*, **38** (4): 416-419, London.

Chaudhari, R.P., K.D. Verma, V.K. Chandla and S.M.A. Rizvi, 1977. Epidemiology of aphid vectors of potato virus diseases in Jammu and Kashmir. *Indian Journal of Plant Protection*, **5** (2): 161-163.

Chaudhuri, R.P., 1953. Control of cutworms in potato fields. *Indian Journal of Entomology*, **15** (3): 203-206.

Chaudhuri, R.P., 1955. Mealy bug infestation on potato in storage. *Science and culture*, **21** (2): 99-100.

Chaudhuri, R.P., 1957. Studies on the methods of protecting stored potatoes from potato tuber moth damage. *Indian Journal of Entomology*, **19** (4): 268-278.

Chelliah, S. 1973. Investigation on damage potential of *Aphis gossypii* G. in eggplant (*Solanum melongena* L.). *Madras Agricultural Journal*, **60** (2): 127-128.

Cheng, C.H., 1967. An observation on ecology of *Stephanitis typica* Distant (Hemiptera: Tinigidae) on banana. *Journal Taiwan Agriculture Research*, **16** (2): 54-69, Taipei.

Cherian, M.C., 1934. Notes on some S. Indian syrphidae. *Journal of Bombay Natural History Society*, **37** (3): 697-699.

Cherian, M.C. and Mohamad Basheer, 1938. A new cecidomyid pest of Moringa. *Madras Agricultural Journal*, **26**(3): 92-95, Madras, Coimbatore.

Cherian, M.C. and Mohamad Basheer, 1939. *Noorda moringae* Tarns, a new Pyralid pest of *Moringa pterygosperma*. *Indian Journal of Entomology*, **1** (3): 77-82.

Cherian, M.C. and Mohamad Basheer, 1940. *Euborellia stall* Dohrn.

(Forficulidae) as a pest of Groundnut in South India. *Indian Journal of Entomology*, **2** (2): 155-158.

Cherian, M.C. and P. Israel, 1939. Notes on *Perina nuda* Fabr. (Layman) and its natural enemies. *Madras Agricultural Journal,* **26** (6): 203-207.

Cherian, M.C. and Krishna Menon, 1940. *Oncocephala tuberculata* Oliv. (Chrysomelidae). *Indian Journal of Entomology*, **2**(1): 97-98.

Cherian, M.C. and B. Rangiam Pillai, 1942. *Selepa docilis* Butl. and its parasite *Euplectrus euplexiae* Roh. *Indian Journal of Entomology*, **4** (2): 236.

Cherian, M.C. and C.V. Sundaram, 1939. Notes on the life history and habits of *Dacus brevistylus* Bezzi (Family Trypetidae) a pest of *Coccina indica* fruits. *Indian Journal of Agricultural Sciences,* **9** (1): 127-131.

Chhibber, R.C., 1975. Know the Insect pests of sugarbeet. *Indian Farmers' Digest,* **8** (8): 21-24, Pantnagar.

Chopra, R.N., 1933. *Indigenous drugs of India*, 483 pp., The Art Press, Calcutta.

Choudhary, J.P. and Pandita, ML., 1974. Population dynamics of aphids, jassids and whiteflies in relation to the production of virus free seed of potato. *Haryana Journal of Horticultural Science,* **3** (1-2): 33-38, Hisar.

Choudhary, R.S., 1982. Control of brinjal shoot and fruit borer *Leucinodes orbonolis* Guen (Pyralidae: Lepidoptera). *Entomon,* 7 (3): 257-259.

Choudhuri, H.C., 1957. A survey of Aphids infesting potatoes in the plains of West Bengal. *American Potato Journal,* **34** (1): 10-19.

Choudhuri, H.C., 1958: *Potato in West Bengal,* 125 pp., Government of West Bengal, Calcutta.

Choudhury, B., 1967. *Vegetables*, 198 pp., National Book Trust, India, New Delhi.

Choudhury, B., T.S. Kalda and P.M. Bhagchandani, 1971. *Pusa Kranti* a new name in brinjal. *Indian Horticulture,* **16**(1): 11-12, New Delhi.

Chowdhuri, A.N., 1965. Control of *Epiachna* Beetles attacking potato in Simla Hills. *Indian Potato Journal,* 7 (2) ; 108, New Delhi.

Chowdhury, S. and S. Majid, 1954. *Handbook of Plant Protection*, 117 pp., department of Agriculture, Assam, Shillong.

Cleghorn, James, 1914. Melon culture in Peshin Baluchistan and some account of the melon-fly pest. *Agricultural Journal of India,* **9** (2): 124-140, Calcutta.

Clement, Peter, B.V. David, R. Sundrarajan and V. Govindarajalu, 1989. evaluation of Cartap for the control of diamond back moth, *Plutella xylostella* (L.) on cabbage. *Pestology,* **13** (12): 21-24.

Clement Peter and V. Govindarajalu, 1994. Efficacy and persistent toxicity of certain new insecticides against brinjal pest complex. *Pestology,* **18** (1): 27-30.

COPR, 1978. *Pest control in Tropical Root Crops.* Pans Manual No. 4, 235 pp., Centre for Overseas Pest Research, London.

COPR, 1981. *Pest control in Tropical Grain Legumes.* 206 pp., Centre for Overseas Pest Research, London.

Coquerell, Ch., 1859. Note Sur Quelques Insectes De Madagascar Et De Bourbon. *Annales De La Societe Entomologique, France* (3) **7**: 239-260, Paris.

Cramer, Pierre, 1779. *Papillons Exotiques,* **3**: 250, Amsterdam.

Cruz, Y.P. and E.N. Bkhnardo, 1971. The Biology and feeding behaviour of the melon aphid. *Aphis gossypii* Glover (Aphididae: Hemiptera) on four host plants. *Philippine Entomologist,* **2** (2): 155-156, Manila.

CSIR, 1948(1950). *Wealth of India -raw materials,* **1**,253 pp., Council of Scientific and Industrial Research, New Delhi.

CSIR, 1950. *Wealth of India - raw materials,* **2**, 427 pp., Council of Scientific and Industrial Research, New Delhi.

CSIR, 1952. *Wealth of India - raw materials,* **3**, 236 pp., Council of Scientific and Industrial Research, New Delhi.

CSIR, 1959. *Wealth of India - raw materials,* **5**, 332 pp., Council of Scientific and Industrial Research, New Delhi.

Das Bashambar, 1918. The aphididae of Lahore. *Memoirs of Indian Museum,* **6**: 135-274, Calcutta.

Das, N.M. and M.R.G.K. Nair, 1966. Control of the sweet potato weevil *Cylas formicarius* Fb. with some of the newer synthetic insecticides. *Agricultural research Journal,* Kerala, **4** (1): 78-81.

Das, N.M., K.S. Remamony and M.R.G.K. Nair, 1968. On the control of melonfly *Dacus cucurbitae* Coquillett with some newer synthetic insecticides applied as cover sprays. *Agricultural Research Journal, Kerala,* **6** (1): 43-45.

Davey, M.R. M.S. Mccabe, U. Mohapatra and J.B. Power. 2002. *In transgenic Plants and Corps.* G.G. Khachatourians, A. Mc

Mughen, R.scorza, Wai-kh Nip and Y.H. Hui (Eds.) pp. 613-635. Morcel Dekker Inc., New York, NY-10016, U.S.A.

David, A. Leela, 1960. Insecticidal trial for the control of the sweet potato weevil. *Madras Agricultural Journal,* **47** (6): 247-251, Madras, Coimbatore.

David, A. Leela, 1963. A new and safer Insecticide for the control of *Epilachna vigintiociopunctata* F. and fruit borer *Leucinodes orbonalis* G. on brinjal. *Madras Agricultural Journal,* **50** (2): 103.

David, A. Leela, 1964. Insecticidal control of *bhindi* leaf hoppers. *Madras Agricultural Journal,* **51** (1): 21-23.

David, B. Vasantharaj, 1960. *Plusia (Phytomera) (Trichoplusia) ni* Hb. (Noctuidae) as a pest of cabbage, *Brassica oleracea* in South India. *Journal of Bombay Natural History Society,* **57** (3): 679-681.

David, B. Vasantharaj, 1961 A. Some new hosts of *Sylepta lunalis* Gn. in South India. *Indian Journal of Entomology,* **23** (4): 301.

David, B. Vasantharaj, 1961 B. A diaspidine scale on Moringa in South India. *Madras Agricultural Journal.* **48** (6): 225.

David, B. Vasantharaj and T. Kumaraswami, 1975. *Elements of Economic Entomology*, 507 pp. Popular Book Depot, Madras.

David, B. Vasantharaj, N.V. Radha and K.A. Sheshu, 1973. Influence of weather factors on the population of castor Aleyrodid *Trialeurodes rara* Singh. *Madras Agricultural Journal,* **60** (7): 496-499.

David, S. Kanakaraj, 1956. Additional notes on some aphids in the Madras State. *Madras Agricultural Journal,* **43** (3): 103-107.

David, B. Vasantharaj and T.N. Ananthakrishnan. 2004. *General and Applied Entomology,* 1184 pp. Tata Mcgraw-hill Publishing company Limited, New Delhi-110002.

David, S. Kanakaraj, 1958. Aphids capable of infesting potato in India and their relationship to the crop in South India. *South Indian Horticulturist,* **6** (2): 67-74.

Deshmukh, S.N. and Rattan Lal, 1969. Investigations on carbaryl residues in/on brinjal plants. *Indian Journal of Entomology,* **31** (3): 222-234.

Deshmukh, S.N. and A.S. Udeaan, 1972. Dissipation of malathion, carbaryl and endosulfan on brinjal fruits. *Proceedings of 3rd International Symposium on Subtropical and Tropical Horticulture,* (Abstracts): Bangalore.

Deshmukh, S.N. and J. Singh, 1975. Dissipation of carbaryl and malathion from okra fruit. *Indian Journal of Entomology,* **37** (1): 64-68.

Dethier, V.G., 1970. Some general consideration of insects' responses to the chemicals in food Plants. In: *Control of Insect Behaviour by Natural Products,* D.L. Wood, R.M. Silvrstein and M. Nakajima [Eds.]: 21-28. Academic Press, New York, U.S.A.

Dewan, R.S. and Y.P. Beri, 1965. Persistence of DDT and BHC residues on brinjal, *Solanum melongena* L. (Var. *Pusa Purple Long*) Fruit. *Indian Journal of Entomology,* **27** (2): 202-204.

Dewan, R.S., S.S. Misra, S.K. Handa and Y.P. Beri, 1967. Persistence of DDT, Gamma BHC and carbaryl residues on *bhindi, Abelmoschus esculentus* fruits. *Indian Journal of Entomology,* **29** (3): 225-228.

Dhaliwal, G.S. and Ramesh Arora, 1996. *Principles of Insect Pest Management:* 374 pp., National Agricultural Technology Information Centre, Ludhiana.

Dhaliwal, G.S. and Ram Singh 2004. *Host Plant resistance to Insets.* 578 pp. Panima Publishing Corporation, New Delhi-110002.

Dhamdhere, S.V., J. Bahadur and U.S. Misra, 1985. Relative efficacy of Granular insecticides for the control of *Amrasca biguttula biguttula* Ishida on okra. *P.K. V. Research Journal,* **9** (2): 37-40.

Dhankhar, B.S., V.P. Gupta and K. Singh 1977. Screening and variability Studies for relative susceptibility to shoot and fruit borer *(Leucinodes orbonalis* G.) In normal and ratoon crop of brinjal (*Solanum melognena* L.). *Haryana Journal of Horticultural Sciences,* **6** (1-2): 50-58, Hisar.

Dhingra, S. and Shashi Verma, 1984. Relative toxicity of insecticides to the adults of *Oxycarenus lactus,* Kirby a pest of okra. *Journal Entomological Research,* **8** (1): 111-113.

Dichson, R.C.,J. E Swift, L.D. Anderson and J.T. Middleton 1940. Insect vectors of cantalvke mosaic in california's desert vallay. Journal of Economic Entomolvgy. 4 (5): 778-774.

Dikshit, A.K., S.S. Misra and L. Lal, 1985. Persistence of chlorpyrifos residues in Potato and the effect of processing. *Potato Research,* **28** (4): 461-467.

Dikshit, A.K., S.K. Handa and Shashi Verma, 1986 Residues of methamidopho and effect of washing and cooking in cauliflower, cabbage and Indian colza. *Indian Journal of Agricultural Sciences,* **56** (9): 661-666, New Delhi.

Dikshit, A.K., S.K. Handa and Shashi Verma, 1987. Residue of. metasystox in and on cabbage and lablab bean. *Indian Journal of Agricultural Sciences,* **57** (4): 284-286.

Dingra, M.R., S.V. Dhandere and R.R. Rawat, 1989. Natural enemies of *Epilachna ocellata* at Jabalpur (M.P.) India, *Bulletin of Entomology,* **27**, (2): 184-185.

Dixit, A.K., M.D. Awasthi, S.K. Handa, Shashi Verma and R.S. Dewan, 1975. Residues and residual toxicity of malathion and endosulfan on Okra fruits. *Indian Journal of Entomology,* **37** (3): 251-254.

Doane, J.F. and R.K. Chapman, 1962. Control of root maggots on radish, Turnip and Rutabaga in Wisconsin. *Journal of Economic Entomology.* **55** (2): 160-164.

Donocaster John P., 1946. The Shallot Aphis, *Myzus ascalonicus* Sp. N. (Hemiptera, Aphididae). *Proceedings of Royal Entomological Society,* **15** (B): 27-31.

Dureja P. and V.T. Gajbhiya 2004. Insecticides registered under section 9 (3) of the Insecticide Act. 1968 as on 24th August, 2004. *Pesticide Research Journal.* **16** (2): 84-96.

Dutt, H.L., 1914. The Potato tingid (*Recaredus* sp.). *Agricultural Journal,* **2**: 48-68, Department of Bihar and Orissa, Patna.

Dutta, S., 1966. *Horticulture in the Eastern Region of India,* 264 pp., Directorate of Extension, Ministry of food and Agriculture, New Delhi.

Edwards, C.A., G.K. Veeresh and H.R. Krueger (Eds.), 1980. *Pesticide Residues in the Environment in India,* 524 pp. University of Agricultural Sciences, Hebbal, Bangalore.

Edwards, W.H., 1937. Horn-worms which Defoliate Sweet potato vines. *Journal Jamaica Agricultural Society,* **41** (9):515, Kingston.

Egwuatu, R.I., 1984. Field trials with Systemic contact insecticides for the control of *Podogrica unifarma* and *Podagrica sjostedti* (Coleoptera: Chrysomelidae) on okra. *Tropical Pest Management,* **28** (2): 115-121.

Endersby, N.M., W.C. Morgan, B.C. Stevenson and C.T. Waters, 1992. Alternative to regular insecticide applications for the control of Lepidopterous pests of *Brassica oleracea* var. *capitata. Biological, Agriculture & Horticulture,* **88**(3): 189-203.

Eswaramoorthy, S., S. Cuelliah and S. Uthamasamy, 1978. Efficacy of certain insecticides against the sucking pests of *Bhindi, Abelmoschus esculentus* (L.) Moench. *Madras Agricultural Journal,* **63** (4): 254-256.

Fabricius, Johann Christian, 1794. *Entomologia Systematica,* **3** (2): 276, Hafniae.

Feakin, S.D. (Ed.), 1978. *Pest control in Tropical Root crops.* Pans Manual No. 4, 235 pp., Centre for Overseas Pest Research, London.

Feakin. S.D. (Ed.), 1983. *Pest control in Tropical Tomatoes,* Pans Manual, 130 PP., Centre for Overseas Pest Research, London.

Fernando, H.F., 1957. the Biology and control of *Leptoglosus membranaceus,* Fabricius (Fam. Coreidae, Ord. Hemiptera). *Tropical Agriculturist,* **113** (2): 107-118, Paradeniya.

Fletcher, T. Bainbrigge, 1914. *Some South Indian Insects,* 565 pp., Superintendent Government Press, Madras.

Fletcher, T. Bainbrigge, 1917. Amaranthus Pests. *Report of Proceedings of 2nd Entomological Meeting, February 1917, Pusa (Bihar):* 296-297, Calcutta.

Fletcher, T. Bainbrigge, 1920 A: Life-histories of Indian Insects Microlepidoptera: I - Pterophoridae. *Memoirs of Department of Agriculture in India, Entomological Series,* **6** (1): 9-13 & 26-28, Calcutta.

Fletcher, T. Bainbrigge, 1920 B: Life-histories of Indian Insects-microlepidoptera: III - Gelechiadae. *Memoirs of department of Agriculture in India, Entomological series,* **6** (3): 73-74, Calcutta.

Fletcher, T. Bainbrigge, 1920 C. Life-histories of Indian Insects-Microlepidoptera: IV - Cosmopterygidae, Oecophoridae, Physoptilidae, Xyloryctidae, Stenomidae and Orneodidiae. *Memoirs of department of Agriculture in India, Entomological Series,* **6** (4): 97-99 & 102-104, Calcutta.

Fletcher, T. Bainbrigge, 1920 D. Life-histories of Indian Insects-Microlepidoptera: IX - Appendix. *Memoirs of department of Agriculture in India, Entomological Series,* **6** (9): 203, Calcutta.

Fletcher, T. Bainbrigge, 1920 E. Annotated List of Indian Crop Pests. *Report of Proceedings of 3rd Entomological Meeting, February 1919, Pusa (Bihar),* **1:** 33-314, Calcutta.

Fletcher, T. Bainbrigge, 1921. Annotated List of Indian Crop Pests. *Agricultural Research Institute, Pusa, Bulletin No. 100,* 246 pp., Calcutta.

Fotedar, K.R. and Z. Singh, 1980. Effect of some systemic Insecticides on *Tetranychus cinnabarinus* on okra. *Acarology Newsletter,* **10**(3): 22-24.

Franclemont, J.G., 1951. The species of the *Leucania unipuncfa* group, with a discussion of the generic names for the various segregates of

Leucania in North America (Lepidoptera, Phalaenidae, Hadeninae). *Proceedings of Entomological Society of Washington,* **53** (1): 57-85.

Franssen, C.J.H., 1934. Insecten, Schadelijkann Hatbataten-gewas on Java. *Mededelington Van Het Instituut Pla-ziekt,* **92**: 20-21, Korte.

Frohlich, G. and W. Rodewald, 1970. *General Pests and diseases of Tropical crops and their control*, 366 pp., Pergamon Press, London.

Frey, G., 1971: Keys to the Indian and Ceylonese species of the Genus *Holotrichia* Hope (Coleoptera: Melolonthidae). *Ent. Arb. Museum, G. Frey,* **22:** 206-225, Munich (Munchen).

Froggatt, W.W., 1898. Biology of bean Agromyza. *Agricultural Gazette of New South Wales,* **9**: 560-561, Sydney.

Fukuda, K., 1933. Insect Pests of sweet potato in Formosa, Part 1. The Sweet potato weevil (In Japanese). *Report, Government Research Institute, Formosa,* **62**: L-35.

Gajbhiye, V.T., N.P. Agnihotri, H.K. Jain and Samarjit Rai, 1981. Persistence of chlorpyrifos residues on cabbage and potato tubers. *Annals of Agricultural Research,* **2**(1-2): 49- 52.

Gajbhiye, V.T., N.P. Agnihotri, P.S. Sirohi, T. George and R.K. Gupta, 1995: effect of formulation on the persistence and degradation of quinalphos on tomato, *Lycopersicon escnlentus* Mill. *Journal of Entomological Research,* **19** (1): 13-17.

Gajbhiye, V.T., G.T. Gujar, N.P. Agnihotri and H.K. Jain, 1985: Efficacy of some newer synthetic pyrethroids on okra. *Pestology,* **9** (1): 13-18.

Gandhale, D.N., G.N. Salunkhe and L.M. Naik, 1982. Effectiveness of insecticides for the control of aphids (*Brevicoryne brassicae* L.) on cabbage. *Indian Journal of Plant Protection,* **10** (1&2): 85-86.

Gangopadhyay, S. and J. Chandra, 1975. Effect of new fungicides on turnip Aphid (*Hyodophis pseudobrossicae* Davis). *Science and Culture,* **41** (1): 35-36.

Gera, ,S.S. and K.N. Bhatnagar, 1992. Seasonal incidence of pest complex of cabage and their control in a semi arid region. *Pestology,* **16** (3): 38-45.

Gera, R., H. Singh and Sucheta, 1978. Comparative efficacy of various insecticides against cotton Jassid. *Sundaptoryx biguttula biguttula* on *Bhindi. Madras Agricultural Journal,* **65***(12)*: 836-838.

Ghai, Swaraj, V.V. Ramamurthy and S.L. Gupta, 1979. Lepidopterous insects associated with rice crop in India. *Indian Journal of Entomology,* **41** (1): 65-90.

Ghulam Ullah, 1940. Food plants of *Creontiodes pallidifer* Walker. *Indian Journal of Entomology,* **2** (2): 242.

Ghulam Ullah, 1941. *Phthorimaea operculella*(zell.) as a pest of potato leaves. *Indian Journal of Entomology,* **3** (2): 336.

Gibson, A., 1913. Flea-beetles and their control. *Canada Department of Agriculture, Division of Entomology, Circular No. 2*, 11pp., Ottawa.

Gill, H.S., A.S. Kataria, K.A. Balkrishnan and Boo Pal Singh, 1988. Vegetable research highlights. Division of Vegetable Crops. Indian Agricultural Research Institute. New Delhi-74, *Pdvr Technical Bulletin.* **2**: 49 PP., New Delhi.

Goot, P. Van Der, 1930. Biological notes on bean stem Agromyza. *Mededlington Van Het Instituut Pla-ziekt, Buitenz,* **78**: 97, Batavia.

Gopalan, M., U. Venkata-narayanan, S. Uthamasamy And T.R. Subramanian, 1974. Control of major pests of *bhindi* with systemic granular insecticides and foliar sprays. *South Indian Horticulture,* **12:** 101-105.

Gowda, J.V.N., U.N. Sulladmath and K.P.G. Rao, 1977. Phenthoate - a new effective phosphatic insecticide against cabbage diamondback moth, brinjal borer and onion thrip. *Current Research,* **6** (3): 42-43.

Green, E.E., 1903. Notes on *Dorylus orientcdis* West. *Indian Museum Notes,* **5** (2): 39.

Green, E.E., 1933. Notes on some Coccidae from Surinam, Dutch Guiana with description of new species. *Stylops*, **11** (3): 49-58, London.

Greenwood, D.F. and R.N. Hofmaster, 1950. The efficacy of several new insecticides for the control of *H. recurvcdis* F. on fall Spinach. *Journal of Economic Entomology,* **43** (1): 108.

Grewal, J.S., G.S. Shekhawat, S.K. Bhattacharyya and R.A. Singh, [Eds.], 1990. Current facets in potato research. (*Proceedings of National Seminar held at Central Potato Research Station,* Modipuram U.P. 13-15 December, 1991). 216 pp., Indian Potato Association, Shimla.

Gubbaiah Davaiah, M.C., 1978. Occurrence and biology of *Aspidomorpha sanctaecrucis* (F.) on *Ipomoea* sp. *Current Research,* 7 (9): 156-157.

Gupta, D.S. (Ed.), 1980. *Residue Analysis of Insecticides*, 374 pp., Haryana Agricultural University, Hisar.

Gupta, G.K. 1977. Properties and Insect transmission of cauliflower mosaic virus in Himachal Pradesh. *Indian Phytopathology,* **30**: 117-118.

Gupta, H.C.L. and B.L. Pareek, 1978. Removal of carbaryl residues from onion. *Journal of Food and Science Technology,* **15** (5): 215-216.

Gupta, K.M., 1945. *Phyllotreta cruciferae* Goeze (Halticinae-Chrysomelidae) a pest of cultivated Cruciferous crops in the United Provinces. *Indian Journal of Entomology,* 7 (1-2): 239-240.

Gupta, R.L., 1956. A Note on Life History of *Euzophera perticella Rag. Proceedings of 43rd Indian Science Congress, Agra, Iii:* 294, Calcutta.

Gupta, R.L. and R.R. Rawat, 1954. Life-history of *Hypolixus truncatulus (Boh.)* [*Lixus brachyrhinus* (Boh.)] - The Rajgira weevil. *Indian Journal of Entomology,* **16** (2): 142- 144.

Guram, M.S. and G.S. Sohi, 1961 A: Trials of some Insecticides of vegetable origin against *Aphis gossypii* Glover. *Proceedings of 48th Indian Science Congress, Roorkee, Iii:* 490, Calcutta.

Guram, M.S. and G.S. Sohi, 1961 B. Trails of some local plant products against, red pumpking beetle (*Aulacophora foveicollis* Luc.) *Proceedings of 48th Indian Science Congress,* Roorkee, III: 491, Calcutta.

Hami, M.A., 1955. Effect of Borer attack on the Vitamin C contents of brinjals. *Pakistan Journal of Health,* **4**(4): 223-224, Karachi.

Hampson, G.F., 1892. *The Fauna of British India including Ceylon and Burma. Moths,* **1**, 527 pp., Taylor and Francis, London.

Hampson, G.F., 1894. *The Fauna of British India including Ceylon and Burma.* Moths, **2**, 609 pp., Taylor and Francis, London.

Hampson, G.F., 1895. *The Fauna of British India including Ceylon and Burma. Moths,* **3**, 546 pp., Taylor And Francis, London.

Hampson, G.F., 1896. *The Fauna of British India including Ceylon and Burma. Moths,* **4**, 594 PP., Taylor and Francis, London.

Handa, S.K., A.K. Dikshit and Shashi Verma, 1989. Residue of fenitrothion on cauliflower *Brassica oleracea botrytis* and cabbage *Brassica oleracea capitata, Indian Journal of Agricultural Sciences,* **59** (1): 25-27.

Harding, J.A., 1961. Studies on the control of thrips attacking onions. *Journal of Economic Entomology,* **54** (6): 1254-1255, Baltimore, Maryland, USA.

Haseeb Masarrat 2007. Influence of host plants on susceptibility of *Helicoverpa armigera, Spodoptera litura* and *Spilarctia (Spilosoma) obliqua* to *Beauveria bassiam. Annals of Plant Protection Sciences.* **15** (1): 30-33.

Hayes, W.B.1966. Fruits growing in India, 512 pp. Kitabistan, Allahabad.

Hemalatha, B. and T. Uma Maheswari. 2004. Evaludion of tomato varieties against serpentine leaf miner, *Liriomyza trifolii* Burgess. *Indian Journal of Plant Protection* **32** (1): 138-139.

Herklots, G.A.C., 1972. *Vegetables in South-east Asia*, 525 pp., George Allen and Unwin Ltd., London.

Hill, Dennis S. and Jeremy D. Hill, 1994: *Agricultural Entomology*, 635pp. Timber Press Inc., Porland, Oregon.

Hill, Dennis S., 1975/1983. *Agricultural Insect Pests of the Tropics and their Control,* 516 pp. 746 pp. Cambridge University Press, Cambridge.

Hirst, S., 1923. On some new little known species of Acari. *Proceedings of Zoological Society of London,* **4:** 971-1000, London.

Hole, R.C. and M.H. Shorey, 1966. Systemic Insecticides for control of Western flower thrips on bulb onions. *Journal of Economic Entomology,* **58** (4): 793-794, Baltimore, Maryland, USA.

Hole, U.B., A.R. Walunj and A.S. Tambe, 1997. Efficacy hostathion against shoot and fruit borer of brinjal. *Pestology,* **21** (6): 14-16.

Hofsvang, T., 1991. The influence of intercropping and weeds on the oviposition of *Brassica* root flies. *Norwegian Journal Agricultural Science,* **5** (4): 349-356.

Hosamani, A.C., K. Thulasiram, B.V. Patil, M. Bheemanna and S.G. Hanchinal. 2005. Fenpropation (Meothrin) 30 Ec: Anideal Insecticide for chilli (*Capsicum annuum* L.) pest management. *Pestology,* **XXIX**(2): 21-24.

Hunter, W.B. and D.E. Ullman, 1992. Effect of the neem product Rd-repelin on settling behaviour and transmission of zuechini yellow mosaic virus by pea aphid, *Acyrthosiphum pisum* (Harris). *Annals of Applied Biology,* **120**: 9-15, Cambridge.

Husain, M. Afzal and M.R. Bhalla, 1937. The bird enemies of the cotton leaf roller, *Sylepta derogata* Fb., At Khanewala Multan (Punjab). *Indian Journalo of Agricultural Sciences,* 7 (5): 785-792.

Hutson, J.C., 1936. Entomological Notes. *Tropical Agriculturist,* **87** (5): 289-295, Paradeniya.

Hyde, W.C., 1931. *Phthorimaea* moth attacking tomato plants. *New Zealand Journal of Agriculture,* **42** (3): 175-211.

Isaac P.V., 1946. Report of the Imperial Entomologist. *Scientific Report Imperial Agricultural Research Institute, New Delhi,* 1944-45: 73-79, New Delhi.

Isahaque, N. MD. M., 1978, Observations on the bionomics of *Phthorimaea operculella* (Zell.) (Lepidoptera: Gelechiidae) a pest of potato in Assam. *Pesticides,* **12** (4): 57-60.

Iyer, T.V. Subramaniam, 1922. Notes on the more important insect pests of crops in Mysore State. III-diptera, IV- Rhynchota. *Indian Journal of Mysore Agricultural and Experimental Union,* **4** (1): 16-24.

Jacob, Sosama, 1982. Residues and residual toxicity of insecticides methamidophos, quinalphos and malathion on vegetable crops okra and cauliflower. Ph.D. Thesis, division of Entomology, Indian Agricultural research Institute, New Delhi.

Jacob, S. and Shashi Verma, 1984 A. Dissipation of methamidophos on cauliflower. *Indian Journal of Plant of Protection,* **12**: 97-100.

Jacob, S. and Shashi Verma, 1984 B. Persistance of malathion on okra. *Indian Journal of Agricultural Sciences,* **54** (11): 993-994.

Jacob, S. and Shashi Verma, 1985 A. Dissiption of methamidophos on okra. *Pestology,* **9** (10): 8-14.

Jacob, S. and Shashi Verma, 1985 B. Decontamination of okra (*Abelmoschus esculentus* Moe) treated with quinalphos. *Agricultural Research Journal, Kerala,* **23** (1): 15-19.

Jacob, S. and Shashi Verma, 1985 C. Biological Efficacy of methamidophos, quinalphos and malathion against jassid, *Amrasca biguttula biguttula,* Ishida on okra. *Indiann Journal of Plant Protection,* **13** (1): 15-19.

Jacob, S. and Shashi Verma, 1985 D. Quinalphos residues on cauliflower. *Indian Journal of Entomology,* **47** (3): 304-309.

Jacob, S. and Shashi Verma, 1987. Efficacy of Insecticides against the Population of aphid, *Brevicoryne brassicae* on cauliflower, *Indian Journal of Agricultural Sciences,* **57** (11): 857-861.

Jadhav, L.D., D.S. Ajri, M.V. Kadam and S.K. Dorge, 1976. Leaf-footed Plant bug on pomegranate in Maharashtra. *Entomologists' Newsletter,* **6** (10): 57.

Jadhav, L.D., M.V. Kadam, D.S. Ajri and R.N. Pokharkar, 1979. Host Preference of leaf-footed plant bug, *Lepioglossus membronoceus* Fabricius. *Indian Journal of Entomology,* **41** (3): 299-300.

Jagadish, K.S. and Gavi Gowda, 1994. Efficacy of certain insecticides as seed treatment in the management of cowpea stemfly. *Ophiomyia phaseoli* (Tryon) (Diptera: Agromyzidae). *Seed Research,* **22** (2): 156-159.

Jagan Mohan, N. and N.K. Krishna Kumar, 1980. Chemical control of onion thrips, *Thrlps tabaci* Lind. *Pesticides,* **14** *(5)*: 28-29.

Jagan Mohan, N., N.K. Krishna Kumar and V.G. Prasad, 1980. Control of leaf hopper (*Amrasca biguttula biguttula* Ishida) and fruit borer (*Leucinodes orbonalis* Guen.) on brinjal. *Pesticides,* **14** (7): 19-21.

Jagan Mohan, N., K. Krishnaiah and N.K. Krishna Kumar, 1981. Chemical control of mustard aphid, *Lipaphis erysimi* Kalt. And leaf webber, *Crocidolomia binotalis* Zell. on cabbage. *Pesticides,* **15** (2): 29-32.

Jagan Mohan, N., K. Krishnaiah and V.G. Prasad, 1983. Chemical control of insect pests of okra, *Abelmoschus esculentus* Moench. *Indian Journal of Entomology,* **45** (2): 152-158.

Jagan Mohan, N. and V.G. Prasad, 1984. Role of synthetic pyrethroids in the control of brinjal pests. *Indian Journal of Entomology,* **46** (2): 179-182.

Jaglan, R.S. and P.R. Yadav, 1986. Persistence and degration of endosulfan in and on unprocessed and processed cauliflower curds. *Indian Journal of Entomology,* **48** (4): 441-447.

Jagtap, B. and M.V. Kadam, 1981. Studies on chemical control of mustard sawfly on radish. *Pesticides,* **15:** 9-10.

Jain, H.K. and N.P. Agnihotri, 1980. Residues of fenvalerate (Sumicidin) on vegetables, In: *Pesticide Residues in the environment in India:* 171-176, University of Agricultural Sciences, Hebbal, Bangalore.

Jain, H.K., N.P. Agnihotri, S. Rai and V.T. Gajbhiye, 1983. Residues of synthetic pyrethroids on brinjal and tomato fruits. *Annals of Agriculture Research,* **4**: 35-39.

Jain, H.K., N.P. Agnihotri and K.P. Srivastava, 1979. Toxicity of fenvalerate and estimation of its residues on some vegetables. *Journal of Entomological Research,* **3** (2): 212-216.

Jain, S.P., 1957. Control of *Epilachna* beetle. *Allahabad Farmer,* **31** (1): 20-22.

Jakhmola, S.S. and R.R. Rawat, 1969. Chemical control of *bhindi* jassid by insecticidal granules. Department of Entomology, Annual Report, *1968-69:* 67-68. Jawaharlal Nehru Krishi Vishwa Vidyalaya, Jabalpur.

James, Kelsa T. and R. Varatharajan, 1995. Efficacy of two neem products in the field control of *Scirtothrips dorsalis* Hood (Thysanoptera) on *Capsicum annum. Indian Journal of Plant Protection,* **23** (2): 166-168.

Jat, N.R. and B.P. Srivastava, 1973. Extent of malathion residues in and on okra fruits. *Indian Journal of Entomology,* **35** (2): 143-147.

Jayaraj, S., A. Abdul Kareem and P.P. Vasudeva Menon, 1960. *Myllocerus viridenus* F. (Curculoinidse: Coleoptera), a new Pest of papaya in South India. *Current Science,* **29** (10): 413.

Jestin, C.G.L., R.J. Rabindra and S. Jayaraj, 1990. *Bacillus thuringiensis* Berliner and some insecticides against diamondback moth, *Plulella xylostella* (L.) on Cauliflower. *Journal of Biological control,* **4** (1): 40-43.

Johnson, D.E., 1930. The relation of the cabbage maggot and other insects to the spread and development of soft rot of Crucifers. *Phytopathology,* **20** (2): 857-872.

Jones, F.G.W. 1961. The potato root eelworm *Heterodera rostochiensis* Woll. in India. *Current Science,* **30** (5): 187.

Jones, F.G.W. and M. Jones, 1974. *Pests of field crops,* 2nd Edition, 448 pp., Edward Arnold, London.

Jones, H.A., S.R. Daily and S.L. Emsweller, 1934. Thrips resistance in the onion. *Hilgardia,* **8**: 215-232.

Jones, W.W., 1931. Tomato vines injured by a Mirid (*Engytatus geniculatus* Reuter). *Journal of Economic Entomology,* **24** (1): 327-328.

Jotwani, M.G. and Dhamo K. Butani, 1977. Insect pests of leguminous vegetables and their control. *Pesticides,* **11** (10): 35-38.

Jotwani, M.G. and Dhamo K. Butani, 1980 A. Insect pests of crops. in: *Handbook of Agriculture:* 417-551, Indian council of Agricultural Research, New Delhi.

Jotwani, M.G. and Dhamo K. Butani, 1980 B: Insect pests of onion and their control. *Pestology,* **4** (10): 1-4.

Jotwani, M.G. and Dhamo K. Butani, 1981: Insect pest problems of National importance and Government Schemes. In: *Indian Pesticide Industry*: 300-310. Vishvas Publications, Vile Parle, Bombay.

Jotwani, M.G. and Prakash Sarup, 1963. Evolution of control schedule for brinjal, *Solanum melongena* Linn, (Variety *Pusa Purple Long*) particularly Against the fruit borer, *Leucinodes orbonalis* Guen. *Indian Journal of Entomology,* **25** (4): 275-291.

Jotwani, M.G. and Prakash Sarup, 1966. Control of jassids on okra seedlings by treatment of seeds with different insecticides. *Indian Journal of Entomology,* **29**(4): 144-149.

Jotwani, M.G., Prakash Sarup and S. Pradhan, 1961. Relative toxicity of some important insecticides to the adults of lacewing bug, *Urentius echinus* distant (Tingidae: Heteroptera), a pest of brinjal. *Indian Journal of Horticulture,* **18** (1): 81-84, New Delhi.

Jotwani, M.G., Prakash Sarup and S. Pradhan, 1962. Comparative toxicity of insecticides to the grubs and adults of *Epilachna vigintioctopunctata* F. (Coccinellidae: Coleoptera). *Indian Journal of Entomology,* **24** (4): 223-228.

Jotwani, M.G., Prakash Sarup, P. Sircar and D.S. Singh, 1966. Control of jassid (*Emposca devastans* Dist.) on okra (*Ablemoschus esculentus*) seedlings by treatment of seed with different insecticides. *Indian Journal of Entomology,* **28** (4): 477-481.

Jotwani, M.G. and T.D. Yadav, 1965. Effect of some common fumigants on the germinability of vegetable seeds - I. *Indian Journal of Entomology,* **27** (4): 472-475.

Jotwani, M.G. and T.D. Yadav, 1966. Effect of some common fumigants on the germinability of vegetable seeds - II. *Indian Journal of Entomology,* **28** (1): 111-113.

Kakar, K.L. and J.P. Sharma, 1991. Role of natural enemies on the mortality of cabbage butterfly, *Pieris brassicae* (L.) in cauliflower seed crop. *Indian Journal of Plant Protection,* **19**(2): 210-213.

Kale, P.B., U.V. Mohod, V.N. Dod and H.S. Thakare, 1986. Screening of brinjal germplasm (*Solanum* spp.) for resistance to shoot and fruit borer (*Leucinodes orbonalis* Guen.) under field conditions. *Vegetable Science,* **13** (2): 376-383.

Kalloo G. and Kirti Singh 2001 [Eds.]. *Emerging Scenario in Vegetable Research and development.* 379 pp. Research Periodicals and Book Publishing House Houston, Texas 77272, U.S.A.

Kalra, R.L. and R.P. Chawla, 1981. Monitoring of pesticide residue in the Indian environment. In: *Indian Pesticide Industry.* 251-285, Vishvas Publications, Mumbai.

Kalra, V.K. and R. Singh, 1979. Biology of bottle-gourd plume moth *Sphenarches anisadoctylus* Walker. *Bulletin of Entomology,* **20** (1-2): 34-39.

Kandasamy, C.M. Mohanasundaram and P. Karuppuchamy, 1990: Evaluation of insecticides for the control of thrips, *Scirtothrips dorsalis* Hood on chillies (*Capsicum annum* L.). *Madras Agricultural Journal,* **77** (3-4): 169-172.

Kao, C.H., C.S. Chiu and E.Y. Cheng, 1990. Field evaluation of microbial and chemical insecticides for diamondback moth and other lepidopterous pests control on cabbage. *Journal of Agricultural Research of China,* **39** (3): 221-227, Taichung.

Kapoor, V.C., 1970. Indian Tephritidae with their recorded hosts. *Oriental Insects,* **4** (2): 207-25 L.

Kapoor, V.C. 1972. Identification of common fruit flies (Tephritidae: Diptera) of India. *Entomologist Record,* **84**: 165-169.

Kapoor, A.P., 1950. The biology and external morphology of the Larvae of *Epilachninae* (Coleoptera, Coccinellidae). *Bulletin of Entomological Research,* **40** (1): 161-208.

Kapoor, A.P. and T.G. Vazirani, 1956. The identity and geographical distribution of the Indian species of the Genus *Dysdercus* Boisduval (Hemiptera: Pyrrhocoridae). *Records of Indian Museum,* **54** (3&4): 159-175.

Kareem, P.C. 1926 Chilli leaf-curf disease Bengal Agricultural. 6(3) : 118-119.

Kareem, A. Abdul and M. Basheer, 1965. Correlation studies on the effects of weather factors on the possible dispersal of aphids. *Indian Journal of Entomology,* **27** (4): 455-459.

Kashyap, R.K., M.L. Pandita and A.N. Verma, 1979. Effect of soil application of Granular systemic insecticides on aphid aontrol and yield of seed Potato. *Haryana Journal of Horticultural Science,* **8** (3-4): 152-155.

Kathpal, T.S., Gulab Singh, PR. Yadav, R.K. Kashyap and A.N. Verma, 1983: Persistence of phorate in soil and its translocation into potato seed tubers. *Pesticides,* **17** (12): 39-43.

Katiyar, R.N. and V.N. Bodade, 1967. Effect of dimecron on the population of jassid and other cotton pests and possibilities to include it in an effective control schedule of cotton pests. *Proceedings of 54th Indian Science Congress, Hyderabad, III:* 552, Calcutta.

Kaushik, Satish K. and Gulab Singh, 1982. A short note on the incidence of blue butterfly *Lampides boeticus* L. *Indian Journal of Plant Protection,* **10** (1&2): 94.

Kawchuk, L.M. 2002. Potato Transformation Value-added Traits. In *Transgenic Plants and Crops* G.G. Khachatounans, A. Mc Mughen, R. Scorza, Wai-kit Nip, Y.H. Hui [Eds.] PP. 637-673-697. Morced Dekker Inc., New York, N.y-110016, U.S.A.

Kendappa, G.N., G. Shankar, S. Mallikarjunappa and M.S. Mithyantha, 2005. Evaluation of certain incecticides against serpentine leaf miner

Liriomyza trifolii (Burgess) (Agromyzidae: Diptera) on french bean (*Phaseolus vulgaris* L.) *Pestology*, ***XXIX*** (3): 36-38.

Kevan, D. Keith McE., 1954. A study of the Genus *Chrotogonus* Aduient-surville (Orthoptera: Acrididae). ILL - A review of available information on its economic importance, biology, etc. *Indian Journal of Entomology*, **16**(2): 145-172.

Khan, M. Haroon and M. Abdul Ghani, 1944. Studies on *Earias* Spp. (The spotted bollworm of cotton) In the Punjab, Pt. II – The crossing of *E. Fabia* Stoll and *E. insuiana* Boisd. *Indian Journal of Entomology*, **6** (1-2): 133-138.

Khan, Quadruddin and V.P. Rao, 1960. Insect and mite pests. In: *Cotton in India* - A Monograph, **2**: 217-301, Indian Central Cotton Committee, Mumbai.

Khurana, S.M. Paul, 1978. Yield losses due to PVX and PVY. *Annual Scientific Report, central Potato Research Institute*, 87 pp., Shimla.

Kimura, A. 1989. Resistance of the diamondback moth (Lepidoptesa: Yponomeutidae) to insecticides. *Plant Protection of North Japan* **40**: 145-148.

Kiran, S. Ravi, P. Sita Devi and K. Janardhan Reddy, 2007. Bioactivity of essential oils and Sesquiterpenses of *Chloroxylon suietenia* Dc against *Helicoverpa armigera*, *Current Science*, 91 (4): 544-548.

Kittur, S.U. and G.A. Patel, 1953. Difference in the injury caused by two species of pumpkin beetles during feeding. *Indian Journal of Entomology*, **15** (3): 262-263.

Knott, James E. and Rose R. Deanon, Jr., 1967. *Vegetable production in South-east Asia*, 366 pp., University of Philippines, Laguna.

Kring, J.B. and D.J. Schuster, 1992. Management of insects on Peooer and Tomato with UV-reflective Mulches. *Florida Entomologist*, **75** (1): 119-129.

Krishna Naik, L., S.G. Rayar and R.S. Giraddi, 1997. Efficacy of Teflubenzuron in the management of diamopdback moth *Plutella xylostella* L. on cabbage. *Pestology*, **21** (7): 36-38.

Krishnaiah, K., 1980. Methodology for assessing crop losses due to pests of vegetables. In: *Assessment of Crop Losses due to Pests and diseases*. technical series, *No.* **33**: 259-267, University of Agricultural Sciences, Bangalore.

Krishnaiah, K. and N. Jagan Mohan, 1977. Control of cabbage pests. *Second oriental Entomology Symposium, Madras:* 88-89, Madras.

Krishnaiah, K. and N. Jagan Mohan, 1983. Control of cabbage pests and New insecticides. *Indian Journal of Entomology,* **45** (3): 222-228.

Krishnaiah, K. and V.G. Prasad, 1978: Note on the persistence of monocrotophos residues in okra. *Indian Journal of Agricultural Sciences,* **48** (1): 56-58.

Krishnaiah, K. and P.L. Tandon, 1975. Chemical control of *Tetranychus neoadedonicus* Andre (*A. ciicwbitoe* Rahman and Sapra) infesting okra and brinjal. *Current Research,* **4**: 87-89.

Krishnaiah, K., N. Jagan Mohan and B.G. Bagle, 1978. Control of major pests of cabbage by new insecticides. *Indian Journal of Plant Protection,* **6** (1): 44-47.

Krishnaiah, K., N. Jagan Mohan and V.G. Prasad, 1975. Dissipation of carbaryl Residues on okra. *Indian Journal of Plant Protection,* **3** (3): 223-224.

Krishnaiah, K., N. Jagan Mohan and V.G. Prasad, 1978. Control of major insect pests of cabbage by new insecticide. *Indian Journal of Plant Protection,* **6**: 44-47.

Krishnaiah, K., N. Jagan Mohan and V.G. Prasad, 1979. Studies on the dissipation of monocrotophos in sweet pepper [*Capsicum annum* var. *grossum* (L.) Sendt.]. *Pesticides,* **13** (5): 39 & 47.

Krishnaiah, K., N. Jagan Mohan and V.G. Prasad, 1981. Efficacy of *Bacillus thuringiensis* Ber. for the control of lepidopterous pests of vegetable crops. *Entomon.,* **6**: 87- 93.

Krishnaiah, K., V.G. Prasad and N. Jagan Mohan, 1980. Carbaryl residues on brinjal and peas. In: *Pesticide Residues in the wnvironment in India:* 229-231, University of Agricultural Sciences, Hebbal, Bangalore.

Krishnaiah, K., P.L. Tandon and N. Jagan Mohan, 1976. Control of shoot and fruit borer in brinjal (*Leucinodes orbonolis* Guen.) with new insecticides. *Pesticides,* **10** (6): 41-42.

Krishnaiah, K., P.L. Tandon, A.S. Mathur and N. Jagan Mohan, 1976. Evaluation of insecticides for the control of major pests of okra. *Indian Journal of Agricultural Sciences,* **46** (4): 178-186.

Krishnamoorthy, C., 1956. A preliminary note on *Epilachna vigintiocto-punctata* on brinjal and bitter-gourd. *Andhra Agricultural Journal,* **3** (1): 35.

Krishna Moorthy, P.N. and N.K. Krishna Prasad. 2004. Integraled pest management in vegetable crops. In: *Integrated Pest Management in*

Indian Agriculture (Proceedings No. 11) Pratap S. Birthal and O.P. Sharma [Eds.], 265pp. National centre for Agricultural Economics and Policy Research (Ncar) and *National Centre for Integrated Pest Management.* (Ncpm) Pusa Complex, N. Delhi-110012.

Krishnamurti, B., 1932. The Potato *Epilachna* beetle, *Epilachna vigintioctopunctata* (Fabr.). *Department of Agriculture Mysore, Entomological Series, Bulletin No. 8,* 16pp., Bangalore.

Krishnaswami, S., 1954. Studies on the insecticidal and adverse effects of DDT and BHC on vegetables. *Indian Journal of Entomology,* **16** (3): 271-281.

Kulkarni, G.S., 1921. The 'Murdo' disease of Chillies capsicum, Journal and Proccedings of Asiatic Society of Bengal. 17(4) : 89.

Kulkarny, H.L., 1952. Insect pests of cruciferous and other allied vegetables. *Indian Farming,* **2** (11): 24-25.

Kulkarny, H.L., 1953. Studies on the pea aphid (*Macrosiphum pisi*) control, *Proceedings of 40th Indian Science Congress, Lucknow, III:* 195-196. Kolkata.

Kumar, D., G.S. Mavi and Joginder Singh, 1991. Impact of agro-technical practice on the incidence of peaminer, *Chromatomyia horticola* (Gourean) on Peas. *Journal of Insect Science,* **4** (2): 180-181, Ludhiana.

Kumar, K.K. and K.C. Devraj Urs, 1988. Effect of fertilizers on incidence of shoot and fruit borer in okra. *Indian Journal of Agricultural Sciences,* **58** (9): 732-734.

Kumar, Rabinder, 1965. Termites: a new pest of potato in India. *Indian Potato Journal,* 7 (1): 49.

Kundu, G.G. and N.C. Pant, 1967. Studies on *Lipaphis erysimi* (Kalt.) with special reference to insect plant relationship. I - susceptibility of different varieties of *Brassica* and *Eruca sativa* to the mustard aphid Infestation. *Indian Journal of Entomology,* **29** (3): 241-251.

Kundu, G.G., V.K. Sharma, R.K. Anand and Samarjit Rai, 1965. New Record of *Dioeretiella rapae* (Curtis) as a parasite of mustard aphid, *Lipaphis erysimi* (Kalt.) (Hemiptera: Aphididae). *Indian Journal of Entomology,* **27** (4): 497-498.

Kushwaha, K.S., 1960. A note on infestation of termites (Insecta: Isoptera) around Udaipur (Rajasthan). *Science and Culture,* **26** (1): 39-40.

Kushwaha, K.S.,.J.S. Sharma, L.S. Sharma and P.K. Pathak, 1961. A note on *Cosmopteryx phaeogastra* Meyr. (Insecta: Lepidoptera) infesting bean creepers in Rajasthan. *Indian Journal of Entomology,* **23** (4): 279-281.

Kushwaha, K.S. and L.S. Sharma, 1961. Some common insect pests of vegetable crops in the State of Rajasthan with brief notes on their infestation. *Proceedings of 48th Indian Science Congress,* Roorkee, III: 428, Kolkata.

Lal, K.B. and K.M. Gupta, 1951. Potato tuber moth and its control. Agricultural and Animal Husbandry, Uttar Pradesh, 1 (Supplement): 54-57, Lucknow.

Lal, K.B. and Chhotey Singh, 1951. The Biology and field Ecology of cotton leaf roller (*Sylepta derogata* Fabr.) In Uttar Pradesh. *Indian Cotton Growing Review,* **5** (3): 137-147.

Lal, L., 1991. Effect of intercropping on the incidence of potato tuber moth, *Phthorimoea opercalella* (Zell.). *Agriculture Ecosystem and environment,* **36** (3-4): 185-186, Amsterdam.

Lal, O.P., R.K. Sharma, T.S. Verma, P.M. Bhagchandani and J. Chandra, 1976. Resistance in brinjal to shoot and fruit borer, *Leucinodes orbonalis* Guen. (Pyralididae: Lepidoptera). *Vegetable Science,* **3**(2): 111-115.

Lal, O.P., 1972. Sterility in black bean aphid through teramycin, Likuden and flavomycin. *Current Science,* **41** (2): 74-76.

Lal, O.P., 1973 A, Brinjal and fruit borer. *Indian Horticulture,* **18**(1): 21.

Lal, O.P., 1973 B: control of cabbage crop pests. *Plant Protection Reporter,* **3** (2): 407, New Delhi.

Lal, O.P., 1973 C, Record of *Epicauta* sp. (Coleoptera: Meloidae) on soybean and eggplant from Himachal Pradesh. *Indian Journal of Entomology,* **35** (4): 342-43.

Lal, O.P., 1974. Occurrence of *Cauliops fallax* Scott. (Hemiptera: Lygaeidae) on french beans and Horse gram in Himachal Pradesh. *Indian Journal of Entomology,* **36** (1): 67-68.

Lal, O.P., 1974. Insect Pests of cucurbits and their control. *Plant Protection Reporter,* **4** (1): 9-21.

Lal, O.P., 1975 A. Insecticidal spraying causing pollen sterility in Chinese cabbage. *Acta Agronomica,* **24** (1-2): 145-47, Hungary.

Lal, O.P., 1975 B. Field trials of some Insecticides against *Epilachna* beetle on eggplant and Tomato. *Agrochimica,* **19** (3-4):306-310, Pisa.

Lal, O.P., 1976 A. Effect of certain insecticidal granules to control the pea leaf-miner, *Phytomyza atricornis.* Meign, in Kulu Valley. *Proceedings of the National Academy of Sciences, India,* **46(b)**(L-2): 302-304, Allahabad.

Lal, O.P., 1976 B. Effect of different levels if nitrogen on the development and reproductivity of *Brevicoryne brassicae* Linn, on cauliflower plants. *Proceedings of the National Academy of Sciences, India,* **46(b)**(L-2): 305-307, Allahabad.

Lal, O.P., 1976 C. Don't feast the pests on vegetables. *Intensive Agriculture,* **14**(10): 12-15.

Lal, O.P., 1978. Relative susceptibility of some cucumber and squash varieties to melon aphid, *Aphis gossypii* Glov. *Indian Journal of Plant Protection.* **5**(2): 208-10.

Lal, O.P., 1980. Field observations on the susceptibility of some cabbage varieties against cabbage worm, *Pieris rapa* L. (Lepidoptera: Pieridae) in Tripoli, Libya. *Arab development Journal of Science & Technology,* **3***(1-2)*: 39-42, Libya.

Lal, O.P., 1981. Control of brinjal flea beetle, *Psyilliodes tenebrousus* Jacoby with certain organo-phosphorous Insecticides. *Indian Journal of Entomology,* **43**(1): 59-63.

Lal, O.P., 1990 A. Response of some Chinese cabbage germplasms to the development, survival and reproductivity of cabbage aphid, *Brevicoryne brassicae* Linn. *Journal of Entomological Research,* **14**(1): 35-38.

Lal, O.P., 1990 B. Host preference of *Epilachna ocellata* Redt. (Coccinellidae: Coleoptera) among different vegetable crops. *Journal of Entomological Research,* 14 (1): 39-43.

Lal, O.P., 1991a: Varietal Resistance In cabbage against The cabbage aphid, *Brevicoryne brassicae* L. (Homoptera: Aphididae) In Kulu Valley, India. *Zeitschrift Fur Planzenkrank-heltan Pflanzensch Utz.* **98**(1): 84-91, Berlin.

Lal, O.P., 1991b: Studies on field resistance in different cultivars and crosses of cabbage against cabbage caterpillar, *Pieris brassicae* L. (Lepidoptera: Pieridae). *Zeitschrift Fur Planzenkrank Ankhettan Und Pflanzensch Utz.* **98** (3): 293-300, Berlin.

Lal, O.P., 1991 Ic: Varietal resistance in eggplant (*Solarium melongena* L.) against the shoot and fruit borer, *Leucinodes orbonalis* Guen. (Lepidoptera: Pyralididae). *Zeitschrift Fur Planzenkrank-ankhettan Und Pflanzensch Utz.* **98**(4): 405-410, Berlin.

Lal, O.P., 1992 A: Evaluation of certain insecticides against the pest complex of french bean seed crop under field conditions. *Journal of Entomological Research,* **16**(1): 57- 61.

Lal, O.P., 1992 B: Pest management in vegetable crops. *Farmer and Parliament,* **27**(7): 17 & 24.

Lal, O.P. and V.D. Agarwal, 1973. Insecticidal trial against lygus bug *Chaulips fallax* Scott. (Hemiptera: Lygaeidae) on french bean. *Indian Journal of Plant Protection,* **1**(1): 32-34.

Lal, O.P. and J. Chandra, 1975. Sex differentiation in the Pupae of cabbage butterfly. *Pieris brassicae* Linn. (Lepidoptera: Pieridae). *Indian Journal of Entomology,* **37**(3): 310-311,.

Lal, O.P. and J. Chandra, 1976. Some parasites of cabbage worm *Pieris brassicae* Linn. (Pieridae: Lepidoptera) in Kulu Valley. *Current Science,* **45**(21): 766-767, Bangalore.

Lal, O.P., H.S. Gill, S.R. Sharma and R. Singh, 1991. Field resistance in different cultivars of cauliflower against head borer, *Hellula undalis* Fab. *Journal of Entomological Research,* **15** (4): 277-28 L.

Lal, O.P. and A.H. Naii, 1986. Field trials on the control of insect pests infesting seed crop of radish. *Haryana Journal of Horticultural Science,* **15**(1-2): 65-70.

Lal, O.P., S.R. Sharma and Ram Singh, 1994. Field resistance in different hybrids of cauliflower against Bihar hairy caterpillar, *Spilosoma obliqua* walker (Lepidoptera: Arctiidae). *Journal of Entomological Research,* **18**(1): 45- 48.

Lalitha, P. and V.G. Prasad, 1980. A review of work done in India on insecticide residues on/in vegetables. In: *Pesticide Residues in the Environment in India*: 120-146, University of Agricultural Sciences, Hebbal, Bangalore.

Lall, Bishambar, 1939. *Pieris brassicae* Linn. *Indian Journal of Entomology,* **1** (1-2): 114.

Lall, B.S., 1946. Biological notes on *Pleurotropis foveicollis* Crawford - a larval parasite of *Epilachna vigintioctopunctata* Fabr. *Current Science,* **15** (5): 138-139.

Lall, B.S., 1949: Preliminary observations on the bionomics of potato tuber moth (*Gnorimoschema operculella* Zell.) and its control in Bihar, India. *Indian Journal of Agricultural Sciences,* **19** (2): 295-306.

Lall, B.S., 1964. Vegetable pests. In: *Entomology In India:* 187-211, Silver Jubilee Number, Entomological Society of India, New Delhi.

Lall, B.S., 1977: Pests in tropical crops - Fruitflies, *Dacus* spp. In *Diseases,*

Pests and Weeds in Tropical Crops (J. Kranz, H. Schmutterer and W. Koch) [Eds.]: 524-526, Verlaug Paul Parey, Berlin.

Lall, B.S. and L.M. Singh, 1960. On biology of onion thrip, *Thrips tabaci* (Lind.) (Thysanoptera: Thripidae): *Current Science,* **29** (8): 314-316,

Lall, B.S. and L.M. Singh, 1968. biology and control of onion thrips in India. *Journal of Economic Entomology,* **61** (3): 676-679.

Lall, B.S. and R.D. Singh, 1960. On the Biology and morphology of *Pleurotropis foveicollis* Craw. (Hymenoptera: Eulophidae). *Science and Culture,* **26** (4): 189.

Lefroy, H. Maxwell, 1906. Bombay Locust. Memoirs of *Department of Agriculture in India, Entomological Series,* **1** (1): 1-112, Kolkata.

Lefroy, H. Maxwell, 1910. Life-history of Indian insects. Coleoptera - I. *Memoirs of department of Agriculture in India, Entomological Series,* **2**(8): 139-164, Kolkata.

Lefroy, H. Maxwell and C.C. Ghosh, 1908. The mustard sawfly (*Athalia proxima* Klug.). *Memoirs of department of Agriculture in India, Entomological Series,* **1** (6): 357-370, Kolkata.

Lefroy, H. Maxwell and F.M. Howlett, 1909. *Indian Insect Life*, 764 pp., Thacker, Spink & Co., Kolkata.

Linden, A. Van Der, 1991. Biological control of the leaf-miner, *Liriomyza huidobrenis* (Blanchard) in Dutch glasshouse tomatoes. *Mededelingen Van Der Faculteit Land Bouwetenschappen Rijksuniversiteit Gent,* **56** (2a): 265-271.

Linnaeus, Caroli A., 1758. Hemiptera. Systemanoturae, Editio Decima, *Heformato,* **1:** 453, Holmiae.

Liu, M.Y., Y.J. Tzeng and C.N. Sun, 1982. Insecticide resistance in the diamondback moth. *Journal of Economic Entomology,* **75**: 153-155.

Logiswaran, G. and S. Sambandamurth, 1994. Biological efficacy of the new formulation Ekalux 20 Af (quinalphos) against tomato fruit borer, *Helicoverpa (Heliothis) armigera* Hubn. *Pestology,* **18 (7): 14-16.**

Lovelock, Yann, 1972. *The Vegetable Book*, 383 pp., George Allen and Unwin Ltd., London.

Lu, F.M., 1990. Colour preference and using silver mulches to control onion thrips, *Thrips tabaci* Lindemann. *Chinese Journal of Entomology,* **10** (3): 337-342, Beijing.

Madden, A.H. and F.S. Chamberlin, 1938. Biological studies on the tomato worm on tabbacco in Florida, 1936 and 1937. *Journal of Economic*

Entomology, **31** (6): 703-706.

Madhava Rao, V.N., 1965. The Jackfruit in India. *Farm Bulletin* (New Series), **34**: 18 pp., Indian Council of Agricultural Research, New Delhi.

Mahal, M.S., H. Lal, R. Singh and B. Singh, 1993. Field resistance on okra to cotton jassid, *Amrasca biguttula* (Ishida) in relation to crop age. *Journal of Insect Science*, **6** (1) 60-63.

Maheswariah, B.M. and M. Puttarudriah, 1956. Some observations on life-history and habits of *Anoploenemis phasianci* Fabricius. *Mysore Agricultural Journal*, **31** (4): Bangalore.

Maini, Nirmal Singh, Harbans Lal Nijhawan and Dhamo K. Butani, 1964. Physiological factors governing resistance of *Brassica* Crops to aphid attack. *Indian Oilseeds Journal*, **8** (2): 142-145.

Majumder, S.K., 1960. Pesticides and health hazards. *Annual Review of Food Technology*, **1**: 36-72.

Mammen, K.V. and K.V. Joseph, 1965: Biology of the pea leaf roller, *Nacoleia vulgaris* Guenee. *Agricultural Research Journal, Kerala*, **3** (1): 51-54.

Mane, Anita and A.G. Chandele, 1995. Comparative efficacy of diafenthiuron with other insecticides against diamond back moth in cabbage. *Pestology*, **19** (4): 19-20.

Mani, M.S., 1973. *Plant Gall of India*, 354 pp., Macmillan India Ltd., Madras.

Mangal, J.L. and S. Siddiqui 2001. Post-harvest technology of vegetables: present status and future strategies. In: *Emerging Scenario in Vegetable Research and development*. G. Kalloo and Kriti Singh [Eds.] pp 227-241. Periodical and Book Pulishing House. Houston, Texas 77272, U.S.A.

Marshall, G.A.K., 1936. Curculionidae attacking cultivated plants. *Bulletin of Entomological Research*, **27**: 256.

Marshall, Guy, A.K., 1916. *The Fauna of British India including Ceylon and Burma. Coleoptera: Curculionidae*, 307 pp., Taylor and Francis, London.

Martin, P.M. and R. Ruberte, 1978. Vegetables for the hot humid Tropics. Part 3, Chaya - C. *Chayomanso*, 11pp., Institute of Tropical Agriculture, Mayaguez, Puerto Rico.

Mathur, A.C., K. Krishnaiah and P.L. Tandon, 1974. Control of tomato fruit borer (*Heliothis armigera* Hub.). *Pesticides*, **8** (11): 34-35.

Mathur, A.C. and J.B. Srivastava, 1964. *Epilachna vigintioctopunctata* as a defoliator of some Solanaceous medicinal plants. *Indian Journal of Entomology,* **26** (2): 240.

Mathur, Y.K., S.V. Singh and R.S. Singh, 1987. Key pests of rapeseed and mustard and their management. In: *Recent Advances in Entomology* (Y.K. Mathur, A.K. Bhattacharya, N.D. Pandey, K.D. Upadhyay and J.P. Srivastava) [Eds.]: 335-352, Gopalan Prakasan, Kanpur.

Maulik, S., 1926. *The Fauna of British India including Ceylon and* Burma. *Coleoptera: Chrysomelidae,* 442 pp., Taylor and Francis, London.

May, A.W.S., 1946. Pests of cucurbit crops. *Queensland Agricultural Journal,* **62** (3): 137-150, Brisbane.

Meher, H.C. and N.P. Agnihotri, 1991. Persistence of sebuphos residues in soil and brinjal. *Pesticide Research Journal,* **3**:73-78.

Meher, H.C., N.P. Agnihotri and C.L. Sethi, 1985. Persistence of carbofuran in soil and pea crop and its bioefficacy against soil nematodes. *Indian Journal of Entomology,* **15**: 155-158.

Meher, H.C., N.P. Agnihotri and C.L. Sethi, 1984. Persistence of aldicarb residues in cowpea and soil following foliar application and its nematicidal activity. *Indian Journal of Entomology,* **14**: 160-165.

Meher, H.C., N.P. Agnihotri and C.L. Sethi, 1989. Persistence of aldicarb residues in cowpea *(Vigna unguiculata)* and soil under Tropical condition. *Indian Journal of Agricultural Sciences,* **59:** 771-777.

Meher, H.C., C.L. Sethi and N.P. Agnihotri, 1985. Persistence and basipetal movement of UC 54229, A carbofuran analog in pea. *Indian Journal of Nematology,* **15**: 71-74.

Meher, H.C., C.L. Sethi and N.P. Agnihotri, 1987a. Evaluation of analytical technique for estimation of phenamiphos residues from soil and plant. *Indian Journal of Entomology,* **49**: 213-219.

Meher, H.C., C.L. Sethi and N.P. Agnihotri, 1987b. Uptake of aldicarb residues by pea and okra. *Indian Journal of Entomology,* **49**: 247-252.

Meher, H.C., C.L. Sethi and N.P. Agnihotri, 1990. Persistence and basipetal translocation of phenamiphos in cowpea and pea. *Pestology.* **14:** 8-11.

Mehrotra, K.N. and A. Phokela, 1995. Resistance on Pest, In *Kothari's Desk, Series on Pesticide Industry* Pp. 349-360.

Mehta, P.N. and Z.M. Nyiira, 1973. An Evaluation of five Insecticides for use against pests of cowpea [*Vigna. unguiculata* (L.) Walp.] with special reference to green pod yield. *East African Agricultural and Forestry Journal,* **39** (2): 99-104.

Mehta, P.R. and B.K. Varma, 1968. *Plant Protection,* 587 pp.. Directorate of Extension, Ministry of Food, Agriculture, Community development and Cooperation, New Delhi.

Mehto, D.N. and B.S. Lall, 1981. A note on fertilization response against brinjal fruit and shoot borer. *Indian Journal of Entomology,* **43** (2): 106-107.

Menon, K.P.V. and K.M. Pandalai, 1958. The coconut palm - A. Monograph:251-292. Indian central Coconut Committee, Ernakulam.

Menon, M.G. Ramdas and N.S. Bhattacherjee, 1960. *Cyrtozemia cognata* Marshall (Curculionidae: Coleoptera), A Pest of vegetables and flowering plants of Delhi. *Indian Journal of Entomology,* **22** (4): 297.

Menon, M.G. Ramdas, K.N. Katiyar and V.C. Kapoor, 1972. Taxonomic Identities of some Indian species of the Genus *Aulacophora* Chevralot associated with Cucurbitaceous plants (Coleoptera: Chrysomelidae: Galerucinae). *Journal Natural History,* **6** (1): 17-20, London.

Meyrick, Edward, 1894: On a collection of Lepidoptera from upper Burma. *Transactions of Entomological Society of London:* **15**, London.

Meyrick, Edward, 1897. Description of Australian Microlepidoptera. *Proceedings of Linnean Society of New South Wales, Series Ii,* **22:** 297-435, Sydney.

Meyrick, Edward, 1917. Two New Indian species of *Cosmopteryx. Entomologist's Monthly Magazine,* **53**: 257- 258, London.

Mishra, S.S., 1995. Evaluation of Insecticides against white grubs, *Holotrichia* Spp. damaging potatoes. *Indian Journal of Plant Protection,* **23** (I): 55-57.

Misra, S.S., 1995. Evaluation of insecticides for the management, of cut worm, *Agrotis segetum* Schiff.-mull damaging potatoes in North-western Hills. *Indian Journal of Plant Protection,* **23** (2): 222-223.

Misra, S.S. and H.O. Agarwal, 1989. Persistence of aldicarb and carbofuran residues in potato grown in North-western Plains of India. *Tropical Pest Management,* **35** (4): 370-370.

Misra, S.S. and H.O. Agarwal, 1991. Potato tuber moth, *Phthorimoea operculella* (Zell.) and its Integrated management. *Pesticides Information,* **17** (2): 3-7.

Misra, S.S. and H.O. Agarwal, 1992. Persistence of phorate and disulfoton residues in potato grown in North-western Hills. *Indian Journal of*

Plant Protection, **20** (1): 138-143.

Misra, S.S., M.D. Awasthi and R.S. Dewan, 1977. Residue of some contact insecticides in potato. *Journal of Food and Science Technology,* **14** (1): 247-258.

Misra, S.S. and V.K. Chandla, 1989. White grubs infesting potatoes and their management. *Journal Indian Potato Association,* **16** (1-2): 29-33.

Misra, S.S. and Lakshman Lal, 1981. Evaluation of foliar systemic insecticides against leaf hopper, *Amrasca devastans* (Dist.) on Potato crop. *Entomon.,* **6** (3): 201-205.

Misra, S.S., L. Lal and M.D. Awasthi, 1982. Persistence of aldrin and dieldrin residues in potatoes: *Journal of Food Science and Technology,* **19** (1): 11-14.

Misra, S.S., H.C. Sharmaand H.O. Agarwal, 1992. Economic threshold level of cutworm, *Agrotis segetum* Schiff-mull. on potato crop grown in North-western Higher Hills of Himachal Pradesh (India). *Journal of Entomological Research,* **16** (4): 283-286.

Mitra, P.K., 1954. A Note on *Heliothis armigera* Hubn., as a pest of pea *(Pisum sativum). Current Science,* **13** (12): 312-313.

Modawal, C.N., 1941. A biological note on *Chilomenes sexmaculata* Fabr. *Indian Journal of Entomology,* **3** (1): 139-140.

Mogal, B.H., A.R. Mali and S.G. Rajput, 1982. Efficacy of granular systemic insecticides against onion thrip (*Thrips tabaci* Lindeman). *Pestology,* **6** (5): 13-15.

Mohan, S., N. Vardharajan and M. Balasubramanian, 1989. Development and testing of pesticide bait-pallets for tobacco cutworm, *Spodoptera litura* Fab. *Pestology,* **13** (4): 22-23.

Mohite, P.B., V.S. Teli and P.R. Moholkar, 1992. Efficacy of organic synthetic pesticides against onion thrips, *Thrips tabaci* Lind. *Pestology,* **16** (7): 8-10.

Momshita M., K. Azuma and S. Yang, 1992. Changes in insecticide susceptibility of the diamondback moth in Wakayama, Japan, *Japan Agricultural Research Quarterly,* **26**: 139-143, Tokyo.

Mookherjee, P.B. and S.R. Wadhi, 1956. Further studies on the phytotoxicity of modern organic insecticides. *Indian Journal of Entomology,* **18** (4): 332-335.

Morgan, W.L., 1931. The tomato stem caterpillar (*Phthorimoea plaesiosema* Turner). *Agricultural Gazette Of New South Wales,* **42** (12): 919-921,

Sydney.

Morgan, W.L., 1938. Biological aspects of some Agromyzids. *Agricultural Gazette of New South Wales,* **49**: 501-503, Sydney.

Mosley, Edward, 1841. Aphides, *Gardeners' Chronicle,* **1**: 684, London.

Mote, U.N., 1976. Control of onion thrips, *Thrips tabaci* Lind. *Pesticides,* **10** (7): 42-43.

Mote, U.N., 1977: Evaluation of different modern insecticides against cabbage aphids. *Vegetable Science,* **4** (1): 33-36.

Mote, U.N., 1978 A. Chemical control of brinjal jassid (*Amrasca devastans* Dist.) and shoot and fruit borer (*Leucinodes orbonalis* Guen.). *Pesticides,* **12** (7): 20-23.

Mote, U.N., 1978 B. Control schedule of okra pests. *Journal of Maharashtra Agricultural Universities,* **3**: 197-199.

Mote, U.N., 1978 C. Studies on the varietal resistance of brinjal (*Salanum melongena* L.) to jassid (*Amrasca biguttula biguttula* Ishida) under field condition. *Vegetable Science,* **5** (2): 107-110.

Mote, U.N., 1979. Varietal resistance of brinjal to fruit borer, *Leucinodes orbonalis* Guen. *Bulletin of Entomology,* **20** (1 - 2): 75-77.

Mote, U.N., 1979. Varietal resistance of okra *Abelmoschus esculentus* L. to jassid, *Amrasca biguttula biguttula* Ishida. *Bulletin of Entomology,* **20** (1-2): 78-81.

Mote, U.N., 1982. Varietal susceptibility to brinjal (*Solanum melongena* L.) to jassid (*Amrasca biguttula biguttula* Ishida). *Journal of Maharashtra Agricultural Universities,* 7 (1): 59-60.

Mote, U.N. and R.B. Pawar, 1994. Effect of new insecticides as seed dressers on the infestation of sucking pests and plant growth characters of okra. *Pestology,* **18** (11): 5-9.

Mound, L.A. and S.H. Halsey, 1977. *Whiteflies of the World,* 340 pp., British Museum (National History) and John Wiley and Sons, London.

Muesebeck, C.F.W. and B.R. Subba Rao, 1958. A New Braconid parasite of *Hymenia recurvalis* (Fabricius). *Indian Journal of Entomology,* **20** (1): 27-28.

Mukhopadhyay, A. and A. Mandal, 1994. Screening of brinjal (*Solanum melongena*) for resistance to major insect pests. *Indian Journal of Agricultural Sciences,* **64** (11): 798-803.

Murakami, S., 1933. Results of studies on *Cylas formicarius* Fabr. I. (In Japanese]. *Journal of Plant Protection,* **20**: 979-980, Tokyo.

Murthy, D.V., 1958. A Short note on the banana leaf thrip, *Caliothrips (Heliothrips) kadaliphila* Ramk. & M. *Mysore Agricultural Journal,* **33** (1): 2-4.

Murthy, D.V., 1959. A preliminary note on the control of red-ant (*Solenopis geminata* Fbr.) on some vegetable crops. *Mysore Agricultural Journal,* **34** (1): 9-14.

Muthukrishnan, T.S. and K.R. Nagarajarao, 1959. A first record of two root mealy bugs in Madras State. *Indian Journal of Veterinary Science,* **29** (2-3): 135-136.

Muthukrishnan, T.S., K.R. Nagaraja Rao, T.R. Subramanian, I.P. Janaki and E.V. Abraham, 1958. Brief notes on a few crop pests noted for the first time in Madras. *Madras Agricultural Journal,* **45** (9): 363-364.

Nagaich, B.B. 1977. The potato production in India- Diversification, its uses and export potential. *Journal of Indian Potato Association,* **4**: 30 35.

Nagaraja Rao, K.R. and T.R. Subramaniah, 1958. *Glyphodes bivitrolis* Guen. as a new pest of jack. *Indian Journal of Agricultural Sciences,* **28** (4): 561-562.

Nagaraja Rao, K.R., 1955. Farmers try new ways of controlling chilli thrips. *Indian Farming,* **5** (2): 7, New Delhi.

Nagaraja Rao, K.R., 1959. Studies on Phytotoxicity of certain Systemic pesticides. *Indian Journal of Entomology,* **21** (2): 89-92.

Nagpal, H.D., 1948. *Insect Pests of Cotton in India,* 81 pp., Indian Central Cotton Committee, Mumbai.

Naik, B.G., Shashi Verma and K.G. Phandke, 1993. Occurrence of pests in relation to degradation of insecticides in brinjal crop during Summer and Kharif Seasons. *Pesticides Residue Journal,* **5** (1): 94-104.

Nair, M.R.G.K., 1975. *Insects and Mites of Crops in India,* 404 pp., Indian Council of Agricultural Research, New Delhi.

Nakazawa, H. and R. Akanura, 1993. Insect pests and diseases control of cabbages in Japan. *Agrochemicals Japan,* **63**: 8-10, Tokyo.

Narasimha Rao Ch. V., K.C. Punnaiah, V. Deva Prasad and P.V. Krishnaiah, 1996. Studies on the Efficacy of certain newer insecticides against brinjal whitefly, *Bemisia tabaci* Gennadius. *Pestology,* **20** (7): 17-19.

Narayanan, E.S., 1941. Notes on some Indian parasitic Hymenoptera, with a description of a new Cynipid. *Indian Journal of Entomology,* **3** (1): 59-63.

Narayanan, E.S., 1953 A: The red pumpkin beetle and its control. *Indian*

Farming, **3** (2): 8-9.

Narayanan, E.S., 1953 B. The mustard sawfly, *Athalia proxima.* Klug., a pest in the Nurseries of the cabbage family. *Indian Farming,* **3** (8): 8-9.

Narayanan, E.S., 1954. The mustard aphid. *Indian Farming,* **3** (11): 8-9.

Narayanan, E.S. and H.N. Batra, 1960. *Fruitflies and their control,* 68 pp., Indian Council of Agricultural Research, New Delhi.

Narayanan, E.S., B.R. Subba Rao, M. Ramachandra Rao and A.K. Sharma, 1960. Biology and morphology of the immature stages of *Microctonus indicus,* New Species (Braconidae: Hymenoptera), a parasite of *Phyllotreta crucijerae* Goeze (Chrysomelidae: Coleoptera). *Proceedings of Indian Academy of Sciences,* **B-51** (6): 280-287, Bangalore.

Narayanan, K., S. Jayaraj and T.R. Subramaniam, 1973. Control of diamondback moth caterpillar, *Pluella xylostella* (L.) and the green aphid, *Myzus persicae* (Sulzer) with insecticides and *Bacillus thuringiensis* Berliner. *Madras Agricultural Journal,* **62**: 498-503.

Narke, C.G. and O.S. Suryawanshi, 1987. Chemical control of major pests of okra. *Pesticides,* **21** (1): 37-39.

Nasir, M. Maqsud, 1946. *Chrysopa cymbela* Banks and its two new varieties. *Indian Journal of Entomology,* **8**(1): 119- 120.

Nasir, M. Maqsud, 1947. Biology of *Chrysopa scelestes* Banks. *Indian Journal of Entomology,* **9** (1): 177-189.

Nassar, O.A. and Swaraj Ghai, 1981. Taxonomic studies on Tetranychoid Mite infesting vegetables and fruit crops in Delhi and surrounding areas. *Oriental Insects,* **15** (4): 333-396.

Nath, G. and S.A. Srtvastava, 1984. Effect of some processings on the removal of endosulfan from cauliflowers. *Indian Journal of Entomology,* **46** (1): 24-32.

Nath, G., N.R. Jat and B.P. Srtvastava, 1975. Effect of washing, cooking and dehydration on removal of some insecticides from okra. *Journal of Food and Science Technology,* **12**: 127.

Nath, Prem, 1964 A. Resistance of cucurbits to the red pumpkin beetle. *Indian Journal of Horticulture,* **21** (1): 77-78.

Nath, Prem, 1964 B. Observations on resistance of cucurbits to the fruitfly. *Indian Journal of Horticulture,* **21** (2): 173-175.

Nath, Prem, 1971. Breeding cucurbitaceous crops for resistance to insect pests. *Sabrao Newsletter,* **3** (2): 127- 134.

Nath, Prem, 1976. *Vegetables for the Tropical Region.* 109 pp., Indian

Council of Agricultural Research, New Delhi.

Nayar, K.K., T.N. Ananthakrishnan and B.V. David, 1976. *General and Applied Entomology*, 589 pp., Tata Mcgraw-hill - Publication Co. Ltd., New Delhi.

Neeraja, G., M. Vijaya, Ch Chiranjeevi and B. Gautham. 2004. Screening of okra hybrids against pest and diseases. *Indian Journal of Plant Protection.* **32** (1) 129-131.

Netekar, M.G., S. Raj and N.P. Agnihotri, 1988. Bioefficacy of synthetic pyrethroids and their residues on brinjal fruit. *Pestology,* **12** (11): 18-21.

Nikkoo, F. and M. Moiini, 1991. Study of cutworm biology in the vegetable growing Areas around Tehran and determination of its control measures. *Applied Entomology and Phytopathology,* **58** (1-2): 1-3.

Nirmal, B. and T.V.K. Singh 2004. Insecticide resistance in dimondback moth, *Plutella xylostella* Linn. *Indian Journal of Plant Protection,* **32** (2): 17-21.

Nirula, K.K., 1961. Insecticidal control of *Agrotis ipsilon* Rott. *Indian Potato Journal,* **3** (1): 42-47.

Nirula, K.K., J. Anthony and K.P.V. Menon, 1952. A new pest of coconut palm in India. *Indian Coconut Journal,* **5** (3): 137-140.

Nirula, K.K. and Rabinder Kumar, 1963. Studies on the rate and mode of application of heptachlor and aldrin for the control of *Agrotis ipsilon* Rott. *Indian Potato Journal,* **5**(1): 38-43.

Nirula, K.K. and Rabinder Kumar, 1969. Soil application of systemic insecticides for the control of aphid vectors, leaf roll and Y virus in potato. *Indian Journal of Agricultural Sciences*, **39** (7): 699-702, New Delhi.

Nuttonson, M.Y., 1951. *Ecological Crop Geography and Field Practices in Japan.* 213 pp., American Institute of Crop Ecology, Washington, Dc.

Oka, I.N., 1957. Pertjobann Laboratorium Dalam Pemberantasan Ulat Kubis *Plutella rnaculipennis* Curt. Dengan *Bacillus thuringiensis* Berl. *Tehaick Perkaman,* **6** (4): 113-114, Bogor, Indonesia.

Osburn, M.R., 1937. Termites infesting turnips. *Journal of Economic Entomology,* **30** (6): 697.

Pagden, H.T., 1928. *Leptoglossus membranaceus* F., A Pest of cucurbitaceae. *Malaya Agricultural Journal,* **16** (2): 387- 403, Kuala Lumpur.

Panchabhavi, K.S. and M. Sudhindra, 1994. *Bacillus thuringiensis* in the Ipm of cabbage diamondback moth. *Pestology,* **18** (9): 28-30.

Panda, N., A. Mahapatra and M. Sahoo, 1971. Field evaluation of some brinjal varieties for resistance to shoot and fruit borer (*Leucinodes orbonalis* Guen.). *Indian Journal of Agricultural Sciences,* **41** (7): 597-601.

Pande, Y.D., D. Ghosh and S. Guha, 1987. Toxicity of leaf extract of *Ageratum conzydoides* L. (Compositae) to red pumpkin beetle, *Raphidopalpa foveicollis* (Lucas) (Coleoptera: Chrysometidae) in Tripura. *Indian Agriculturist,* **31** (4): 251-255.

Pandey, N.D. and U. Shankar, 1975. Studies on host preference of *Honosepilochna viginitioctopuntata* Fabr. *Indian Journal of Entomology,* **37** (3): 321-323.

Pandey, U.K., M. Pandey and S.P.S. Chauhan, 1981. Irisecticidal properties of some plant material extracts against painted bug. *Bagrada cruciferarum* Kirk. *Indian Journal of Entomology,* **43** (4): 238-242.

Patel, G.A. and M.V. Kadam, 1957. Pests of crucifers, chillies, onion, cucurbits and *bhendi.* In: *Crop Pests and How to Fight Them*: 91-100. Directorate of Publicity, Government of Bombay, Bombay. Also in *Farmer Bombay,* **5** (5): 31-36.

Patel, R.C. and H.L. Kulkarny, 1955. Bionomics of *Urentius echinus* Dist. (Hemiptera: Heteroptera: Tingidae) an important pest of brinjal (*Solanum melongena* L.) in North Gujarat. *Journal of Bombay Natural History Society,* **53** (1): 86-96.

Patel, R.C. and H.L. Kulkarny, 1956. Bionomics of pumpkin caterpillar, *Margaronia indica* Saund. (Pyralidae: Lepidoptera). *Journal of Bombay Natural History Society,* **54** (1): 118-127.

Patel, R.M., 1958. Some Notes on the biology of cabbage gall weevil. *Plant Pathology,* **7** (4): 137-141, London.

Patil, B.V., M. Bheemanna and Tulsi Ram, 1996. Bioefficacy of cypermethrin dust against brinjal shoot and fruit borer, *Leucinodes orbonalis* G. *Pestology,* **20**(4): 13-15.

Patil, C.S., R.G. Jadhav, U.N. Mote and V.M. Khaire, 1991. Evaluation of insecticides dusts against brinjal fruit borer. *Annals of applied Biology,* **118**: 6-7.

Paul, Anderson, S.R. Sharma, T.V. S. Stresty, Shantibala Devi, Suman Bala, P.S. Kumar, P. Pardha Saradhi, Roger Frutos, I. Altosaar, and P. Kumar.

2005. Transgenic Cabbage (*Brassica oleracea* var. *capitata)* Resistant to diamondback moth) *Plutella xylostella*) *Indian Journal of Biotechnology.* **4** (1): 72-77.

Pawar D.B., U.N. Mote and K.E. Lawande, 1991. Monitoring of fruitfly population in bittergourd crop with the help of lure traps. *Journal* of *Maharashtra Agriculture Universities,* **16**(2): 281.

Pawar, D.B. and K.E. Lawande, 1993. Threshold level of jassid and effect of sprays on fruit borer in okra. *Journal of Maharashtra Agricultural Universities,* **18** (1): 112-113.

Pawar, J.G., 1964. Banish the *Epilachna* beetle. *Indian Farming,* **13** (10) 25 & 27.

Peregrine, W.T.H., 1983. Chaya (*Cnidoscolus aconitifolius*) - A potential new vegetable crop for Brunei. *Tropical Pest Management,* **29** (1): 39-41, London.

Perron, J.P., J.J. Jasmin and J. Lafrance, 1958. Varietal resistance of seeded onions to the onion maggot *Hylemya antiqua* (Meig.) (Diptera: Anthomyiidae). *Canadian Entomologist,* **90**: 653-656, Toronto.

Peswani, K.M. and Rattan Lal, 1964. Estimation of loss of brinjal fruits caused by shoot and fruit borer, *Leucinodes orbonolis* Guen. *Indian Journal of Entomology,* **26** (1): 112-123.

Peter, C. and R. Sundarajan, 1991. Field tests with two new insect growth regulators for the control of diamondback moth, *Plutella xylostella* L. *Indian Journal of Plant Protection,* **19** (2): 161-166.

Peter, K.Y. and V.S. Devadas, 1989. Leafy vegetables. *Indian Horticulture,* **33** (4) & **34** (1): 8-11.

Pingale, S.V., 1956 A. Insecticide residues on fruit and vegetable products. fruit and vegetable preservation Industry, India: 190, Mysore.

Pingale, S.V., 1956 B. Prevention of insect contamination to the fruit and vegetable products. fruit and vegetable preservation Industry, India: 194, Mysore.

Poe, S.L., 1973. Tomato pinworm, *Keiferia lycopersicella* (Walsingham) (Lepidoptera: Gelechiidae) in Florida. Florida Department of Agriculture, Entomology Circular No. 131, 2pp., Florida.

Pradhan, S., 1969. *Insect Pests of Crops,* 208 pp., National Book Trust, India, New Delhi. Revised by M.G. Jotwani, 1983.

Pradhan, S., H.S. Bhambhani and S.R. Wadhi, 1960. Fumigation of infested potatoes. *Indian Journal of Entomology,* **22** (3): 181-189.

Pradhan, S., M.G. Jotwani and B.K. Rai, 1958 A. Bioassay of insecticides -

VIII: Relative toxicity of some insecticides to red pumpkin beetle, *Aulacophora foveicollis* Lucas (Coleoptera: Chrysomelidae). *Indian Journal of Entomology,* 20 (2): 104-107.

Pradhan, S., M.G. Jotwani and B.K. Rai, 1958 B. Bioassay of insecticides - X: Comparative toxicity of some insecticides to *Cylas formicarius* Fabricius and *Myllocerus maculosus* Desbrochers (curculionidae: Coleoptera). *Indian Journal of Entomology,* **20** (3): 171-174.

Pradhan, S. and M.G. Ramdas Menon, 1942. *Antilochus cocquerberte* (Fabr.), a Predator of *Dysdercus cingulatus. Indian Journal of Entomology,* **4** (1): 91.

Prakash, Om, Zile Singh and M.Z. Panditan, 1980. Efficacy of different spray schedules against insect pests of okra (*Abelmoschus esculentus* (L.) Moench). *Haryana Journal of Horticultural Science,* **9** (3): 201-206.

Prasad, J. and K.S. Kushwaha, 1990. Role of pathogenic microbes in management of tobacco caterpillar, *Spodoptera litura* (Fab.) in cabbage and cauliflower. *Indian Journal of Entomology,* **52** (4): 641-652.

Prasad, S.K., 1959. The association between the different pests of cabbage. *Indian Journal of Entomology,* **21** (3): 171-174.

Prasad, S.K., 1960. The nature and extent of damage to potatoes in relation to density of potato flea beetles. *Indian Journal of Entomology,* **22** (3): 231.

Prasad, S.K., 1961 A. Quantitative estimation of the damage to cabbage by cabbage worm, *Pieris rapae* (Linn.). *Indian Journal of Entomology,* **23** (1): 54-61.

Prasad, S.K., 1961 B. Quantitative estimation of damage to potatoes caused by potato leaf hopper, *Empoasca fabae* (Harris). *Indian Potato Journal,* **3** (2): 105-117.

Prem Chand, 1995. *Agricultural and Forest Pests and their Management,* 374 pp., Oxford and IBH Publishing Co. Pvt. Ltd., New Delhi.

Prem Nath, Sundari Velayudhon and D.P. Singh, 1987. *Vegetables for the Tropical Region.* 195pp. Publication and Information Division, Indian Council of Agricultural Research, Krishi Anusandhan Bhavan, Pusa, New Delhi.

Pruthi, Hem Singh, 1921. Morphology and biology of red cotton bug, *Dysdercus cingulatus* Fabr. *Journal and Proceedings Asiatic Society of Bengal,* **17** (4): 147-149, Kolkata.

Pruthi, Hem Singh, 1923. on the Anatomy and bionomics of the red'cotton Bug, *Dysdercus cingulatus* (Fabr.). *Journal and Proceedings Asiatic*

Society of Bengal, **19** (1): 15-42. Kolkata.

Pruthi, Hem Singh, 1940. Descriptions of some new species of *Empoosca* Walsh (Eupterygidae, Jassoidea) from North India. *Indian Journal Of Entomology*, **2**(1): L-10.

Pruthi, Hem Singh, 1946. Report of the Imperial Entomologist. *Scientific Report, Imperial Agricultural Research Institute,* 1941-44: 64-67, New Delhi.

Pruthi, Hem Singh and H.N. Batra, 1960. *Important Fruit Pests of Nort-West India.* Bulletin No. 80, 113pp., Indian Council of Agricultural Research, New Delhi.

Pruthi, Hem Singh and L.N. Nigam, 1939. The Bionomics, life history and control of the grasshopper *Poekilocerus pictus* (Fab.) a new pest of cultivated Crops in North India. *Indian Journal of Agricultural Sciences,* **9** (4): 626-641.

Purewall, S.S., 1954. Growing onions. *Indian Farming*, **4** (6): 26-32, New Delhi.

Puri, S.N., K.S. Murthy and O.P. Sharma. 2001. Integrated Pest Management in Vegetables: issues and Strategies In: *Emerging Scenario in Vegetable Research and development* G. Kalloo and Kirti Singh [Eds.] Pp 293-303. Research House, Houston, Texas 772 72, U.S.A.

Puttarudariah, M. and G.P. Channa Basavanna, 1958: Some insect predators of aphids (Aphididae) in Mysore. *Mysore Agricultural Journal,* **32** (3-4): 158-161.

Puttarudariah, M. and B. Krishnamurti, 1954. Problem of *Epilachna* control in Mysore: Insecticidal control found inadvisable when natural incidence of parasites is high. *Indian Journal of Entomology,* **16** (2): 137-141.

Puttarudariah M. and B.M. Maheswariah, 1958. Some observations on the life-history and habits of *Anomis flava* (Fabricius) (Noctuidae: Lepidoptera). *Indian Cotton Growing Review,* **12** (5): 337-34 L.

Quesales, F.O.V., 1918. Leaf mining insects in Philippines. *Philippines Agriculturist,* 7: 3-32, Los Banos.

Rabindra Prasad 2007. Note on phytophagous mites associated with common vegetables on Ranchi (Jharkhand). *Journal of Insect Scienc* **20** (1): 101.

Raghapathy, A.B. Habeebullah and M. Balasubramaniam, 1985. Dissipation of insecticides applied to control *Plutella xylostella* Curtis and

Spodoptera litura Fabr. in cauliflower. *Pesticides,* **19** (7): 53-57.

Rahman, Khan A., 1940. Insect pest number, *Punjab Agricultural College Magazine,* 7 (5-7): 43, Lyallpur.

Rahman, Khan A. and Anshi Lal Batra, 1944. The onion thrips, *Thrips tabaci* (Lind.) (Thripidae: Terebrantis: Thysanoptera). *Indian Journal of Agricultural Sciences,* **14** (4): 308-310.

Rai, N. and D.S. Yadav, 2005. *Advances in Vegetable Production* 996 pp. Researchco Book Centre, New Delhi-110005.

Rai, S., N.P. Agnihotri and H.K. Jain, 1986. Persistence of residues of synthetic pyrethroids on cauliflower and their residual toxicity against aphids. *Indian Journal of Agricultural Sciences,* **56**: 667-670.

Rai, Samarjit, H.K. Jain and N.P. Agnihotri, 1980. Bioefficacy and dissipation of synthetic pyrethroids in okra. *Indian Journal of Entomology,* **42** (4): 657-660.

Rai, Samarjit, M.G. Jotwani and R.P. Gupta, 1979. Occurrence of *Helicoverpa. armigera armigera* (Hubner) as a serious pest of onion (*Allium cepa* Linn.). *Bulletin of Entomology,* **20** (1-2): 109-110.

Raizada, Usha, 1965. The Life-history of *Scirtothrips* Hood with detailed external morphology of its immature stages. *Bulletin of Entomology, Loyalla College,* **6**: 30-49.

Raja Rao, S.A., 1954. Bionomics and life-history of *Aphidius* Sp., A parasite of *Aphis gossypii* Glov. on brinjal (*Solanum melongena*). *Indian Journal of Entomology,* **16** (4): 362-371.

Rajavel, D.S. and D. Veeraraghavathatham, 1989. Efficacy of certain plant extracts on major pests of cabbage. *South Indian Horticulture,* **37** (3): 186-188.

Rajavel, D.S. and M. Gopalan, 1991. Control of cabbage diamondback moth, *Plutella xylostella* (L) with synthetic pyrethroids. *Madras Agricultural Journal,* **78** (1-4): 98-99.

Rajavel, D.S., M. Gopalan and G. Balasubramanian, 1989. Evaluation of synthetic pyrethroids in the control of brinjal shoot and fruit borer, *Leucinodes orbonalis G. Madras Agricultural Journal,* 76 (2): 114-116.

Rajpurohit, T.S. and R.D. Singh, 1991. Control of Cercospora leaf spot of sugarbeet by fungicidal seed treatment. *Pestology,* **15** (2): 36-37.

Rajukkannu, K., R. Raghurai, K. Asaf Ali and K.K. Krishnamurthi, 1976. Residues of endrin, parathion, carbaryl and endosulfan in vegetables. *Pesticides,* **10** (12): 19-20.

Rajukkannu, K., K. Saivaraj, P. Vasudevan and M. Balasubramaniam, 1980. Residues of certain newer insecticides in brinjal. *Pesticides,* **14** (1 1): 15-17.

Rakshpal, R. , 1949: Notes on the biology of *Bagrada cruciferarum* Kirk. *Indian Journal of Entomology,* **11** (1): 11- 16.

Ram Kishore and S.S. Misra, 1984: Impact of different planting dates on the incidence of cutworms in potato crop. *Indian Journal of Entomology,* **46** (3): 366-368.

Ramakrishnan, C. and P.S. Naryanaswamy, I960. Studies on the control of the sweet potato weevil *Cylas formicarius* F. *Madras Agriculture Journal,* **47** (1): 33-36.

Ramamani. S. and B.R. Subba Rao, 1965. On the identity and nomenclature of paddy cutworm commonly referred to as *Cirphis unipuncta* Haworth. *Indian Journal of Entomology.* **27** (3): 363-364.

Ramaseshiah G., 1967. The fungus *Entomophthora coronata* parasitic on three species of aphids infesting cucurbits in India. *Journal of Invertebrate Pathology,* **9**: 126-128.

Ramaseshiah, G. and P.R. Dharamadhikari, 1968. Occurrence of fungus *Cepholosporium aphidicola* on some aphids in India: *Technical Bulletin No. 10:* 150-155, Commonwealth Institute of Biological Control, Trinidad.

Ramasubbaiah, K., 1980. Toxicity and residues of phosphamidon against pests of brinjal and bhendi. In: *Pesticide Residues in the Environment in India:* 231-237, University of Agricultural Sciences, Hebbal, Bangalore.

Ramasubbaiah, K. and Rattan Lal, 1975. Studies on residues of phosphamidon in cowpea crop. *Indian Journal of Entomology,* **37** (2): 179-184.

Ramaswamy, G., 1952. Some insect pests of my vegetable garden. *Presidency College Zoology Magazine,* **7**: 14-18, Madras.

Ramchandra Rao, Y., 1917. *Adisuna atkinsoni* Moore. *Madras Agricultural department Yearbook:* 87-91, Madras.

Ramzan, M. and Darshan Singh, 1980. Chemical control of the tobacco caterpillar, *Spodoptera litura* Fabricius on cauliflower. *Journal of Research, P.A.U.,* **17** (2): 236-239, Ludhiana.

Ranga Rao, P.V. and K.L. Narayana, 1957. Effect of insecticidal spray treatment on the quality of onion. *Andhra Agricultural Journal,* **4** (4): 100-103 .

Ranganathan, T. and K. Govindan, 1996. Efficacy of three new chemicals over diamondback moth *Plutella xylostella* L. on cauliflower. *Pestology,* **20** (2): 16-18.

Rao, N.S. and R.N. Barwal, 1984. Effectiveness of various systemic insecticides against pea aphid, *Acyrthosiphon pisum* (Harris). *Pestology,* **8** (2): 14-16.

Rathi, Y.P.S. and Y.L. Nene, 1974. Some aspects of the relationship between mung bean yellow mosaic virus and its vector, *Bemisia tabaci. Indian Phytopathology,* **27**: 459-462.

Rattan Lal, 1946. Biological observations on *Empoosca kerri* var. *motti* Pruthi on potato plant. *Indian Journal of Entomology,* **8**(1): 195-201.

Rattan Lal, 1950. Biology and control of *Myzus persicae* Sulzer as a pest of potato at Delhi: *Indian Journal of Entomology,* **12** (1): 87-100.

Rattan Lal, and Swaraj Ghaj, 1958. Physiology of digestion of the adult red pumpkin beetle *Aulacophora foveicollis* (Lucas). *Indian Journal of Entomology,* **20** (1): 37-45.

Rattan Lal and Shashi Verma, 1975. Waiting period after insecticidal treatments for the harvest of some vegetables crops. *Entomologists 'Newsletter*, **5** (10-11): 48-49, New Delhi.

Rattan Lal and Shashi Verma, 1982. Toxic residues of pesticides. In: *Agricultural Entomology*, **1** ; 185-208. All India Scientific Workers' Society, New Delhi.

Rawat, R.R. and B.N. Modi, 1969. Studies on biology of *Ferrisia virgata* Ckll. (Pseudococcidae: Homoptera) in Madhya Pradesh. *Indian Journal of Agricultural Sciences,* **39** (3): 274-281.

Rawat, R.R. and H.R. Sahu, 1973. Estimation of losses in growth and yield of okra due to *Empoasca devastans* distant and *Earias* species. *Indian Journal of Entomology,* **35**(3): 252-254.

Rawlins, W.A., M.J. Sloan, R.G. Strong and J.J. Ellington, 1960. The onion maggots and its control. *Cornell Agricultural Experiment Station, Bulletin No. 947,* 18 pp., Cornell.

Reddy, D. Bap, 1961. Bean-fly a Serious pest of French beans. *Current Science,* **30** (5): 192-193 , Bangalore.

Reddy, D. Bap, 1968: *Plant Protection In India,* 267 pp. Allied Publishers, Kolkata.

Reddy, D.N.R. and A. Jagdish, 1980. Chemical control of onion thrip *Thrips tabaci* Lindeman (Thysanoptera: Thripidae). *Pesticides,* **14** (7): 25-26.

Reddy, K.C.S. and G.C. Joshi, 1990. Effect of insecticides and plant growth regulators on plant growth, incidence and yield of brinjal. *Journal of Research,* APAU, 18: 141-145, Hyderabad.

Regupathy, A., B. Habeebuuah and M. Balasubramanian, 1985. Dissipation of insecticides applied to control *Plutella xylostella* Curtis and *Spodoptera litura* in cauliflower. *Pesticides,* **19** (9): 53-56.

Rengarajan, Sreelathat. and B.J. Diwakar, 1995. Residue of alphamethrin, fluvalinate and carbaryl in/on okra. *Indian Journal of Plant Protection,* **23** (2): 206-207.

Rishi, N.D., 1967. Studies on insect pest of Kashmir: Part II - Vegetables: major pests with their life history and control measures. *Kashmir Science,* **4** (1-2): 62-78, Srinagar.

Rizvi, S.M.A. and V.K. Chandla, 1977. Insecticidal control of cutworms in potatoes. *Indian Journal of Plant Protection,* **5**: 105-107.

Rizvi, S.M.A., V.K. Chandlaand B.S. Bist, 1976. Evaluation of some systemic granular insecticides against *Myzus persicae* Sulzer on potato. *Potato Research,* **19** (2): 183-185

Rizvi, S.M.A., Lakshman Lal and S.M.P. Khurana, 1976. Host plants of *Myzus persicae* Sulzer in Jullunder and Simla. *Indian Journal of Plant Protection,* **4** (1): 32-34.

Rizvi, S.M.A. and P.K. Sanyal, 1977. *Euzophera perticella* Rag. (Lepidoptera: Pyralidae) as a root borer of brinjal. *Indian Journal of Plant Protection,* **5** (2): 213.

Roberts, B.W. and B. Cartwright, 1991. Alternative soil and pest management practices for sustainable production of fresh market cabbage. *Journal of Sustainable Agriculture,* **1** (3): 21-38, Binghamton, New York, USA.

Roe, Robbins Anne, 1974. *25 Vegetables anyone can grow.* 216 pp., Dover Publications Inc., New York.

Rotino, G.L., F. Suneri, N. Acciarii, S. Arpaia and M. Zottini. 2002. Transgenic parthenocarpic and insect-resistant eggplant. In: *Transgenic Plants and crops.* G.G. Khachatourians, A. Mcmughen, R. Scorza, Wai-kit Nip, Y.h. Hui [Eds.], Pp. 587-601 Marcel Dekker Inc., New York, Ny-10016, U.S.A.

Sahai, B. and Y.R. Mehta, 1954. A new insect pest of cauliflower in India. *Indian Journal of Horticulture,* **11** (1): 32, New Delhi.

Sakimura, K., 1937. The life and seasonal histories of *Thrips tabaci* Lind. In vicinity of Tokyo, Japan. *Ova-dobuteuoaku-zaoshi,* **9**: 1-24, Tokyo.

Sakurai, H.R. Sirashi, Y. Ito, T. Teruva and S. Takeda, 1989. Effect of Gamma radiation on the spermato-genesis of melonfly, *Dacus cucurbitae, Research Bulletin of Agriculture Gifu University,* Japan, **54:** 59-60, Tokyo.

Sandhu, S.S., 1978. Red pumpkin beetle - A serious pest of cucurbits. *Progressive Farming,* **14:** 22, Ludhiana.

Sannaveerappanavar, V.T. 1995. Studies on insecticidal resistance in the diamondback Moth. *Plutella xylostella* (Lepidoptera: Yponomoutidae) and strategies for its management. Ph.D. Thesis submitted to University of Agricultural Sciences, Banglore.

Sanyal, A.K. 1953. Pumpkin beetle menance in U.P. *Indian Journal of Entomology,* **15** (4): 381-382.

Sardana, H.R. and Shashi Verma, 1986. Preliminary studies on the prevalence of insect pests and their natural enemies on cowpea crop in relation to weather factors at Delhi. *Indian Journal of Entomology,* **48** (4): 448-458.

Sardana, H.R. and Shashi Verma, 1988 A. Dissipation of aldicarb residues in soil and green gram and cowpea crops. *Indian Journal of Plant Protection,* **16:** 163-169.

Sardana, H.R. and Shashi Verma, 1988 B: Dissipation of dimethoate residues in/on green gram and cowpea. *Proceedings National Academy of Sciences, India,* **58-b** (2): 307-312, Allahabad.

Sardana, H.R. and Shashi Verma, 1988 C: Combined influence of insecticides and fertilisers on the growth and yield of cowpea, *Vigna unguiculata,* L. Walp. *Proceedings National Academy of Sciences, India,* **58-b** (3): 463-468, Allahabad.

Sardana, H.R. Sumitra Arora, D.K. Singh and L.N. Kadu. 2004. Development and validation of adaptable IPM in egg plant, *Solanum melongena* L. A farmers participatory approach. *Indian Journal of Plant Protection.* **32** (1): 123-128.

Sardar Singh and M.S. Guram, 1961. Retreatment of soil against cutworms on potato. *Indian Journal of Horticulture,* **18** (2): 156-161.

Sardar Singh and G.S. Sohi, 1961. Field key for the determination of insects infesting vegetables. *Punjab Horticulture Journal,* **1**: 48-57, Patiala.

Sarode, S.V. and K.K. Kumar, 1983. Efficacy of some granular insecticides against major pests of cabbage. *Pesticides,* **17** (4): 26-27.

Sarup, Prakash, M.G. Jotwani and S. Pradhan, 1960. Relative toxicity of aome important insecticides to the bean aphid, *Aphis craccivora* Koch. (Aphididae: Homoptera). *Indian Journal of Entomology,* **22**

(2): 105-108.

Sarup, Prakash, D.S. Singh, Prasad Sircar, Swaran Amarpuri, Rattan Lal, V.S. Saxena and V.S. Srivastava, 1971. Relative toxicity of some important pesticides to the adults of *Bagrada cruciferarum* Kirk. (Pentatomidae: Hemiptera). *Indian Journal of Entomology,* **33** (4): 452-456.

Sarup, Prakash, Prasad Sircar, D.N. Sharma, D.S. Singh, Swaran Dhingra, R.S. Dewan and Rattan Lal, 1974. Evaluation of biological efficacy of insecticidal granular formulations against some important Predators/ Pests of Pea Crop. *Indian Journal of Entomology,* **36** (2): 153-159.

Sarup, Prakash, S. Wadhwa, Daya Shankar Singh and Rattan Lal, 1966. Testing of Pesticides against cabbage aphid, *Brevicoryne brassicae* Linn. *Indian Journal Of Entomology,* **28** (3): 369-374.

Sastry, K.S.M. and S.J. Singh, 1971. Effect of different Insecticides on the control of whitefly (*Bemisia tabaci* Gen.) population in Tomato Crop and the Incidence of Tomato Leaf Curl Virus. *Indian Journal of Horticulture,* **28**: 304-309.

Sastry, K.S.M. and S.J. Singh, 1978. Field Evaluation of Insecticides for the control of whitefly (*Bemisia tabaci*) in relation to the Incidence of yellow Vein Mosaic of Okra. (*Abelmoschus esculentus*). *Indian Phytopathology,* **26:** 129-138.

Satpathy, S., S. Rai and K.S. Kapoor 2002. *Souvenir, AICRP on Vegetables Crops.* IIVR, Varanasi. pp 123-130.

Satpathy, S., Samarjit Rai, Nirmal De and A.P. Singh 2004, effect of Insecticides on Leaf Net Carbon Assimilation rate and Pest Incidence in Okra. *Indian Journal of Plant Protection.* **32** (2): 22-25.

Saxena, A.P. and S.S. Misra, 1980. Potato Pests in Hilly tracts and their control. *Seeds & Farms,* **6** (10): 43-50.

Saxena, H.P., 1976. Pest problem in Pulse Crops. *Proceedings of National Academy of Sciences,* **B-46** (1 & 2): 173-176, Allahabad.

Saxena, H.P., Sunil Kumar and S.K. Prasad, 1971. Efficacy of some systemic granular insecticides against Galerucid beetle *Madurasia obscurella* Jacoby Infesting *Kharif* Pulses. *Indian Journal of Entomology,* **33** (4): 470-471.

Saxena, J.D., S. Rai, K.M. Srivastava and S.R. Singh, 1989. Resistance in the field population of the Diamondback moth to some commonly used Synthetic Pyrethroids. *Indian Journal of Entomology,* **51:** 265-

268.

Saxena, K.N., 1955. Studies on the passage of food hydrogenion concentration and enzymes in the gut and salivary glands of *Dysdercus koenigii* Fabr. (Pyrrhocoridae: Heteroptera). *Journal of Zoological Society of India,* 7 (2): 145: 154.

Saxena, R.C., 1971. Bionomics of *Caliothrips indicus* (Bagnall) infesting onion (Thripidae: Thysanoptera). *indian Journal of Entomology,* **33** (3): 342-345.

Sen, P.C., 1921. The large brown cricket, *Brachytrypes portentosus* Licht. *Bengal Agricultural Journal,* **1** (4): 111-112, Dacca, Calcutta.

Sen, P.C., 1923. Leaf roller. *Bengal Agricultural Journal,* **3** (1): 23, Dacca, Calcutta.

Sengupta, G.C., 1957. On the biology of *Lema proeusta* Fab. *Journal of Economic Entomology,* **50** (4): 471-474.

Sengupta, G.C. and B.K. Behura, 1956. On the life-history of , *Lema semiregularis* Jac. *Journal of Bombay Natural History Society,* **53** (3): 484-485.

Sengupta, G.C. and N. Panda, 1958. Insecticidal control of *Epilachna vigintioctopunctata* (F.). *Journal of Economic Entomology.* **51** (6): 749.

Sengupta, G.C., N. Panda and N. Shi, 1959. Tests of insecticides for control of *Epilachna vigintioctopunctata* Fabr.(Coccinelidae) *Proceedings of 46th Indian Science Congress, Delhi,* III: 505, Calcutta.

Sengupta, G.C. and N. Shi, 1960. Chemical control of sweet potato weevil *Cylas formicarius* Fabr. (Curculionidae: Coleoptera) attacking roots and stems of sweet potatoes in the field. *Proceedings of 47th Indian Science Congress, Bombay, III:* 264, Calcutta.

Seshagiri Rao, D., 1957. A serious infestation of potato tubers by the red ant, *Solenopsis geminota* F., in the field. *Mysore Agricultural Journal,* **32** (2): 114.

Seshagiri Rao, D., 1958. Control of fruitflies attacking fruit and vegetable Crops. *Mysore Agricultural Calendar and Yearbook,* 1957-58: 171 - 173.

Shah, A.H., 1979. Field evaluation of some new insecticides against the brinjal fruit and shoot borer, *Leucinodes orboncdis* Guen. *Indian Journal of Entomology,* **41** (1): 195- 196.

Shah, A.H., C.B. Patel and M.B. Patel, 1994. Field evaluation of some new insecticides for the control of Diamondback moth, *Plutella xylostella* (L.). *Pestology,* **8**: 20-22.

Shah, S.A., 1939. Mode of damage by *Aulacophora foveicollis* (Lucas). *Indian Journal of Entomology,* **1** (3): 97.

Sharangapani, M.V. and S.V. Pingale, 1955. Hazards of DDT treatment of potato. *Farmer Bombay,* **6** (11): 54-55, Bombay. also in *Bulletin of Central Food Technology Research Institute,* **4** (3): 57.

Sharma, A.K. and B.R. Subbarao, 1964. A further contribution to the knowledge of taxonomy and biology of Aphidinae (Ichneumonidae: Hymenoptera) with particular reference to Indian forms. *Indian Journal of Entomology,* **26** (2): 238-239.

Sharma, K.C. and O.P. Bhalla, 1991. Predatory potential of Syrphid species on different aphids of cruciferous crops in the Mid Hills region of H.P. *Indian Journal of Plant Protection,* **19.** (1): 73-75.

Sharma, M.L., T.R. Sharma, R.C. Sharma and D.N. Nandkar, 1990. Field evaluation of tomato cultivars to fruit borer (*Helicoverpa armigera*) in Satpur Plateau Region of M.P. *Orissa Journal of Horticulture,* **18** (1-2): 22-25, Bhubaneswar.

Sharma, S.L., H.S. Sohi, B.R. Verma and K.B. Sachdeva, 1966. Chemical control of defoliation of tomatoes. *Indian Journal of Horticulture,* **23** (3 and 4): 172-176, New Delhi, Bangalore.

Sharma, T.R., A.K. Datta Gupta and H.L. Kundu, 1959. Occurrence of the indian Pea-leaf miner, *Phytomyza atricornis* Meigen (Agromyzidae: Diptera) at Pilani (Rajasthan). *Proceedings of 46th Indian Science Congress, Delhi, III:* 397, Calcutta.

Sheeba Jasmine, R., S. Chandrasekaran, S. Kuttalam, and R. Jayakumar. 2005: Bioefficacy of carbosulfan (Marshal 25ec) against leaf hopper and harvest time residues of carbosulfan in/on brinjal. *Pestology,* **XXIX** (4): 28-31.

Shetgar, S.S., V.M. Pawar and S.N. Puri, 1983. Residues of monocrotophos in/on okra. *Indian Journal of Entomology,* **45** (3): 237-243.

Shetgar, S.S., V.M. Pawar, S.N. Puri, A.K. Raodeo and V.K. Patil, 1982. residues of monocrotophos in/or okra (*Hibiscus esculenlus*). *Indian Journal of Plant Protection,* **10** (1 & 2): 57-63.

Shi, N. and B.N. Satpathy, 1960. Comparative toxicity of different insecticides to control *Epilachna vigintioctopunctata* Fabr. (Coccinellidae: Coleoptera) attacking brinjal crop. *Proceedings of 47th Indian Science Congress, Bombay,* III: 554, Calcutta.

Shi, N. and B.N. Satpathy, 1960. Tests of insecticides for control of jassid,

Empoosca devastans Distant attacking Lady's finger (*Hibiscus esculentus*). *Proceedings of 47th Indian Science Congress, Bombay, III:* 555, Calcutta.

Shi, N., G.C. Sengupta and B.N. Satpathy, 1960. Relative toxicity of some important insecticides to *Epilachna vigintioctopunctata* F. (Coccinellidae: Coleoptera). *Indian Journal of Entomology,* **22** (2): 87-91.

Shylesha, A.N., N.S. Azad Thakur and Mahesh Kumar, 1996. Insect-pest complex of jack-bean *Canavcdia ensiformis* Linn, and sword-bean *C. gladiata* Jacq. vegetable crops of North-East India. *Pestology,* **20** (7): 11-14.

Sidhu, H.S. and S. Singh, 1964. Biology of mustard aphid *Lipaphis erysimi* Kalt. In Punjab. *Indian Oilseeds Journal,* **8** (4): 348-349.

Simonet, D.E., 1981. Carrot weevil management in Ohio vegetables. *Ohio Report,* **66** (6): 83-85, Ohio Agricultural Research and Development Centre, Wooster (Ohio), USA.

Singh, America, T.P. Trivedi. H.R. Sardana, O.P. Sharma and Naved Sabir, 2004: *Recent Advances in Integrated Pest Management. National centre for Integrated Pest Management,* Lal Bahadur Shastri Bhawan, Pusa Campus, New Delhi-110012. 290 pp.

Singh, C., 1953. *Lamprosema (Nacoleia)* species as a Pest of *Lal* Sag, *Amaranthus* Sp. *Indian Journal of Entomology,* **15** (1): 38.

Singh, D. and B.S. Chahal, 1978. Control of tomato fruit borer *(Heliothis armigera* Hubner*)* in Punjab. *Haryana Journal of Horticultural Science,* **7** (3-4): 182-186.

Singh, D., G.C. Varma and M. Ramzan, 1976. A field evaluation of some new insecticides against diamondback moth *Plutella xylostella* Linn. *Indian Journal of Plant Protection,* **4** (1): 17-19.

Singh, D.S. and Prasad Sircar, 1980. Relative toxicity of pyrethroids to *Lipaphis erysimi* Kalt. and *Aphis craccivora* Koch. *Indian Journal of Entomology,* **42** (4): 597-605.

Singh, Darshan, M. Ramzan and G.S. Mann, 1977. Record of *Monomoriwn* Sp. infesting okra fruits (*Abelmoschus esculentus* Moench) At Ludhiana (Punjab). *Science and Culture,* **43** (7): 323

Singh, Dilbagh, 1983. Control of brinjal fruit borer, *Leucinodes orbonolis* Guen. with synthetic pyrethroids, *Pesticides,* **17** (9): 14-15.

Singh, G. and A. Mukherjee, 1987. On the Oligophagous nature of

Henosepilochna dodecastigma (Wiedemann) and *Henosepilacha vigintioctopuctata* (Fabr.) (Coleoptera: Coccinethidee). *Indian Journal of Entomology,* **49** (1): 118-126.

Singh, G. and P.N. Misra, 1978. Effect of some insecticides on aphid population and yield of potato crop. *Indian Journal of Agricultural Sciences,* **48** (1): 12-16.

Singh, H. and G. Singh, 1970. Evaluation of some important pesticides for the control of fruit borer (*Heliothis armigera* Hubner) Infesting Tomato. *Indian Journal of Entomology,* **32** (3): 205-210.

Singh, H., 1984. *Household and Kitchen Garden Pest-principles and Practices,* 430 pp., Kalyani Publishers, New Delhi.

Singh, H.B. and R.K. Arora, 1978: *Wild Edible Plants of India*, 89 pp., Indian Council of Agricultural Research, New Delhi.

Singh, H.B., 1963: Pusa Kesar and Pusa Kanchan carrot and turnip varieties. *Indian Horticulture,* **7** (4): 21.

Singh, Harcharan and Manjit Singh Dhooria, 1971: Bionomics of the pea pod borer, *Etiella zinckenella* (Treitschke). *Indian Journal of Entomology,* **33** (2): 123-130.

Singh, I.P., A. Ipe and M.I. Ipe, 1991. Control of *Liromyza brassicae* Riley By flooding. *Entomon.,* **16** (1): 83-85.

Singh, Jagdish, Sanjev Kumar And G. Kalloo. 2003: Nutritional improvement in vegetable crops using biotechnological approaches pp. 280-312 *In: Plant Genetic Engineering* Vol **5** Rana P. Singh and Pawan K. Jaiwal [Eds.], Sci. Tech. Publising Lll, Houston, Texas - 77036, U.S.A.

Singh, J.P., 1966. Insect pests of cucurbits. *U.P. Agricultural University Magazine, 1965-66*: 72-79, Kanpur.

Singh, J.P., 1970: *Elements of Vegetable Pests,* 275 pp., Vora & Co. Publishers Pvt. Ltd., Bombay.

Singh, K.P., G.D. Kapadia and A.B. Paralikar, 1983. Aldicarb residues in potato. *Indian Journal of Entomology*, **54** (1): 60-63, New Delhi.

Singh, K.P., Y.S. Malik and M.L. Pandiste, 1983. Control of fruit and shoot borer, *Leucinodes orbonalis* Guen. through synthetic pyrethroids and their effect on the yield of seed and fruits of brinjal, *Pesticides,* **17** (9): 28-29.

Singh, M.N., S.M. Paul Khurana, B.B. Nagaich and H.O. Agarwal, 1982. Strains of potato leaf role virus and their aphid transmission. *Journal of Indian Potato Association,* **8**: 121.

Singh, M.P., M.S. Khuman and J.S. Salam, 1995. Efficacy of certain plant extracts against *Myzus persicae* Sulz. (Homoptera: Aphididae) on cabbage in Manipur. *Indian Journal of Plant Protection,* **23** (2): 139-141.

Singh, Pritam and Rattan Lal, 1966. Dissipation of malathion residues on tomato and peas. *Indian Journal of Entomology,* **28** (3): 332-338.

Singh, R., H.C.L. Gupta and B.L. Pareek, 1976: Estimation of carbaryl residues in/on sponge gourd. *Indian Journal of Environmental Health,* **18** (3): 212-218. Nagpur.

Singh, Rajwinder and G.S. Battu. 2007. Laboratory evalution of selected phagostimulente-adjuvants for tank-mix formulation of HEARNPV (PAV-1 Strain) against *Helicoverpa armigera* (Hubner), *Journal of Insect Science,* **20**(1): 58-64.

Singh Rana P. and Pavan K. Jaiwal. 2003. *Plant Genetics Engineering.* Vol. **5** , *Improvement of Vegetables,* 315 p. Sci. Tech. Publishing, Llc, Houstan, Texes 77036, U.S.A.

Singh, Ranjit, 1969. Fruits. 213pp. National Book Trust Ltd., New Delhi.

Singh, S.P., 1972. Cotton bollworm *Heliothis armigera* Hb. (Lepidoptera: Noctuidae) on cabbage. *Entomol. Obozr.,* **51**(1): 43-46 (Russian); English Translation *Entomological Review,* **51:** 27-28, Washington (USA).

Singh, S.P., 1974. Observation on the biology of *Trichoplusia orichalcea* Fabr. on Potato. *Indian Journal of Plant Protection,* **2** (1&2): 127-128.|

Singh, S.P., 1978. Occurrence of *Metarrhizinm anisopliae* (Metch.) Sorokin on white grubs in potato fields. *Current Research,* 7 (10): 173.

Singh, S.P., 1982. Some observations on potato cutworms and their natural enemies. *Entomon,* 7 (2): 197-203.

Singh, S.P., 1986. A Note on the adult biology of *Agrotis spinifera* (Hubner) on potato. *Bulletin of Entomology,* **27** (2) 191-193.

Singh, S.P., 1987. Studies on some aspects of the bioecology of potato cutworms in India. *Journal of Soil Biology and Ecology,* **7**: 135-143.

Singh, S.P. and K.D. Verma, 1982. A note on white grubs and wireworms infesting potatoes in Shimla: In: *Potato in developing Countries,* 379-380, 426 Pp., Cpri, Shimla.

Singh, T.V.K., T. Raman Goud and D.D.R. Reddy, 1993. Control of diamondback moth *Plutella xylostella* (L.) with Insecticides on Cabbage. *Pestology,* **17** (5): 12-13.

Singh, Umrao, A.M. Wadhwani and B.M. Johri, 1983. *Dictionary of Economic Plants of India*, 288 pp., Indian Council of Agricultural

Research, New Delhi.

Singh, Y.P., 1995. Persistence and degradation of endosulfan in/on cauliflower (*Brassica oleracea* var. *botrytis* Linn.) At Medium High altitude hills. *Indian Journal of Plant Protection,* **23**(1): 101-104.

Singh, Yogendra and S.L. Bichoo, 1973. Bionomics of *Earias fabia* Stol on okra. *Bulletin of Entomology*, **14** (1-2): 36-40.

Singh, Z. and S.M. Vaishampayan, 1967. Results of trials on the chemical control of *Caliothrips indicus* (Bagnall) on pea. *Indian Journal of Entomology,* **29** (2): 219-220.

Sinha, P.K., R. Singh and R.P. Yadav, 1978. India - *Crioceris impressa* Fabr. on air potato. *FAO Plant Protection Bulletin,* **26** (1): 27, Rome.

Sinha, Purnima S. and S.K. Saxena, 1990. Effect of latex of *Euphorbia hirta* on fruit rot of tomato caused by presence of *Drosophila busckii. Indian Phytopathology,* **43** (3): 462-464.

Sinha, S.N., N.P. Agnihotri, A.K. Chakrabarti, H.K. Jain and H.C. Meher, 1986. Field Evaluation of some chemical control schedules against insect pest of brinjal and their residues on fruit. *Pestology,* **10**: 5-9.

Sinha, S.N., N.P. Agnihotri, H.K. Jain and A.K. Chakrabarti, 1985. Bioefficacy of synthetic pyrethroids against tomato fruit worm *Heliothis armigera* Hubn. And their residues on tomato fruit. *Pestology,* **9** (5): 9-12.

Sinha, S.N. and A.K. Chakrabarti, 1983. Effect of seed treatment with carbofuran on the incidence of red pumpkin beetle *Raphidopalpa foveicollis* Lucas on cucurbits. *Indian Journal of Entomology,* **45** (2): 145-151.

Sinha, S.N., A.K. Chakrabarti, N.P. Agnihotri, H.K. Jain and V.T. Gajbhiye, 1984. Efficacy and residual toxicity of some systemic granular insecticides against *Thrips tabacci* on onion. *Tropical Pest Management,* **30**: 32-35.

Sinha, S.N., A.K. Chakrabarti and KM. Peswani, 1977. Control of Red pumpkin beetle, *Raphidopalpa foveicollis* Lucas infesting cucurbit seedlings under field conditions. *Seed Research,* **5**: 44-48.

Sohi, Gurcharan Singh, 1945. Trials with DDT against vegetable pests, *Indian Journal of Entomology,* 7(1-2): 242.

Sohi, G.S., 1958. Bionomics and control of *Athalia proxima* Klug. (Hymenoptera: Tenthredinidae). *Proceedings of 45th Indian Science Congress, Madras, III:* 470-471, Calcutta.

Sreeramulu, C., 1976. Improved varieties of chillies. *Arecanut and spices*

Bulletin, 7 (3): 65-68, Calicut.

Srinivas, S.V. and Clement Peter, 1993. Efficacy of certain new insepticides to the brinjal shoot and fruit borer, *Leucinodes orbanalis* Guen. *Pestology,* **17** (6): 36-38.

Srinivasan, K. and G.K. Veeresh, 1986. Economic analysis of promising cultural practices in the control of moth pests of cabbage. *Insect Science and its Application,* **7** (4): 559-563, Nairobi (Kenya).

Srinivasan K., C.A. Viraktamath, Meera Gupta and G.C. Tewari, 1995. Current status, host range and parasitoids of serpentine leaf-miner, *Liriomyza trifolii* in South India. *Indian Journal of Plant Protection,* **23** (1): 64-67.

Srinivasan, P.M., 1959 A. Control of fruit borer (*Heliothis armigera* H.). *Indian Journal of Horticulture,* **16** (3): 187- 188.

Srinivasan, P.M., 1959 B: Guard your bitter gourd against fruitfly. *Indian Farming,* **9** (9): 8.

Srinivasan, P.M., 1961. Banish the *Epilachna* beetle from the brinjals. *Indian Farming,* **10** (11): 13-14.

Srinivasan, P.M. and M. Basheer, 1961a. Some borer resistant brinjals. *Indian Farming,* **11** (8): 19.

Srinivasan, P.M. and M. Basheer, 1961 B. A trap crop for the control of fruitfly in cucurbits. *Indian Horticulture,* **5** (4): 8-9.

Srinivasan, P.M., M. Basheer and R.B. Gowder, 1959. Here's good way of controlling the tomato borer. *Indian Horticulture,* **3** (3): 7.

Srinivasan, P.M. and R.B. Gowder, 1957. A note on the preliminary investigations for the control of *bhendi* shoot and fruit borer, *Earias* sp. *Indian Journal of Horticulture,* **14** (4): 241-244.

Srinivasan, P.M. and R.B. Gowder, 1959 A. How to control the brinjal borer. *Indian Horticulture,* **3** (2): 7-8.

Srinivasan, P.M. and R.B. Gowder, 1959 B. A note on the control of the brinjal shoot and fruit borer, *Leucinodes orbonalis* Guen. *Indian Journal of Agricultural Sciences,* **29** (1): 71-73.

Srinivasan, P.M. and R.B. Gowder, 1960. A preliminary note on the control of the *bhindi* shoot and fruit borer, *Earias fabia* and *E. insulana. Indian Journal of Agricultural Sciences,* **30** (1): 55-57.

Srinivasan, P.M. and P.S. Narayanaswamy, 1960 A. A preliminary note on the control of fruitfly *Dacus cucurbitae* Coq., on bitter gourd *Momordica charantis. Madras Agricultural Journal,* **47** (1): 34.

Srjnivasan, P.M. and P.S. Narayanaswamy, 1960 B: Experiment on the control of the shoot and fruit borer, *Earias fabia* Stol. And *Earias insulana* Boisd., on *bhindi, Abelmoschus esculentus* L. *Madras Agricultural Journal,* **47** (1): 34-35.

Srinivasan, P.M. and P.S. Narayanaswamy, 1961a. Bringing the tomato borer to book. *Indian Farming,* 11 (2): 28-29.

Srinivasan, P.M. and P.S. Narayanaswamy, 1961b. A research note on the appropriate time for taking up control measures against fruitflies on ash gourd, *Benincasa cerifera. Madras Agricultural Journal,* **48** (10): 395-396.

Srinivasan, P.M. and P.S. Narayanaswamy, 1961c. Varietal susceptibility in *bhendi* (*Abelmoschus esculentus* L.) to the shoot and fruit borers, *Earias fabia* Stoll and *Earias insulana* Boisd. *Madras Agricultural Journal,* 48 (9): 345-346.

Srivastava, B.K., 1958. A new cutworm pest of potato in Rajasthan, *Current Science,* **27** (12): 504.

Srivastava, B.K., 1959. Studies on Cucurbit Fly. *Kanpur Agriculture College Journal,* **9** (2): 17-21.

Srivastava, B.K., 1964. Pests of Pulse Crops. *In*: *Entomology in India:* 83-91, Silver Jubilee Number, Entomological society of India, New Delhi.

Srivastava, B.K. and S.K. Bhatia, 1957. Occurrence of *Spermophogus* Sp. as pest of okra seed. *Indian Journal of Entomology,* **19** (3): 216-217.

Srivastava, B.K. and R.M. Khan, 1962. Cutworms of potato and their control in Rajasthan. *Indian Potato Journal,* **4** (2): 88-92.

Srivastava, B.K. and R.M. Khan, 1963. Bionomics and control of *Hototrichia insularis* (Melolonthidae: Coleoptera). *Indian Journal of Entomology,* **25** (4): 347-354.

Srtvastava, B.P., 1957. Morphology of the alimentary canal of the moth *Leucinodes orbonolis* Guen. (Lepidoptera: Pyraustidae). *Proceedings of National Academy of Sciences, India,* **27** (2): 106-112, Allahabad.

Srivastava, B.P., 1960 A. Morphology of the reproductive organs of *Leucinodes orbonalis* Guen. (Lepidoptera: Pyraustidae): Part I - the female organs. *Indian Journal of Entomology,* **22** (1): 35-46.

Srivastava, B.P., 1960 B. Morphology of the reproductive organs of *Leucinodes orbonalis* Guen. (Lepidoptera: Pyraustidae): Part II - the male organs. *Indian Journal of Entomology,* **22** (3): 160-171.

Srivastava, B.P., 1961. Biology and control of the brinjal stem borer, *Euzophera perticella* Rag. (Lepidoptera: Pyralidae). *Madras Agricultural Journal,* **48** (11): 429-433.

Srivastava, J.B. and G.L. Bhagat, 1966. Studies on the biology of *Criocerls impressa* Fabr. (Chrysomelidae: Coleoptera) a defoliator pest of *Dioscorea bulbifera* Linn. *Indian Journal of Entomology,* **28** (4): 435-437.

Srivastava, J.B. and B.P. Saxena, 1965. *Euzophera perticella* Rag. (Pyrallidae: Lepidoptera) as a stem borer of *Solanum aviculare* Forst. *Indian Journal of Entomology,* **27** (4): 487.

Srivastava, K.P., 1993. *Textbook of applied Entomology*, 2: 424 pp., Kalyani Publishers, New Delhi.

Srivastava, K.P., N.P. Agnihotri and H.K. Jain, 1987. Efficacy, persistence of Synthetic Pyrethroid residues in Summer grown cowpea and green gram. *Indian Journal of Agricultural Sciences,* **57**: 419-422.

Srivastava, P.D., 1957. Studies on the choice of food-plants and certain aspects of the digestive physiology of the larvae and adults of *Athalia lugens proximo.* (Klug) and *Epilachna vigintiactopunctata* (F.). *Bulletin of Entomological Research,* **48** (2): 289-297.

Srivastava, R.P. and Dhamo K. Butani, 1972. A method to prevent the mango Mealybug (*Drosicha mangiferae* Green) damage. *Entomologists, Newsletter,* **2** (5): 35.

Srivastava, R.P., S.B. Mathur and M.L. Sahni, 1965. An entomogenous fungus *Entomophthora* Sp. (Entophthoraceae: Entomophthorales) pathogenic on *Empoasca devastans* Distant (Jassidae: Hemiptera) a *bhindi* jassid. *Indian Journal of Entomology,* **27** (3): 367.

Srivastava, U.S. and J. Bahadur, 1958. Observations on the life history of red cotton bug, *Dysdercus cingulatus* (Hemiptera: Pyrrhocoridae). *Indian Journal of Entomology*, **20** (3): 228-234.

Stebbing, E.P., 1903. Insect pests of the sugarcane in India. *Indian Museum Notes*, **5** (3): 64-91.

Stilt, L.L., 1953. Insecticide tests for control of maggots in turnip. *Journal of Economic Entomology.* **46** (6): 961-965.

Stollalis, C., 1982. Papillons Exotiques, **4**:355, Amsterdam.

Subba Rami Reddy, S., V.I.V. Prasada Rao and K. Lakshaminarayan, 1981. Control of chilli aphids in Andhra Pradesh. *Pesticides,* **15** (3): 38-39.

Subba Rao, B.R., 1957. Some new species of Indian Eulophidae. *Indian Journal of Entomology,* **19** (1): 50-53.

Subbaratnam, G.V., 1982. Studies on the internal characters of shoot and fruit of brinjal Governing resistence to shoot and fruit borer, *Leucinodes orbonalis* Guen. *South Indian Horticulture,* **30** (3): 217-220, Coimbatore.

Subbaratnam, G. V., N.P. Agnihotri and H.K. Jain, 1984. Dissipation of fenvalerate in/on brinjal fruit. *Indian Journal of Entomology,* **46**: 287-290.

Subbaratnam, G.V. and Dhamo K. Butani, 1981. Screening of egg-plant varieties for resistance to insect pest complex. *Vegetable Science,* **8** (2): 149-153, New Delhi.

Subbaratnam, G.V. and Dhamo K. Butani 1982. Chemical control of insect pest complex of brinjal. *Entomon,* **7** (1): 97- 100.

Subbaratnam, G.V., Dhamo K. Butani and B.H. Krishna Murthy Rao, 1983. Leaf characters of brinjal governing resistance to jassid, *Amrasca biguttula biguttula* (Ishida). *Indian Journal of Entomology,* **45** (2): 171-173.

Subramaniam, T.R., 1957. Description and life-history of a new weevil of the Genus *Protocylos* from Coimbatore. *Indian Journal of Entomology,* **19** (3): 204-213.

Subramaniam, T.R., 1959 A. A note on the life-history of *Apion ampulum* Fst. *Madras Agricultural Journal,* **46** (6): 236.

Subramaniam, T.R., 1959 B. biology of *Colobodes dolichus* Marshall - A pest of *Dolichus lablab* L. In South India. *Indian Journal of Entomology,* **21** (1): 35-45.

Subramaniam, T.R., 1959 C. Observations on the biology of *Cylas formicarius* F. At Coimbatore. *Madras Agricultural Journal,* **46** (8): 293-297.

Subramaniam, T.R., 1960 A. Description and biology of a new species of weevil damaging snake gourd (*Trichosanthes anguina*) in South India (Part I). *Madras agricultural Journal,* **47** (4): 219-220.

Subramaniam, T.R., 1960 B. Description and Biology of a new species of weevil damaging snake gourd (*Trichosanthes anguina*) in South India (Part II). *Madras Agricultural Journal,* **47**(6): 258-264.

Subramaniam, T.R., 1965. A note on weevils damaging *Moringa. Indian Journa! of Entomology,* **27** (4): 485-486,

Subramaniam, T.V., 1920. The Life-history of Moringa stem-borer. *Report of Proceedings of 3rd Entomological Meeting, Pusa (Bihar), February, 1919,* **3**: 922-925, Calcutta.

Sudheer Reddy, V. and B. Murlimohan Reddy, 1994. Effect of seed protectants on susceptibility of eggplant (*Solanum melongena* L.) Seed. *Seed Research,* **22** (2): 181-183, New Delhi.

Suganya Kanna, S., S. Chandra Sekaran, A. Regupathy and J. Stanly, 2005.

Field efficacy of emamectin 5 Sg against tomato fruit boses, *Helicoverpa armigera* (Hiibner) *Pestology,* **XXIX** (4): 21-24.

Sunndita and Divneder Gupta. 2007. A note on Host range of Fruit fly species infesting Summer vegetable crops in mid Hills of Himachal Pradesh. *Journal of Insect Science,* **20** (1): 106-107.

Swank, G.R., 1937. Tomato pinworm, *Gnorimoschema lycopersicella* (Busck.) in Florida. *Florida Entomologist,* **20** (3): 33-43, Gainesville.

Swarup, Vishnu, 1994. Vegetables: export needs thrust: 125-127pp., the Hindu survey of Indian Agriculture, 1994, Chennai.

Swinhoe, C., 1885. On the Lepidoptera of Bombay and the Deccan. Part IV - Heterocera (Continued). *Proceedings of Zoological Society,* 1885. 885.

Tabashnik, B.E., N. Finson and M.W. Johnson, 1991. Managing resistance to *Bacillus thuringiensis:* lesson from diamondback moth (Lepidoptera: Plutellidae). *Journal of Economic Entomology,* **84** (1): 49-55.

Talbot, G., 1939. *The Fauna of British India, including Ceylon and Burma.* Butterflies, **1**: 413-429, Taylor and Francis Ltd., London.

Taylor, T.A., 1964. Studies on Nigerian Yam Beetles. II - Bionomies and control. *Journal West African Science Association,* **9**: 3-31, Achimota, London.

Teotia, T.P.S. and S.N. Sinha, 1971. Relative toxicity of some insecticides to the larvae of brinjal fruit and shoot borer, *Leucinodes orbondlis* Guen. *Indian Journal of Entomology*, **33** (2): 217-218.

Tewari, G.C. and H.R. Sandana, 1990. An unusual heavy parasitisation of brinjal shoot and fruit borer, *Leucinodes orboncalis* Guen., by a new Praconid Parasite. *Indian Journal of Entomology*, **52** (2): 338-341.

Tewari, G.C. and P.N.K. Moorthy, 1983. Effectiveness of Synthetic pyrethroids, against the pest complex of brinjal, *Entomon,* **8**: 165-168.

Tewari,. G.C., N.K.K. Kumar and P.N.K. Moorthy, 1984. Optimising the dose and spray interval of synthetic pyrethroid against brinjal fruit and shoot borer, *Leucinodes orbonalis. Entomon,* **9** (3): 197-200.

Tewari, R.N., D.C. Pachauri and R.K. Sharma, 1996. Tomatoes from IARI. tomato 'Pusa Hybrid-1' and 'Pusa Hybrid-2'. *Indian Horticulture,* **40** (4): 26-29.

Thakur, M.R., 1989. Okra [*Abelmoschus esculentus* (Linn.) Moench.]. *Indian Horticulture,* **33** And **34** (4 and 1): 49-51.

Thanki, K.V. and J.R. Patel, 1991. Field evaluation of some brinjal varieties to insect pests and diseases in North Gujarat conditions. *Gujarat Agricultural University Journal,* **16** (2): 94-97.

Thevasagayam, E.S. and S.C. Gangasingham, 1961. Major insect pests of brinjal (*Solanum melongena*) and their control. *Tropical Agriculturist,* **117** (2): 105-114, Paradeniya.

Thomas, C.A., 1932. The potato pinworm (*Gnorimoschema lycopersicella* Busck.) a new pest in Pennsylvania. *Journal of Economic Entomology,* **35** (1): 137-138.

Thomas, Cyrus, 1878. A list of the species of the Tribe Aphidini Family Aphidae, found in the United States, which have been herefore named, with descriptions of some new species. *Bulletin Illinois State Laboratory Natural History,* **2**: 3-16, Chicago, USA.

Thompson, C.H. and William C. Kelly, 1957. *Vegetable Crops,* 611 pp., McGraw-Hill Book Co. Inc., USA.

Thompson, W.R., 1946. A catalogue of the parasites and predators of insect pests. Section 1 (7): 265-266. the Imperial Parasite Service. Ontario (Canada).

Thusaliram, B.V. Patil M, Bheemanna, A.C. Hosamani, and S.G. Hanchinal. 2005. Bioefficacy of bifenthrusi (Talstar) 10 EC. against chilli insect pests in irrigated ecosystem *Pestology,* **XXIX** (3): 17-19.

Tirumala Prasad, R., G.P.V. Reddy, M.M.K. Murthy and V. Devaprasad, 1993. Efficacy and residues of certain newer insecticides on okra. *Indian Journal of Plant Protection,* **21** (1): 47-50. *Madras Agricultural Journal,* **77** (3-4): 169-172.

Tirumala Rao, V. and M. Koteswara Rao, 1955. Incidence and control of brinjal pests. *Andhra Agricultural Journal,* **2** (2): 76-85.

Tirumala Rao, V., A. Perraju and P.V. Ranga Rao, 1954. Chemical control of brinjal tingid (*Urentius echinus* D.). *Andhra Agricultural Journal,* **1** (5): 312-313.

Tomar, B.S., T.S. Kalda, K.A. Balakrishnan and S.S. Gupta, 1996. Exploring nutritive potential of brinjal, *Indian Horticulture,* **40** (4): 32, New Delhi.

Trehan, K.N., 1957. Brief notes on crop pests and their control in the Punjab (India). *Journal of Bombay Natural History Society,* **54** (3): 581-625.

Trehan, K.N. and S.R. Bagal, 1949. Life-history, bionomics and control of sweet potato weevil, *Cylas formicarius* in Bombay Province. *Current Science,* **18** (4): 126-127.

Trehan, K.N. and S.R. Bagal, 1957. Life-history, bionomics and control of sweet potato weevil (*Cylas formicarnius* F.) with short notes on other pests of sweet potato in Bombay State. *Indian Journal of Entomology,* **19** (4): 245-252.

Trehan, K.N. and H.R. Pajni, 1960. Bioassay of insecticides -i. Relative toxicity of some insecticides as stomach poison against red pumpkin beetle *Aulocophora foveicollis* Lucas (Chrysomelidae: Coleoptera). *Research Bulletin, East Punjab University,* **11**: 25-27, Hoshiarpur.

Trehan, K.N. and V.K. Sehgal, 1963. Range of host plants and larval feeding in *Phytomyza atricornis* Meigen (Diptera: Agromyzidae). *Entomologists Monthly Magazine,* **99:** 1-3, London.

Tripathi, R.L., 1963. Biology of *Athalia proximo* (Klug), the mustard saw-fly. *Indian Journal of Entomology,* **25** (2): 179-181.

Twine, P.H., 1971. Cannibalistic behaviour of *Heliothis armigem* (Hubn.). *Queensland Journal of Agricultural and Animal Sciences,* **28** (5): 153-157, Brisbane.

Umapathy, G. and P. Baskaran, 1991. Insecticidal control of major pests of radish. *Madras Agricultural Journal,* **78** (1-4): 96-97.

Unni, K.K., 1980. Synthetic pyrethroids - the new generation pesticides. *Andhra Agricultural Journal,* **27** (1 & 2): 72-73.

Urs, K.C. Devaraj and V.L.V. Prasad Rao, 1980. Residues of quinalphos in vegetables. In: *Pesticide Residues in the environment in India*: 179-192, University of Agricultural Sciences, Hebbal, Bangalore.

Urs, K.C. Devaraj and V.L.V. Prasad Rao, 1980. Residues of monocrotophos in cabbage, cauliflower and knol-khol. In: *Pesticide Residues in the environment in India* .'219-229, University of Agricultural Sciences, Hebbal, Bangalore.

Usman, S., 1956. Some field parasites of potato tuber moth *Gnorimoschema operculella* Zell. In Mysore. *Indian Journal of Entomology*, **18** (4): 463-468.

Uthamasamy, S. and M. Balasubramanian, 1978. Efficacy of some insecticides in controlling the pests of *bhendi, Ablemoschus esculentus* (L.) Moench. *Pesticides,* **12**: 39-41.

Uttyet, G.C., 1943. Some aspects of parasitism in field populations of *Plutella maculipennis* Curt. *Journal of Entomological Society, Southern Africa,* **6:** 65-80, Pretoria.

Vankova, J., 1962. Application of a bacterial preparation of *Bacillus thuringiensis* against some pests of agricultural crops: I - Application against the large white butterfly (*Pieris brassicae* L.). *Rostl. Vyroba,*

8 (4): 571-576, Czechoslovakia.

Varela Ochoa, G. and F. Guharay, 1988. The use of multiple cropping (cabbage-carrot), as a component of integrated pest management of cabbage defoliators. *Revista Nicaraguense De Entomologia,* **4**: 49-50, Managua, Nicaragua.

Varma, B.K., 1954. Notes on *Cassida circumdata* Hbst., *Cassida tndicola* Duv. And *Glyphocosis trilineata* Hope (Coleoptera: Chrysomelidae: Cassidinae) as pests of sweet potato (*Ipomoea batatas*) at Kanpur. *Indian Journal of Agricultural Sciences,* **24** (3): 261-263.

Varma, B.K., 1961. Bionomics of *Phyllotreta cruciferae* Goeze (Coleoptera: Chrysomelidae) reared on radish in India. *Indian Journal of Agricultural Sciences,* **31** (!): 59-63.

Varma, K.D. and S.P. Singh, 1974. Slug damage to potato in Himachal Pradesh. *Indian Journal of Plant Protection,* **1** (2): 71-72.

Veeravel, R. and P. Baskaran, 1977. Control of insect pests of brinjal (Solanum melongena L). *Pesticides,* **11** (3): 31-32.

Veeresh, G.K., 1977. *Studies on the root grubs in Karnataka.* UAS Monograph Series No. 2, 87 pp., University of Agricultural Sciences, Hebbal, Bangalore.

Veire, M. Van De, 1991. Control of leafminer flies in glass house tomatoes and lettuce with the selective insecticide cyromazine. *Revue De I'agriculture,* **44** (3): 923-927, Bruxelles.

Venugopal, M.S., B. Habeeb Ullah and M. Balasubramanian, 1980. Residues of insecticides in and on tomato. In: *Pesticide Residues in the environment in India:* 258-260, University of Agricultural Sciences, Hebbal, Bangalore.

Venugopal, S., 1958. Studies on *Hyposidra successaria* Walker, a Geometrid pest of *Sesbania speciosa* in South India. *Indian Journal of entomology,* **20** (4): 283-290.

Venugopal, S. and K.S. Venkataramani, 1954. An agromyzid pest of bhindi. *Journal of Madras University, B*-**24** (3): 335-340.

Verma, A.M., P.O. Sharma and P.U. Saramma, 1969. Relative toxicity of some contact insecticides against cabbage caterpillar, *Pieris brassicae. Journal of Research,* P.A.U., **6** (1): 197-199.

Verma, A.N., J.S. Balan and S.M. Singhvi, 1970. Relative efficacy of diferent insecticides for the control of *Ak* Grasshopper, *Poekilocerus pictus* F. on Castor, *Telhan Patrika,* **2** (2): 1-5, Hyderabad.

Verma, G.C., O.S. Bindra and D. Singh, 1974. Comparative efficacy of

Bacillus thuringiensis Berliner formulations against *Pieris brassicae* Linnaeus. *Current Science,* **43** (22): 734-735.

Verma, G.C. and S.S. Gill, 1977. Laboratory studies in the comparative efficacy of biotic insecticide for the control of *Plutella xylostella. Journal of Research, Punjab Agricultural University,* **14**: 304-308.

Verma, K.D. and A.C. Mathur, 1966. First record of two species and males of *Lipaphis pseudobrassicae* Davis and *Protrama penecaeca* Stroyan (Homoptera: Aphididae) from India. *Indian Journal of Entomology,* **28** (2): 277-278.

Verma, K.D. and S.S. Mishra, 1975. Chemical control of *Epilachna oullata* infesting potato crop. *Indian Journal of Plant Protection,* **3** (2): 214-215.

Verma, S. and Lal, R., 1976. Residues and residual toxicity of endosulfan on cauliflower. *Indian Journal of Agricultural Sciences,* **46** (3): 125-129.

Verma, S.K., 1966a. Control of onion thrips, *Thrips tabaci* Lindeman and viability of onion seeds. *Indian Journal of Entomology,* **28** (2): 268-269.

Verma, S.K., 1966b. Studies on the host preference of the onion thrips, *Thrips tabaci* Lindeman to the varieties of onion. *Indian Journal of Entomology,* **28** (3): 396-398.

Verma, Shashi, 1981. Degradation of insecticides on brinjal. *Seminar on Strategies of Pest Management Held at Iari, New Delhi, 21st to 23rd December, 1981 (Abstracts):* 77, New Delhi.

Verma, Shashi, 1985. Efficacy of insecticides and dislodgable residues on okra, *Abelmoschus esculentus. Indian Journal of Entomology,* **47** (3): 336-345.

Verma, V., 1980. *A Textbook of Economic Botany*, 336+6 pp., Emkay Publications, Delhi.

Vevai, E.J., 1942. Note on *Empoosca punjabensis* Pr., the causal agent of hopper-burn in potato. *Indian Journal of Entomology,* **4** (1): 89.

Vevai, E.J., 1969 A. Know your crop, its pest problems and control - 6: potato. *Pesticides,* **3** (4): 33-37.

Vevai, E.J., 1969 B. Know your crop, its pest problems and control - 7: chillies. *Pesticides,* **3** (5): 29-35.

Vevai, E.J., 1969 C. Know your crop, its pest problems and control - 12: cabbage. *Pesticides,* **3** (9): 29-34.

Vevai, E.J., 1969 D. Know your crop, its pest problems and control - 13:

Lady's Finger, *Bhindi. Pesticides,* **3**(10): 29-33.

Vevai, E.J., 1970 A. Know your crop, its pest problems and control - 16: cucurbits. *Pesticides,* **4**(1): 27-33.

Vevai, E.J., 1970 B: Know your crop, its pest problems and control - 25: brinjal. *Pesticides,* **4**(10): 26-35.

Vevai, E.j., 1971 A. Know your crop, its pest problems and control - 28: tomato. *Pesticides,* **5**(1): 21-34.

Vevai, E.J., 1971 B. Know your crop, its pest problems and control - 30: onion & garlic. *Pesticides,* **5** (3): 17-25.

Vijaychandra, P.V. and G.P.V. Reddy, 1980. Persistence of malathion in tomato, *bhindi* and brinjal crops. In: *Pesticide Residues in the environment in India:* 264-269, University of Agricultural Sciences, Hebbal, Bangalore.

Visalakshi, A. and M.R.G.K. Nair, 1978. Biology of *Lema lacordairei* Baly (Coleoptera: Chrysomelidae: Criocerinae) a pest of yam *Dioscorea alata* in Kerala, *Entomon,* **3**(1): 129-131.

Visalakshi, A., S. Naseema Beevi and C.P.R. Nair, 1981. Relative contact toxicity of some newer insecticides to adults of banana aphid, *Pentalonia nigronervosa* Coq. *Entomon,* **6** (2): 109-110.

Visalakshi, A., S. Naseema Beevi, T. Premkumar and M.R.G.K. Nair, 1980: Biology of *Leptoglossus australis* (Fabr.) (Coreidae: Hemiptera) a pest of snake gourd. *Entomon,* **5** (1): 77-79.

Visalakshi, A., K., Santhakumari, George Kosi and M.R.G.K. Nair, 1980. Biological studies on *Aspidomorpha furcata*thunb. (Chrysomelidae: Cassidinae: Coleoptera). *Entomon,* **5** (3): 167-169.

Vyas, H.N. and H.P. Saxena, 1982. Field evaluation of some insecticides against stemfly *Melanogromyza phaseoli* (Diptera: Agromyzidae) infesting greengram *Vigna radiata* (L.). *Gujarat Agriculture University, Research Journal,* **7**: 102- 10 4.

Waghray, R.N. and Shiv Raj Singh, 1965. Influence of host plant nutrition on *Aphis craccivora* Koch. *Indian Journal of Entomology,* **27** (2): 196-201.

Walker, Francis, 1859. Catalogue lepidoptera, *British Museum,* **17**: 499 Pp., London.

Walsingham, Thomas De Grey (Lord), 1855. On the lepidoptera of Bombay and the Deccan. Part IV - Heterocera (Continued) by C. Swinhoe. *Proceedings of Zoological Society, 1885*: 885, London.

Walsingham, Thomas De Grey (Lord), 1891. African Micro- lepidoptera.

Transactions of Entomological Society of London, 1891: 63-131, London.

Walunj, A.R. and M.D. Dethe, 1996 A. Efficacy of Thiodicarb 75wp Larvin and methomyl 40 Sp lannate against shoot and fruit borer of brinjal. *Pestology*, **20** (10): 28-30.

Walunj, A.R. and M.D. Dethe, 1996 B. Dissipation of thiodicarb 75 Wp and methomyl 40 Sp residues on brinjal fruits. *Pestology*, **20** (11): 12-13.

Walunj, A.R. and M.D. Dethe, 1997. Persistence of thiodicarb 75 Wp (Larvin) and methomyl 40 SP (lannate) against leaf hopper and whitefly of brinjal. *Pestology*, **21**(6): 52-54.

Wang, Ping-yuan, 1963. Notes on Chinese pyralid moths of the Genus *Diphornia* Hubner. *Acta Entomology, Sin.* **12** (3): 358-365, Peking.

Ware, George W. and J.P. Mc Collum, 1968. *Producing Vegetable Crops*, 558 pp., the Interstate Printers and Publishers, Inc. Danville, Illinois.

Watson, J.R., 1944. *Herse cingulatus* Fab. As an armyworm. *Florida Entomologist*, **27** (3): 58, Gainesville, Florida, USA.

Wesley, W.K., 1952. *Atroctomorpha crenulata* Fb., a common destructive grasshopper and its control. *Allahabad Farmer*, **26** (1): L-6.

Wesley, W.K., 1956 A. I. Major Insect Pests of Vegetables in Allahabad, U.P. and their control. *Allahabad Farmer*, **30** (3): 121 -128.

Wesley, W.K., 1956 B. II. Major insect pests of vegetables in Allahabad, U.P. and their control. *Allahabad Farmer*, **30** (4): 135-149.

Wesley, W.K., 1956 C. III. Major insect pests of vegetables in Allahabad, U.P. and their control. *Allahabad Farmer*, **30**

Wesley, W.K., 1957 A. IV. Major insect pests of vegetables in Allahabad and their control. *Allahabad Farmer*, **31** (1): 13-14.

Wesley, W.K., 1957 B. V. Major insect pests of vegetables in Allahabad and their control. *Allahabad Farmer*, **31** (4-5): 91-99.

Wesley, W.K., 1958 A. VI. Major insect pests of vegetables in Allahabad and their control. *Allahabad Farmer*, **32** (1): 5-12.

Wesley, W.K., 1958 B. Control of the brinjal lace bug. *Allahabad Farmer*, **32** (3): 90-92.

Wickramasinghe, N. and Fernando, 1962. Investigations on insecticidal seed dressings, soil treatments and foliar sprays for the control of *Melanagromyza phaseoli* in Ceylon. *Bulletin of Entomological Research*, **52**: 223-240.

Wilkinson, D.S., 1928. A revision of the Indo-australian species of the

Genus *Apanteles* (Mym, Brae.) Part I. *Bulletin of Entomological Research,* **19** (1): 79-105; Part II. *Ibid.*, **19** (2): 109-146.

Yadav, P.R., J.N. Sachan and C.P.S. Yadava, 1974 A. Bionomics of the diamondback moth, *Plutella xylostella* Linn, on cauliflower. *Bulletin of Entomology,* **15** (1-2): 87-91.

Yadav, P.R., J.N. Sachan and C.P.S. Yadava, 1974 B. Bionomics of cabbage semilooper, *Trichoplusia ni* Hub. *Bulletin of Entomology*, **15** (1-2): 92-96.

Yadav, P.R., B.P. Srivastava and V.S. Kavadia, 1977. Adsorption of BHC in root crops raised on treated soils. *Indian Journal of Entomology,* **39** (2): 247-258.

Yathom, S., 1956. Biology of spiny bollworm (*Earias insulana* Boisd.). *Katavim*, **7** (I): 43-57, Rehovot (Israel).

Yazdani, S.S., 1971. Note on the comparative efficacies of organic phosphates against the spotted bollworm (*Earias fabia* Stoll). (Noctuidae: Lepidoptera). *Indian Journal of Agricultural Sciences,* **41** (12): 1128-1129.

Yodev, O.C., 1983. Use of pathogen-produced toxic in genetics Engineering of plants and pathogens. In: *Genetic Engineering of Plants* T. Kosuge, C.P. Meredith and A. Hollaender [Eds.]: 335-353, Plenum, New York.

Young, J.R. and L.P. Ditman. 1959. The effectiveness of some insecticides on several vegetable crops. *Journal of Economic Entomology,* **52** (3): 477-481.

Zaitlin, M., P. Day and A. Hollaender, [Eds.], 1985. *Biotechnology in Plant Science: Relevance to Agriculture in the Eighties*. Academic Press, Orlando, Florida.

Zaka-ur-rab. Md. and S.M. Alam, 1959. Mode of feeding in the melonfly *Dacus cucurbitae* Coq. (Diptera: Trypetidae). *Proceedings of 46th Indian Science Congress, Delhi, III:* 494-495, Calcutta.

Zeck, E.H., 1933. Investigations on the green vegetable bug (*Nezara viridula.* Linn.). *Agricultural Gazette of New South Wales,* **44** (8-9): 591-594 & 645-682, Sydney.

Zeller, S.M., 1847. Isis: 851-852. London.